Greening Berlin

Inside Technology

edited by Wiebe E. Bijker, W. Bernard Carlson, and Trevor Pinch

Paul Rosen *Framing Production: Technology, Culture, and Change in the British Bicycle Industry*

Richard Rottenburg *Far-Fetched Facts: A Parable of Development Aid*

Susanne K. Schmidt and Raymund Werle *Coordinating Technology: Studies in the International Standardization of Telecommunications*

Wesley Shrum, Joel Genuth, and Ivan Chompalov *Structures of Scientific Collaboration*

Chikako Takeshita *The Global Politics of the IUD: How Science Constructs Contraceptive Users and Women's Bodies*

Charis Thompson *Making Parents: The Ontological Choreography of Reproductive Technology*

Dominique Vinck, editor *Everyday Engineering: An Ethnography of Design and Innovation*

Greening Berlin

The Co-Production of Science, Politics, and Urban Nature

Jens Lachmund

The MIT Press
Cambridge, Massachusetts
London, England

MIT Press books may be purchased at special quantity discounts for business or sales promotional use. For information, please email special_sales@mitpress.mit.edu or write to Special Sales Department, The MIT Press, 55 Hayward Street, Cambridge, MA 02142.

Set in Stone Sans and Stone Serif by Toppan Best-set Premedia Limited, Hong Kong. Printed and bound in the United States of America.

Library of Congress Cataloging-in-Publication Data

Lachmund, Jens.
Greening Berlin : the co-production of science, politics, and urban nature / Jens Lachmund.
p. cm. — (Inside technology)
Includes bibliographical references and index.
ISBN 978-0-262-01859-3 (hbk. : alk. paper)
1. Urban ecology (Sociology)—Germany—Berlin. 2. Urban wildlife management—Germany—Berlin. 3. City planning—Environmental aspects. 4. Urban policy—Environmental aspects—Germany—Berlin. I. Title.
HT243.G32L33 2013
307.760943'155—dc23
2012022464

10 9 8 7 6 5 4 3 2 1

Contents

Acknowledgments

While researching and writing this book, I was first a fellow at the Max Planck Institute for the History of Science in Berlin and later a member of the Department for Technology and Society Studies at Maastricht University. Each of those institutions provided me with a supportive intellectual environment. Portions of chapters 2 and 5 appeared, in a different format, in "Exploring the City of Rubble: Botanical Fieldwork in Germany after World War II" in *Science and the City*, edited by S. Dierig, J. Lachmund, and J. A. Mendelsohn (*Osiris*, volume 18, University of Chicago Press, 2003) and in "Knowing the Urban Wasteland" in *Earthly Politics: Local and Global in Environmental Governance*, edited by Sheila Jasanoff and Marybeth Long Martello (MIT Press, 2004).

I am thankful to the ecologists, landscape planners, and activists who allowed me to interview them, provided me with ample documents, and in some cases allowed me to accompany them in the course of fieldwork or political or organizational activities. I am especially grateful to Herbert Sukopp, who gave me access to his personal archive and who commented on portions of the manuscript.

I also owe an enormous debt to colleagues who sharpened my thinking in fruitful ways. I want to mention especially Pedar Anker, Karin Bijsterveld, Soraya de Chadarevian, Arthur Daemmrich, Sven Dierig, Michael Hagner, Sabine Höhler, Sheila Jasanoff, Abigail Lustig, Andrew Mendelsohn, Michelle Murphy, Thomas Potthast, Hans-Jörg Rheinberger, Anke te Heesen, and Rosemary Wakeman. I also want to thank Wiebe Bijker, Andreas Fickers, Ruud Hendriks, Margareth Meredith, Alexandru Preda, Geert Somsen, and the two anonymous reviewers who all read some or all of the draft and provided me with valuable feedback. I am also grateful to Paul Bethge and Byron Evans for their help during the process of preparing the manuscript and to Helen Craggs for proofreading. Finally I want to thank Evelyn Hagenah for her encouragement and personal support.

Introduction

In May of 2000, the German capital of Berlin witnessed the festive opening of a "nature park" at the Südgelände, a former railroad yard in the densely populated district of Schöneberg. The yard had been severely damaged during World War II and had been abandoned after the political division of the city. Although material relics from the railway past gave the site some significance, that was not the main reason for the park's creation. Since about 1980, nature conservationists, ecologists, and activists in West Berlin (which was still walled in) had opposed plans for rebuilding the site and instead called for preservation of the abundant flora and fauna that had settled there. Indeed, the area had become a remarkable piece of urban nature. Since 1948, when the railroad yard was abandoned, it had become overgrown with a dense plant cover, which had accrued into an impressive wood-like wilderness. Ecological surveys revealed that species diversity was extremely high at the site and that it contained plant and animal species that were threatened or had been extinguished elsewhere in the region. For many years the Südgelände remained a bone of contention among traffic planners and promoters of the nature park. It was only after the fall of the Berlin Wall and the subsequent reunification of the city in 1990 that the campaign for the nature park progressively achieved success. Partly a nature reserve, partly a green space with artfully designed pathways, the new nature park has allowed citizens to experience a wide amplitude of wildlife in the midst of their city.

Only a few decades earlier, an abandoned lot in the midst of a large metropolis would hardly have become a target of a campaign for nature protection. Traditional nature conservation had focused almost exclusively on the countryside or the outskirts of the city, and on protection of pastoral landscapes and their original flora and fauna. Attempts to "green" a city by creating conventional parks and other open spaces had been more concerned with the shaping of aesthetically pleasing spaces and recreation facilities than with promoting species diversity and natural ecosystems. Sites such as the Südgelände and the "nature" that abounded there generally had been considered "wastelands," and the plant species that settled there "weeds."

In the 1970s, however, protection of urban wildlife species and their biotopes (meaning their natural living spaces) began to figure increasingly on the public agenda of West Berlin and other cities in Germany and elsewhere. In the former West Berlin this was due in part to the rich tradition of ecological fieldwork that had thrived there after World War II. Attracted by the flora and partly also the fauna that had flourished in the empty spaces throughout the city left by the war, and, at the same time being fenced off by the wall from the adjacent countryside, field ecologists began to focus on their own city as their main research object. Notably through the work of the botanist Herbert Sukopp and a group of researchers around him, Berlin soon became widely acknowledged as one of the world's leading centers of urban ecology. Far from being purely academic researchers, these ecologists also ventured into urban politics, confronting traditional planning practices and promoting their own visions of a more harmonious development of city and nature.

Earlier and probably more consequentially in Berlin than in other European cities, this ecological vision of the city had also found resonance among planning practitioners and significant segments of the urban public. By 1980, conflicts over urban land-use planning in West Berlin had become fundamentally ecologized. As with the Südgelände, the impacts of urban development projects on wildlife species, on biotopes, and on other aspects of the "household of nature" had become major concerns of civic activists. Such impacts were probed and evaluated in numerous professional reports, and, as a result, projects sometimes were canceled or modified. At the same time, the Berlin authorities were busy developing a comprehensive master plan that would complement traditional urban planning schemes with a systematic policy of species protection. Although the far-reaching wishes of its advocates were only partly realized, biotope protection has occupied a stable place in Berlin's planning system ever since and has affected the material form of the city in a variety of ways.

This book is an account of how species protection has emerged as a common focus of scientific and political concern in Berlin and how it has materialized in new ways of planning and managing urban space. After a scene-setting chapter on earlier traditions of green planning, it covers a period that stretches roughly from around 1948–49—the time of the political division of Germany as well as of the city of Berlin—nearly to the

present. Since the specific configuration of urban ecology and nature conservation discussed in the book developed in West Berlin, the book focuses
mainly on that part of the divided city. Only when treating the period after
1990, when, along with other political structures, the West German nature-
conservation policy was extended to the East, does the book widen its focus
beyond the former West Berlin. This does not mean that issues of nature,
green planning, and ecology did not also achieve wider attention in the
former East Berlin. The political circumstances and the content of these
policies, however, differed remarkably from the ecological policy in the
West, and although interesting in themselves they will be addressed only
insofar as they are relevant to understanding the trajectory of biotope-
protection policies in the post-unification period.

The first of the book's two main goals is to shed light on the changing
place of nature in the modern city. On the basis of the idea that many
of the moral and physiological ills of the city were due to its alienation
from nature, hygienists, planners, and social reformers have long sought
to remedy these problems through various strategies of greening, including
the preservation of recreation landscapes in the adjacent countryside, the
shaping of parks and greeneries in the city, and even experiments with
new city forms such as garden cities or loosely scattered suburbs. The wave
of environmentalism that has swept through Western countries since the
1970s has given new relevance to these issues but has also caused them
to be perceived and addressed in new ways. Reframed as a matter of species
and biodiversity protection, nature in West Berlin has come to be seen as
a dimension of a broader environmental problematique alongside such
issues as air pollution, chemical pollution, water management, noise abatement, and the reduction of energy consumption and traffic. Like those
other issues, it figures in administrative programs and in citizen initiatives
of cities as well as in the national and international debates and policy
initiatives to make cities more environmentally friendly or "sustainable."
By tracing the new politics of biotope preservation in Berlin, this book
seeks to shed light on how such a policy has taken shape and has become
operative in a local context. At the same time, it also reveals the internal
tensions, dilemmas, and limitations that attempts at nature conservation
are faced with in a city.

The book's second goal is to understand the political use of science in
an important environmental conflict. As with many environmental affairs,
science has played a pivotal role in the definition of the problem and the
means through which it is subsequently tackled. The very notion of the
biotope was originally a scientific concept, used by ecologists to make sense

of the relations that tied plant and animal species to certain conditions of the environment. This book tells the story of how ecologists—both credentialed academics and amateur naturalists who worked with them— carried out fieldwork in the city and thereby created an ecological understanding of urban space on which the later biotope-protection policy was based. Furthermore, it traces how the growing assortment of people who embraced the idea of biotope protection—planners, activists, policy makers—appropriated ecological knowledge, made it fit with their own political goals and agendas, and thereby, to some extent, became embroiled in the formation of the city as an object of ecological knowledge. Such interactions between science and politics are not restricted to the field of urban nature conservation. They have become typical for all kinds of environmental problems. Because of its long duration and its social complexity, urban nature conservation in Berlin can contribute to our more general understanding of the politics of environmental issues in contemporary culture.

Exploring the Nexus of Science, Politics, and Nature

Since the late 1970s, a widespread concern in urban studies has been to unravel the "social production of space" (Lefèbvre 1991). Various authors have traced how social factors—originally conceived in terms of political economy or hegemonic ideologies, later as cultural imaginations and discourses—materialize in the socio-spatial arrangements of cities (Soja 1996; Eade 2000; Gottdiener 1985; King 1996; Harvey 2009; Dear 2000). More recently, scholarship in environmental history and environmental sociology has shown to what extent the evolution of the spaces that we usually consider as the natural environment is also mediated by social processes (Blackbourn 2006; Franklin 2002; Cronon 1995; Macnaghten and Urry 1998; Shields 1991; Matless 1997; Cioc 2002; Lekan 2005).[1] Accordingly, the natural environment is anything but a distinct realm that can be treated in isolation from the worlds of human culture and politics. First, the biophysical character of natural spaces such as rivers, forests, and landscapes is largely molded by the intentional and unintentional effects of human activities. It has even been argued that seemingly pristine sites such as the tropical rainforest show the imprint of active shaping by indigenous inhabitants (Balée 1994). Second, since nature does not "speak directly" to us, as Alan Irwin (2002: ix) has put it, historically contingent acts of meaning-making play a mediating role in how people relate to and engage with nature, including how they perceive, control, and manage their own intended or unintended effects on that nature. Even the very distinction between nature and artifact concerns contingent and negotiable attributes more than it concerns fixed substantial differences (Latour 2004; Reuss and Cutcliffe 2010).[2]

From a similar perspective, recent works of historians, geographers, and sociologists have argued that cities are hybrid socio-natural fabrications, rather than purely man-made artifacts (Kaika 2005; Gandy 2002; Heynen, Kaika, and Swyngedouw 2006; Melosi 2001; Tarr 2002; Franklin 2010; Falck 2010). It is surprising, however, how little attention has been paid to the role of science in the shaping of nature in cities. Science is mostly treated as an explanatory resource to account for the ecological impact of cities on nature, or the effects that environmental degradation has on urban life.[3] What is missing is a systematic exploration of the ways in which the practice of the environmental sciences has become a constitutive part of the dynamics of nature politics and spatial development in cities. This book combines insights and methods from social studies of science, from environmental sociology, from environmental history, and from urban studies to shed light on the nexus of science, politics, and the spaces of the natural environment. My analytical focus is, therefore, on what Sheila Jasanoff (2003), following Bruno Latour (1993), has called *co-production*: the simultaneous constitution of science (including its orderings of nature) and the social worlds in which it is embedded.

I use *nature regime*, or more specifically *biotope-protection regime*, as a shorthand term for the local concatenations of science, politics, and nature that have evolved in Berlin. In contrast to the development anthropologist Arturo Escobar, who has used *nature regime* more broadly to characterize general modes of "incorporating" nature in different cultures, I employ that term here to account for the historically contingent frameworks through which nature is constituted as a focus of public contestation and government.[4] A nature regime, as I understand it here, consists of a dynamic set of relations that include (1) the portions of the non-human world that are claimed to represent valuable forms of nature and that therefore are supposed to require public attention or care, (2) the practices and discourses through which such claims are produced, promoted, and incorporated into public policies, and (3) the individuals, collectives, and institutions that assemble around and actively sponsor these claims.[5] Such regimes are zones of convergence where elements that may be rather diverse in character, and which may have developed largely independent of one another, come together. From this perspective, the environmental problematique, as it has attracted public attention since the 1970s, can be seen as a range of nested nature regimes that, at different geographical scales and with different institutional extensions, have evolved around distinct issues. Besides the urban biotopes that are the topic of this study, one might also think of individual species, water reservoirs, sites at which waste is dumped, or the global climate as foci of nature regimes. The claims that are made on such portions of nature can range from calls for conservation, management, or restoration to calls for control of the industrial risks that are thought to impede their proper functioning. Not all nature regimes are necessarily associated with science; however, the nature regimes of contemporary environmental politics typically are.

This book tells the story of how the practice of ecology became the seedbed of a local nature regime, how that regime branched out and transmogrified, to what extent it became institutionalized in formal structures of governance, and how it lost much of its initial political momentum. The choice of the nature regime as an analytical unit sets this study apart from studies of ecology and environmentalism that have focused on scientific disciplines and institutions (Kwa 1989; Bocking 1997; Kingsland 2005), on regulatory frameworks (Brickman, Jasanoff, and Ilgen 1985; Chaney 2008; Hajer 1995), or on social movements (Bramwell 1989; Engels 2006). It allows me to trace how processes and arrangements in all these spheres have been locally co-produced, and how together they gave form and meaning to a distinct type of nature. Although my main concern here is the urban biotope-protection regime, I will also deal with its genealogical relations to earlier nature regimes, such as the *green-planning regime* or the *traditional nature-conservation regime*. Nature regimes, however, should not be understood as fixed and mutually exclusive structures. They are dynamic constellations that evolve over time, are internally divided, and partly intersect with each other.

Between Science, Politics, and Policy
Whereas studies of science, technology, and society once were mainly concerned with fact construction in classic academic settings, they have recently turned to the broader dynamics through which scientific knowledge is produced and appropriated within the socio-political frameworks in which it is mobilized (Jasanoff 2003). More specifically, my analysis of the biotope-protection regime is based on an understanding of science as local practice as has been developed in many ethnographic and historical studies of scientific work (Knorr Cetina 1981, 1999; Lynch 1985; Latour and Woolgar 1979; Lynch and Woolgar 1990; Shapin and Schaffer 1985). Notwithstanding major conceptual differences among individual authors,

it has been a common aspect of this line of research to probe science through a close attention to the day-to-day work in the actual places in which it is done, such as laboratories, hospitals, and (more recently) fieldwork sites. From this perspective, science is not so much a body of ideas that merely describes or represents the world as it is a socio-material engagement that intervenes in and reconfigures this world. Accordingly, what gets shaped and stabilized as a scientific phenomenon or a knowledge claim depends greatly on the arrangements of objects and situated negotiations through which research problems are made "do-able" (Fujimura 1996) and processed. Likewise, I will treat the knowledge claims made by ecologists and other proponents of nature conservation in Berlin as fundamentally connected to the local practice in which they are created, presented, or appropriated. Notably, I will attend to the day-to-day fieldwork in the city and the negotiating of its results. The knowledge claims in which I am interested here have not all been produced by scientists. Planners, bureaucracies, and activists also have made or appropriated such claims and, to some extent, reconstructed and reshaped their content.

This book, however, aims to be more than just a further analysis of a field of scientific practice. It seeks to understand how these practices have come to be entangled with the practices through which nature-conservation policies are made and executed. In this respect, my understanding of a nature regime has similarities with the work on governing and governmentality that has been inspired by Michel Foucault (Foucault 1991; Burchell, Gordon, and Miller 1991; Dean 1999). Such studies have unraveled the role of "intellectual technologies" (Miller and Rose 1990), such as surveys, accounting systems, and planning schemes, through which domains of human life, including the environment (Oels 2005; Agrawal 2005; Darier 1996; Luke 1995), are formatted into governable phenomena that can be handled in the framework of dominant political practices. Unlike the aforementioned authors, however, I am not concerned here with situating these developments in an overarching trend toward one dominant form of power (biopolitics or neo-liberal governmentality). In contrast, I want to show how the biotope-protection regime has been shaped by, and has shaped, the specific local context in which it evolved. Moreover, governmentality studies often take the political functionality and credibility of the intellectual technologies through which governmentality is executed for granted, and often underestimate the extent to which political strategies are refracted through the practices and concerns of the different social groupings and arenas in which they operate.

The fit between science and policy will be considered here as a locally contingent accomplishment that is constantly negotiated and renegotiated among the participant actors. Not only does this require a transposition of the analytic sensibilities of practice-focused science studies to other, less academic places of knowledge making, such as advisory bodies or protest movements; it also demands attention to the specific logics through which political agendas are formed, to how they are articulated and contested by different constituencies, and to how all this works recursively on the terms under which environmental knowledge claims are produced and stabilized. Much of the book will therefore be concerned with the role of urban ecology in what the environmental sociologist John Hannigan (1995) has called "claims-making processes." This term refers to the public campaigning of experts, media figures, or activists who promote ("assemble," "present," and "contest," as Hannigan distinguishes) claims about the existence of environmental problems and the urgency of policy responses. In addition, I will argue that knowledge practices, in order to become operative in policy, have to be actively aligned with the institutionalized practices of policy making.

Another aspect of the co-production of science and social order in the nature regime concerns its institutional embedding. As I will show, a number of new social collectives, institutional boundaries, and procedures have evolved in close connection to the rise of the urban biotope regime. Together, the institutional structure of academic ecology, professional communities, legal-administrative organizations and procedures, and social movements and activist groups formed what, in chapter 4, I will call the *regime communities* that coalesced within the biotope-protection regime. None of these collectives and institutions were created from scratch (rather, they all built on practices and institutions that had already been in place), and their missions were far from exclusively related to urban nature conservation. As I will show, however, these collectives and institutions, their mutual relations, their internal practices, and the subject positions that were tied to them were profoundly reconfigured through their involvement in the new regime.

Making Space and Place

Spaces are not just neutral containers of social life. They have to be analyzed in terms of the discourses and practices that dwell within them and that give them meaning and structure. Cities are nested spaces that emerge through the operation of various overlapping strands of discourse and practice. They are marked as *urban* through their categorial distinction from the surrounding spaces of the countryside. They also become the anchor of meanings and identities that distinguish them from other cities and thereby constitute them as particular *places*.[6] Moreover, cities are composed of multiple layers of internal spatial divisions, ranging from standardized administrative categories to vernacular definitions of neighborhoods, sites, or buildings. It is a corollary of the complexity of urban life and politics that urban spaces and places have contradictory meanings and that their identity is always contested (Franklin 2010). As Peter Fritsche (1996) showed in his study of everyday life in Berlin around 1900, the ordering of the city—in the case of Fritsche's study, through the production of newspaper narratives—is never complete. Throughout this book, I will show how the biotope-protection regime has participated in the never-ending processes of creating order in the city.

We can see nature regimes as machineries of spatial classification that identify and evaluate existing or projected spaces as nature, and that demarcate them from other (supposedly non-natural, less natural, or differently natural) spaces. Earlier urban nature regimes typically established a clear-cut distinction between the city and the allegedly natural spaces of the countryside. At the same time, they repeated this distinction within the city itself by opposing parks or remains of the pre-urban landscape as islands of nature to the rest of urban space. (See Kaika 2005.) Within the urban biotope-protection regime, nature in the city became reinvented as *wildlife*, *biotopes*, *ecosystems*, or the *household of nature*. Such nature was no longer seen as radically opposed to the city. Urban spaces were, rather, reconceived as a new kind of nature, one that was largely shaped by the *influence* of human culture yet still presented the hallmarks of valuable ecosystems. At the same time, the regime replaced the radical opposition of nature and non-nature within the city that earlier regimes had established with a gradual hierarchy that distinguished between spaces according to their varying degrees of assumed closeness to nature. This had the effect that many sites (such as wastelands or roofs of houses) that hitherto had been regarded as unnatural were re-classified as new and valuable forms of nature in the city.

In chapters 2 and 3, I will analyze how these new spatial orderings emerged from the increasing interaction of ecology with the city. These discursive shifts, however, were anything but purely academic; they were performative of the very order that they described.[7] First, these classifications became extended to and routinized in the practices of planning and administration, in the vocabularies in which citizen activists formulated their political claims, and (at least partly) in the cultural repertoires through which urban dwellers made sense of their immediate environment. They thereby redefined what places and spaces in the city were for these actors, where they began and ended, and how these spaces related to one another. Second, ecological knowledge and planning practices directly affected the material order of these spaces. This was the case when they motivated decision makers to refrain from certain forms of land use and thereby allowed wildlife to develop within these spaces in an unconstrained way. More typically, conservation policies were implemented by technical means that transformed the places in question by making them accessible as recreation sites for the public, or by trying to optimize or newly create positive conditions for wildlife.

Another aspect of this spatializing performance of the urban biotope-protection regime is the establishment of the city as the *scale* at which the issue of biotope loss—an environmental problem that otherwise is also considered as profoundly global—has to be treated. As has been argued in geographical literature on the construction of scale, the local, the regional, the national, and the global should be seen not as pre-given essential layers of social reality but as actively constructed by the political projects that operate at these scales (Masson 2006; Gille 2006). Likewise, there is no natural location for a certain problem at one of these scales. By analyzing how science and politics evolved at the local scale of a city, I will thus also clarify how the local, in conjunction with processes at other scales, was established as a significant space for environmental politics. Although focusing on Berlin as its major object, this book will therefore occasionally seek to relate its findings to what is happening on other scales—upward by tracing how policies and academic debates in national and global contexts have fed into the Berlin nature-conservation policy, and downward by tracing how urban policies and city-wide research endeavors material-ized differentially in site-specific projects within the city.

By viewing urban natures as fabrications of historically contingent nature regimes, my approach diverges from conventional analyses of environmental policy that tend to consider nature as a pre-given objective realm having no direct bearing on the political processes that are related to it. This does not mean, however, that this analysis is ignorant of the causal properties of natural orders or of the environmental damage pro-duced by human action or inaction. Seeing nature as historically consti-tuted does not mean that it has to be reduced to an arbitrary artifact of cultural representation (Irwin 2002: 161–187). By attending to the mundane work practices through which knowledge is produced, and by relating the politics of environmental knowledge to concrete spaces, I seek to grasp the subtle socio-material entanglements through which such meanings are shaped and enacted. In this respect my work has some similarities to recent work on the environment in actor-network theory.[8] In contrast with actor-network theory, however, it is not my intention here to replace the analytical focus on cultural meanings with a radically decentered relational ontology.[9] As we will see, the configurations of material objects and organic processes that assembled in the places that scientists studied, and that conservationists sought to protect, played an active role in the nature regime. Notwithstanding their profound plasticity, they provided constraints that social practices had to adapt to, or offered opportunities on which new trajectories of practice could thrive. Turning points in the environmental history of urban space, such as the destruction of the city during the war and its eventual reconstruction, were therefore also thresholds that spurred or redirected the trajectory of ecological knowledge making. In other words, the biotope-protection regime was as much a product of Berlin's urban environmental history as the latter was a product of the regime.

Putting the Case in Context: From Berlin to West Berlin and Back to Berlin

This study has as its object a city that, within the period under consideration, underwent two dramatic political transformations. Since the late nineteenth century, it had developed from a middle-size residential city to a metropolis and the political capital of Germany. Different political regimes, including the Kaiserreich, the Weimar Republic, and the period of National Socialism, had left their marks on the physical body of the city. The city's development was interrupted dramatically by World War II. Bombardment wiped out large parts of Berlin's built structure, and the fall of the National Socialist regime entailed a radical change of the political order.[10] In 1949, increasing tensions between the Soviet Union and the Western allies led to the constitution of two separate German states—the Bundesrepublik Deutschland (Federal Republic of Germany, FRG), which comprised the territories that French, British, and American forces had occupied after the war, and the Deutsche Demokratische Republik (German Democratic Republic, GDR), which consisted of the Soviet-occupied part of Germany. Although geographically Berlin was in East Germany, since 1945 it had been administered by the four occupying Allied powers, each having jurisdiction over a separate sector. Attempts by the Soviet Union to disentangle its sector of Berlin from the West, and the introduction of separate currencies, were followed in June 1948 by an 11-month military blockade of West Berlin by Soviet troops and the development of separate political and administrative structures in the eastern and western parts of the city. In 1949 East Berlin became the capital of the GDR. West Berlin became a de facto exclave of the FRG. Although formally distinguished from the FRG and under special provisions of the allies, West Berlin was organized like a separate Land (one of the formal units of the FRG). Federal laws, however, became valid there only through formal enactment by Berlin's parliament, the Abgeordnetenhaus. Like the other Länder, West Berlin retained autonomous regulatory competences in some policy fields. In 1949, with about 2 million inhabitants, it was still slightly larger than Hamburg, the biggest city in the FRG. Its regierender Bürgermeister (governing mayor) presided over a Senate that was comparable to the government of a Land.[11] Like the Länder, West Berlin was also represented (though with only consultative competences) in the FRG's second chamber, the Bundesrat, and its citizens elected delegates to the federal parliament, the Bundestag (also with only consultative competences). The status of West Berlin, however, remained a bone of contention between the FRG and the Western allies, on the one hand, and the GDR and the Soviet Union, on the other. The Western powers regarded West Berlin as a bastion against totalitarianism and promoted it as a showcase of democracy. The Eastern powers considered West Berlin a foreign body in their own territory, accepting it as, at best, a political unit, but not as a part of the FRG.[12] They did almost everything they could do to undermine its connections to the FRG and the social and economic relationships that after the war had continued among West Berlin, the GDR, and East Berlin. The Berlin Wall, erected in 1961, became the emblematic materialization of this political division. The confrontation began to attenuate in the early 1970s, when the policy of détente and the tenure of Willy Brandt as chancellor of the FRG led to a significant relaxation or "normalization" of relations between the two German states.

These geopolitical circumstances affected nearly all spheres of social and political life in West Berlin.[13] Its disconnection from the FRG, in economic and social terms as well as in technical infrastructure, was an obstacle for industries located there; they required federal subsidies.[14] Although the federal government offered financial incentives for workers to move to Berlin, the city experienced continuous emigration of its workforce (the only exception being young men, who by moving to Berlin could avoid conscription by the West German army). In comparison to other cities, the composition of the population was unproportionally dominated by the older generation. Partly also because of its marginality, West Berlin became a fertile ground for political protest movements and countercultures, which, as we will see in chapter 4, had an important effect on the emergence of ecological planning policies.[15] Urban planning was torn between a lack of resources and demands that led to a much lower pace of redevelopment than in other cities in the FRG, on the one hand, and (notably as of the late 1960s) expansionist housing and infrastructural projects that were meant to keep or newly attract business and people, on the other. The policies that will be described in this book were, to a large extent, formulated in response to these conditions.

The fall of the wall in November of 1989 and the reunification of Germany in October of 1990 formed the second major turning point in the period considered in this book. In the unification treaty, the GDR formally acceded to the FRG, adding five newly formed Länder.[16] The terms under which the unification was realized meant that the politico-institutional structures as well as the economic structures of the FRG were more or less directly transposed to the former GDR. In the course of those events, the reunited Berlin achieved the status of a Land, separate from the surrounding Land, Brandenburg. Furthermore, in 1991 the German parliament decided to move the major parts of the government and of its own seat from Bonn to Berlin, which thereby was officially re-endorsed as Germany's capital.

Although the reunification removed the pressures that had constrained the walled city, Berlin faced other kinds of challenges. Whereas administrative structures were adopted relatively quickly in the East, economic and structural difference persisted among the former city parts. The extension of a free real-estate market to the East led to an increase in rent in the formerly run-down districts, whose buildings were renovated and which became socially gentrified. The discontinuance of the subsidies that had formerly sustained the West Berlin economy led to the abandonment or migration of many enterprises. Moreover, Berlin suffered from a notorious lack of financial resources, which was further aggravated by the size of public investments that were required for the preparation of the capital moving, as well as the tendencies of firms to settle in Brandenburg, where they enjoyed lower tax rates. On the other hand, the city launched a number of landmark construction projects, among them the rebuilding of the technical infrastructure (new railway facilities, a new airport), the provision of space for the new government institutes and related organizations, and the renovation of run-down neighborhoods in the East. In the years after unification, so much office space was created in the city that it widely exceeded existing demands. Urban policy narratives at this time featured the city as a "new Berlin" and sought to place it in a new historical framework or promoted fantasies of its future role as a national or even global center (Rada 1997).

During the Cold War, West Berlin had always been characterized by a certain marginality—in both the positive and the negative sense of the word—that had allowed the city to go its own way in many respects. After 1990, however, the unified Berlin became increasingly drawn into the vortex of global processes of restructuring. Not only did this result in the visible presence of international firms and international tourism; in addition, the influence of local corporations that hitherto had dominated the real-estate market in West Berlin was increasingly replaced by brokers and international investors. Likewise, the state-centered model of spatial governance of the former West Berlin gave way to new forms of governance such as public-private partnerships and other forms of outsourcing. Urban sociologists have described how the structural problems caused by the unification and the increasingly global economic competition with other cities led to economic and social polarization in Berlin (Häußermann and Kapphan 2000). Although such developments are not peculiar to Berlin, their effects were relatively abrupt, as they followed a situation that had been characterized by considerable isolation.

As I will argue in this book, the biotope-protection regime was largely a product of this particular geopolitical constellation. Though it is not difficult to find parallels of many of its elements in other cities, Berlin's relative isolation and marginality allowed them to become joined in a very dynamic nature regime that, at least in its heyday in the 1980s, was firmly established in the local worlds of science, policy, and civil society. This also meant that the biotope-protection regime lost much of its original momentum when, with the fall of the wall and the subsequent unification, the contextual conditions changed.

Recent Historiography

The last ten years have witnessed a considerable extension of historical scholarship on nature conservation and landscape planning (Landschaftsplanung). As such studies have shown, nature conservationism has been relatively influential in German history since the late nineteenth century. Although largely aesthetically motivated, such strivings were already tied to ecological ways of thinking (Schmoll 2004). Recent scholarship has also placed nature conservation into broader political institutional contexts—those of nation building (Lekan 2004), international relations (Hünemörder 2004), and the political culture and institutional structures of the FRG (Chaney 2008; Engels 2006; Bergmeier 2002; Uekötter 2004). As the afore-mentioned authors have revealed, conservationists—contrary to their own understanding and public self-presentation—have been far from marginal in the political system of the FRG. Attention has also been devoted to the extent to which the ideas of the nature-conservation movement and of Landschaftspflege (landscape care) have remained rooted in certain German peculiarities, notably with respect to their link to the period of National Socialism. In contrast to the thesis promoted by Anna Bramwell (1989), who saw nature conservation as closely associated with the National Social-ist ideology, more recent studies (Radkau and Uekoetter 2003; Brüggemeier, Cioc, and Zeller 2005; Uekötter 2006) plead for a nuanced picture of continuities and differences. Scholarship on the history of German nature conservation and landscape protection provides valuable insights into the broader context of the present work, and I will refer to some of them repeatedly below. Although this literature also touches occasionally on the role of nature around cities (see, e.g., Engels 2005), a systematic analysis of the development of urban nature conservation in one German city is lacking. With the exception of a few insider accounts written by landscape planners or ecologists,[17] the topics addressed in this book have been mainly covered by political science or by geographical accounts of environmental policy in other cities (Hertel 2000; Desfor and Keil 2004; Kaika 2005; Gottlieb 2007). This book not only adds a particularly interesting case to this set of studies; it also extends their perspective toward a systematic exploration of the formative role of ecology in urban policy making.

Sources and Content of the Book

It is not my intention in this book to take stock of the entire spectrum of scientific and political activities of nature conservation in Berlin. Instead I will focus my account on a number of events and topics that I consider particularly well suited to the specific research goals that guide my inquiry. My empirical basis is a qualitative analysis of a broad corpus of historical sources. The main type of data that I consider here are publications in books and journals (including scientific articles), administrative reports, press coverage, activist pamphlets, and a broad spectrum of "gray litera-ture" produced by government agencies and by activist groups. I have also consulted a number of unpublished sources, including the archives of the Berlin Senate departments and districts, the Landesbeauftragter für Naturschutz und Landschaftspflege (Land Appointee for Nature Conservation), the Technische Universität Berlin, and the Berliner Landesarbeitsgemeinschaft Naturschutz (Berlin Land Consortium Nature Conservation).

In addition to written sources, the book is also informed by more than thirty qualitative interviews that I conducted with ecologists, planners, and activists who had been active, or who at the time were still active, in urban nature conservation. The interviews, which lasted from one hour to more than two hours, were taped and, in most cases, transcribed. These interviews were of crucial importance for getting firsthand information and for acquainting myself with the informal components of a field of practices with which I had had no previous personal experience. Often these interviews led me to new textual sources, and many interviewees generously provided me with unpublished documents from their personal archives. In the process of analyzing the data, I decided to base my arguments on written sources to the greatest possible extent, and to refer to interview data only when dealing with biographical information, individuals' points of views, or other more informal aspects that did not leave traces in written documents. In the few places where I base my arguments on interviews, I give the name of the interviewee, the date of the interview, and (where it seems necessary) some information on the interviewee in a note.

Although I did not base this book in any systemic way on ethnographic participant observation, I visited relevant events, such as public planning hearings, meetings of the Berlin botanists' association, and meetings of the nationwide Arbeitsgemeinschaft Biotopkartierung im besiedelten Bereich (Working Group Biotope Mapping in Settled Areas). These observations were related to processes that were underway at the end of the period under consideration, some of which were still unfolding while I was doing the research for this book. Besides providing insights into those processes and events, they also helped me to gain a broader understanding of the informal aspects of the culture and practices whose history I describe in this book. I also learned much from two fieldworkers who allowed me to spend a whole day with them while they were collecting data for ecological surveys.

In structuring the narrative, I have sought to do justice to the chronology as well as to the systematic aspects I am interested in.

Chapter 1 deals with the tradition of open-space planning and nature protection that has existed in Berlin since the late nineteenth century. Only against the background of these former developments will the continuities and differences of the new ecological take on the city that will be analyzed in the subsequent chapters become clear. Chapter 1 provides an overview of the general lines along which these earlier attempts toward greening the city developed. In that chapter, in contrast to the detailed analysis of the interaction of science, policy makers, and activists that is provided in the other chapters of the book, I will restrict myself to the characterization of four main urban nature regimes that I distinguish in this earlier period.

Chapter 2 traces how the fieldwork of ecologists from the 1950s to the 1970s—notably in the rubble fields left by World War II—prepared the way for the biotope-protection regime. By attending to the different strategies of surveying the city (survey networks, site surveys, comprehensive surveys), and to the conceptual developments in scholarly ecology that were tied to them, I will show how a new kind of knowledge of the city and its relation to nature came into being. The city was now represented as a "mosaic of biotopes" (as Sukopp put it) consisting of different patches of space that displayed their own compositions of plant and animal species and which were considered worthy of protection in some cases or eligible for nature-promotion measures in others. At the end of the chapter, I will also examine how the ecological research on Berlin had been tied to the Institute of Ecology created at the Technische Universität in 1973.

In chapter 3, I focus on the development of the West Berlin Species Protection Program, a comprehensive planning scheme through which ecologists translated their views into land-use policy in the early 1980s. As I show, this program evolved through close interaction between ecologists (who conducted surveys, interpreted their political implications, and formulated claims) and administration agencies. Rather than simply being transferred and applied to the political sphere, ecological knowledge had been re-defined and re-represented in new formats that were better attuned to the administrative routines in which they had to be processed. As I show, classificatory practices that delineated the city into standardized parcels of land were critical in constituting urban nature as an object of governing. Moreover, the work on the Species Protection Program implied political negotiations between the relatively far-reaching ambitions to realize conservation claims throughout the entire city and the administration, which had to balance these ambitions with competing land-use claims. It was through this program that the knowledge practices of ecologists coalesced with administrative practices to form an institutionally embedded and politically consequential nature regime.

Chapter 4 looks more systematically at the social groupings and organizations (regime communities, as I call them) that were tied to the evolving

concern of urban biotope protection and that, along with fieldwork practices and planning practices, can be seen as the third fundamental pillar
of the regime. I show how the Institute of Ecology, notably through the
activities of Sukopp as a political entrepreneur as well as through its ties
to a new professional teaching program, was able to position itself as a
major purveyor of expertise on urban nature. Other elements that
co-evolved in this setting were the profession of landscape planning and
a variety of nature-conservation groups and civic activist groups that promoted these issues in the public sphere.

Chapters 5 and 6 focus on the ways in which the discourses and practices
of the biotope-protection regime materialized in particular orderings of
urban space and place. In chapter 5, I focus on one of the most spectacular
results of ecologists' involvement: the creation of nature parks on urban
wastelands, the city's most diverse biotopes. As with the making of the
Species Protection Program, of which the creation of nature parks was one
element, this entailed painstaking negotiations, not only with the opponents of these projects, but also among the various participants of the
epistemo-political setting. Chapter 6 traces a significant evolution that, in
the post-unification period, changed both the politics and the knowledge
formations of the regime. Conflicting land-use interests made it increasingly difficult to realize the goals of the landscape program. Alternative
solutions were now sought by compensating the environmental damage
caused by development projects by creating or improving other biotopes.
The chapter analyzes the conflicts and the uncertainties that were inherent
in this policy and shows how the policy manifested itself in urban space.

The conclusion connects the case study to broader discussions about
the relationship between city and nature in modernity.

1   Traditions of Urban Greening

Since the late nineteenth century, Berlin has witnessed continuous efforts to promote nature and greeneries in the city. This included not only the creation of parks, the planting of trees and other forms of urban horticulture, but also the preservation of extraordinary pieces of the natural landscape in its suburbs and direct surroundings. As I will show, it has been a constant motive of these policies that *nature* or *green* (the two terms were often used interchangeably) was needed as compensation for the negative sides of the modern metropolis. By providing fresh air, light, aesthetic experiences, and quiet recreation ground, urban green spaces were meant to remedy the most pressing of these problems. At the same time, it will be shown that *nature* had no stable meaning throughout this period. Promoters of urban greening considered different features or qualities of the urban landscape as natural, and hence also cherished different models for the shaping of a more livable city. On this basis I distinguish four distinct nature regimes that preceded the regime of biotope protection. These regimes should be considered as ideal types. As we will see, in practice they often blended into one another.

The first two regimes had their roots in the Kaiserreich and in the Weimar years. The regime of *green planning* aimed at the creation of public parks and greeneries and in the Weimar period developed to a systemic program of open-space provision. The regime of *classical nature conservation* revolved around the concept of *natural-monument care* and sought to preserve parcels of nature at the fringe of the city. The other two regimes gained momentum in the postwar period. Motives from the earlier regimes continued to exist, but they became joined with two more comprehensive ways of organizing urban space. The regime of *organic urbanism* considered the creation of urban green spaces as a means to create comprehensive *organic* cities or *urban landscapes*. At the same time, the regime of Landschaftspflege (landscape care) complemented the traditional style of monument conservation. These two new regimes embodied conservative ideas of an organic community and had roots in the period of National Socialism. It was only during the postwar reconstruction of the city that this approach gained practical momentum. Berlin planners took inspiration from, and contributed to, broader specialist discourses among urban green planners, nature conservationists, and promoters of landscape care. As I will show at the end of the chapter, the programmatic framings of Berlin green planners came under increasing pressure in the 1960s and the 1970s, when they clashed with a forceful policy of urban development in West Berlin.

The regime of urban biotope protection that will be analyzed in the subsequent chapters both drew upon and went beyond the discourse and practices of these previous regimes. First, planners were faced with the material manifestations of these policies that they framed in new terms; second, it merged understandings and guiding values of earlier regimes and incorporated them into new problem claims; and third, the new regime developed partly in the framework of the organizational infrastructures that had developed around earlier regimes.

Islands in the Sea of Buildings: Urban Green Planning

As Brian Ladd has pointed out in his study of the evolution of urban planning in Germany, the creation of parks had still been rather exceptional in the first half of the early nineteenth century. It was either a luxury granted to the city by the nobility or the result of contingent efforts of local beautification societies (Ladd 1990: 68). In the subsequent decades, however, planners and urban elites alike began to see the provision of green space as a public necessity and a crucial element of urban policy and social reform. As with the provision of waterworks, sewers, roads, and electricity, as well as with the laying down of general development plans, the creation of parks and other greeneries became an important target of the public involvement of public authorities in city planning.

The formation of the regime of green-space planning was a direct response to the rapid growth that German cities experienced in this period and to its perceived impact on the quality of urban life. From the middle of the nineteenth century on, public health organizations emphasized the importance of "light and air" and called for the restriction of the development density of inner city districts. They also advocated the creation of parks and the planting of trees, which they supposed acted as the "lungs" of the city, regenerating fresh air and binding unhealthy dust and vapors

(Ladd 1990; Rodenstein 1988; Stürmer 1991). At the same time, horticul-turalists and local beautification societies promoted urban greeneries as decorative elements and promenading grounds for urban residents.

Similar views figured prominently in the discourse of urban planning, which in the late nineteenth century became institutionalized in profes-sional organizations, specialist journals, and related administrative prac-tices (Ladd 1990). On top of climatological and aesthetic reasoning, many urbanists considered the physical and moral deprivation of citizens from the experience of "free nature" (Wagner 1915: 1) as a major problem of the city. In an influential 1909 textbook on urban planning, the Viennese architect Camillo Sitte even assumed that city dwellers were driven by an intrinsic urge to flee the "sea of buildings" for the greenery of "free nature" (Sitte [1909] 2001: 187). A crucial function of urban green-space planning was thus to provide the city dweller with the visual "refreshment" of "scat-tered images of nature" (ibid.: 190).

The need for green space was considered particularly pressing in Berlin, a city that had expanded rapidly in only a few decades. In the middle of the nineteenth century, it had still been a medium-size residential city of approximately 415,000 (Richter 1987: 661). Since that time, however, newly developing industries attracted a permanent influx of new inhabit-ants. The city's growth was stimulated by the economic boom that followed the formation of the German Kaiserreich in 1871 and by its newly achieved status as the national capital. The result was an urban agglomeration that reached far beyond the municipal area of Berlin proper, and included a number of adjacent villages and towns, such as Charlottenburg and Schöneberg. It took until 1912 for these independent units to form an administrative association, the Zweckverband Gross-Berlin (Functional Association of Greater Berlin), and until 1920 for them to merge into a single metropolis of about 880 square kilometers and 3.8 million inhabit-ants (Köhler 1987: 749).

By 1920, the villages, fields, and forests that had formerly occupied much of this territory had become a ring of urbanized neighborhoods, industrial areas, and sites of traffic and urban infrastructure. The villages on the outskirts developed into well-to-do garden suburbs; the inner districts became largely dominated by a dense development of tenements with as many as five storeys. This extreme building density was largely the result of the weak planning regulation that only determined the general regular street pattern and therefore allowed developers to achieve maximum revenue from their real estate.[1] Whereas tenements in the more well-to-do districts or street-facing buildings included large prestigious apartments, the less attractive tenements in rear buildings and tiny backyards provided accommodations for workers' families and for the poor.

The development of Mietskasernen, as the tall and densely constructed buildings were dubbed, soon became the crystallization point of critical debate about the negative sides of the modern industrial city (Bodenschatz 1987). Critics bemoaned the lack of aesthetic charm of the buildings and the uniformity of the rectangular street pattern in which they were arranged. The main object of concern, however, was the poor and unhealthy living conditions of the dark and tiny apartments, which tended to be crowded with large numbers of family members and sub-tenants. Social reformers viewed the Mietskasernen as fertile breeding grounds for alcoholism, violence, prostitution, and criminality.[2] Besides a rising concern for the social question, this critique reflected broader romanticist sentiments against the city that were widespread among Germany's educated middle class (Bergmann 1970; Lees 2002).

The first parks in Berlin were gardens of the Prussian court, which since the eighteenth century had also been also open to visitors. The most extraordinary of them was the Große Tiergarten (Great Animal Garden), a large forest and former royal hunting ground on the Western outskirts of the old residential city (Wendland 1993: 100–107). Between 1833 and 1840, the Große Tiergarten had been shaped as a landscape park by the court's garden architect, Peter Joseph Lenné. In the subsequent decades, it had become encircled by the sprawling development in its neighborhood. By 1900, it had become a large central park in the midst of a metropolis. In 1840, Lenné authored a plan in which he proposed the creation of further greeneries in the city. In that year, the city also decided to create a Volksgarten (people's garden) in the Eastern district of Friedrichshain. Opened in 1848, it was the first public park in Berlin that had been created by the municipality. In 1862, the Hobrecht Plan, a general development plan that stipulated the basic street pattern for the extension of the city, set aside some smaller block squares, notably in the more wealthy districts, for decorative greeneries. Eight years later, the city appointed a Gartendirektor (horticultural director) and a municipal park and garden's committee to plan, shape, and maintain greeneries within the expanding city (Stürmer 1991: 12).

Many of these parks were located on sites that hitherto had already been forests and that were deliberately exempted from urban development. Others were sited in areas that were not suitable for development, such as wetlands, which therefore were of little interest for real-estate developers. Under the guidance of a garden architect, massive ground works were carried out that transformed these sites into a cultivated garden environment. In accordance with the dominant ideal of the landscape garden, they were devised with trees, meadows, and ponds. Visitors could promenade along a set of meandering trails. These settings integrated trees, hills, creeks, and other pre-existing landscape features with added attractions such as geological exhibits, flower beds, and children's playgrounds (Stürmer 1991: 16). The Victoriapark even had a machine-driven waterfall. Their deliberate aesthetic design distinguished these parks from the forests on the outskirts, which were also increasingly used as recreation grounds by the urban public. Thus a tamed nature replaced the perceived wilderness of the original areas. Although artistically shaped and located within the urban environment, this nature of the park mimicked the pastoral scenery of the countryside.

Around 1900, the idea of the garden city received much attention in Germany. In a programmatic book originally published in 1902, Ebenezer Howard had envisaged the creation of small towns that were to be smoothly integrated into the countryside and passed through by greenbelts, thereby combining the benefits of the city with the amenities of the countryside. (See Howard 1970.) The concept inspired the foundation of a number of new towns in England (e.g., Letchworth in 1903). In Germany, a Garden City Association was founded in 1902.[3] First closely associated with aspirations of socialist and anarchist lifestyle reform, it soon developed into a pragmatic stakeholder group that advocated land-reform and community building. Hellerau, Germany's first garden city, was built in 1909. During the 1920s, the term *garden city* was broadly applied to all kinds of suburban development projects even if they were only loosely connected with the original concept (Hartmann 1976). Also, landmark modernist public housing projects in Berlin such as the Hufeisensiedlung Britz and the Siedlung Onkel Toms Hütte (both constructed in the 1920s) were inspired by the garden city movement.

A notable shift in the development of Berlin's green-space planning regime around 1900 was the increasing emphasis on the socio-political or "sanitary" role of public greeneries. Garden architects and urban planners increasingly attacked the actual design of the people's gardens as inadequate for the real needs of the urban dweller, and advocated the Volkspark (people's park) as an alternative model. Such parks were not primarily to be seen as aesthetic objects or air-refreshing lungs, but as recreation grounds that would strengthen the physical health of the citizen. The first park project in Berlin that followed this new paradigm was the Schillerpark in the working-class district of Wedding, created between 1909 and 1913

(Wimmer 1992: 47; Henneke 2011). These new parks were meant to provide space for diverse recreation activities such as walking, sports, and children's play, and to appeal to the working class. Although these parks included landscaped scenery and ornamental beds, aesthetic criteria only played a secondary role. As the garden architect Friedrich Bauer (designer of the Schillerpark) put it, it did no harm if, when "the park in its 'use areas' did not provide such a neat impression," it "only achieved its primary goal of benefiting health and life sustaining needs" (quoted in Wimmer 1992: 49). In addition to paths for walking and banks for rest, a Volkspark typically also had special areas for physical exercise, such as a large open lawn, a children's playground, or a swimming pool.

Similar social and political concerns were also at the core of the open-space theory (Freiflächentheorie) that the Berlin architect and planner Martin Wagner expounded in his pathbreaking doctoral thesis (1915). Wagner worked for the newly created Zweckverband and was sympathic to the Social Democratic Party (SPD), which he later joined. In line with the Volkspark movement, he considered the function of urban greeneries mainly in terms of their aptness for the "physical appropriation" through such activities as exercising and walking in fresh air. Wagner thereby advanced the notion of open spaces (Freiflächen), a newly emerging planning category that also included public children's playgrounds, sporting areas, and urban forests. Wagner's concept reflected a preoccupation with the health-promoting function of physical exercise that was widespread in social reform and public health discussions in Germany at the time.[4] He also used physiological theories and statistics showing a negative outcome of military recruitment examinations in cities to argue that physical exercise was crucial to the healthy development of urban young people. On these grounds he criticized "fenced decorative gardens" (ibid.: 10) as not of much actual use to the residents. The open-space theory that Wagner developed in his thesis was an attempt to rationalize urban green planning by quantifying the basic requirement of recreation space in a city.

The surrounding forests aroused much public concern. With the increasing role of public transport, the forests had become significant sites of weekend recreation for Berliners. At the same time, however, the expanding sprawl of the suburbs continuously diminished their size. The Prussian state, which owned the forests and which accrued considerable fiscal benefits from selling the areas to private real-estate developers, had little interest in their preservation. Eventually, in 1915, the newly founded Zweckverband Gross-Berlin purchased a major chunk of these forests from the Prussian state. Within the contract, the Zweckverband committed itself to keeping these areas permanently as forests to serve Berliners' recreational needs (Stürmer 1991).

After World War I, the provision of adequate open spaces remained a central issue in urban planning policy (Stürmer 1991). The realization of individual projects, such as parks, fell into the jurisdiction of the districts, which could thus put their own stamp on the actual development of open spaces. The formation of the unified municipality in 1920 had created a policy infrastructure that also began to launch projects at the comprehensive scale of the Berlin agglomeration. When the focus of green policy shifted from a more aesthetic orientation to the promotion of recreation planning, this was, to a large extent, also a result of the role that the SPD played at all political levels of the new Berlin. The idea of open-space planning, as formulated by Wagner, fit very well with the reform-oriented agenda of the Weimar SPD. In 1926, Wagner was even appointed as the Senate's Baudirektor (Director of Construction), a position from which he influenced the content of urban planning in Berlin. However, public green in Berlin also received support from and was designed by conservative garden architects (Henneke 2011).

After the creation of the municipality of Greater Berlin in 1920, the city took various initiatives to address the lack of open space. In March of 1922, the municipal parliament decided on a plan for seven new green spaces. The realization of ambitious park projects was hampered by massive financial restrictions in the aftermath of the war and by inflation and required private funds. Construction work was done by unemployed people, by students, and by sporting clubs (Stürmer 1991: 134). A considerable range of new parks were created, all of them designed according to the Volkspark idea. Other important park projects followed in the later Weimar period, notably a long greenbelt along a filled canal in the working-class district of Kreuzberg.

The Weimar period also saw a rapid increase in allotment gardening. Allotment gardens, which had originated in the nineteenth century, were used, mainly by members of the working class, for the growing of vegetables and for recreation. These spaces were rented by collective allotment associations, which sublet smaller plots to its members. Although the individual allotments were fenced private plots, they formed larger colonies which were traversed by trails that were also accessible as promenading grounds for the public. According to Stürmer (1991: 68), the acreage of allotment gardens increased from 5,647 to 6,239 between 1921 and 1925.

Another important step toward the provision of urban greeneries was the Construction Order of 1925, which limited the density of development in residential areas (Stürmer 1991: 87) and earmarked portions of the territory as open space. In 1929, the city also commissioned the development of a Generalfreiflächenplan (General Open Space Plan) that provided for different types of green spaces. Three large greenbelts would connect the outer forests with the parks in the center. The realization of the plan, however, was hampered by financial restrictions and political quarrels during the economic crisis (ibid.: 228–230).

The Weimar years were pivotal for the formation and consolidation of a green planning regime in Berlin. Although urban gardening also remained on the administrative agenda during the National Socialist period, this did not lead to a significant extension of green spaces. As we will see later, however, National Socialism was the context in which the concepts of organic urbanism and landscape care that materialized after World War II in two other urban nature regimes evolved.

Protecting the Homeland's Nature: Classical Nature Conservation

Beginning around 1900, Germany witnessed the emergence of nature conservation as a distinct nationwide nature regime (Schmoll 2004; Lekan 2004; Wettengel 1993). Classical nature conservation aimed at the preservation of extraordinary landscapes, including their flora and fauna, which were typically located in the countryside. This, however, brought its advocates into opposition to city-extension projects, notably when they touched on valuable pieces of the surrounding landscape. In Berlin this resulted in the creation of various protected areas in the close vicinity or even within the developed area.

These policies largely followed the concept of Naturdenkmalpflege (care for natural monuments) as it had been formulated in 1904 by the Danzig botanist Hugo Conventz. Conventz had bemoaned the detrimental effects of industrialization and urban development on the flora, the fauna, and the landscape. On these grounds he had proposed that exemplary pieces of nature throughout the country should be set apart and kept in their original state. These natural monuments, as Conventz had called them, were meant to be selected according to their special scientific or cultural-historical significance. The category of natural monuments had been taken in a broad sweep to encompass single sites, objects (e.g., trees, erratic blocks) or even extraordinary plant or animal species. For the protection of larger objects or ensembles of natural monuments, Conventz also proposed the designation of more extended nature reserves.

In the time of the Kaiserreich, the state had embraced the care of natural monuments as an official policy goal. At both the national level and the provincial level, expert committees were created and charged with inventorying natural monuments and proposing preservation measures to the authorities.[5] These committees, which had scientists, amateur naturalists, and administration officials among their volunteer members, operated in close connection with local nature-conservation associations and individual naturalists, who provided information about potential objects of preservation, the dangers to such objects, and the success or failure of protective measures (Wettengel 1993). The committees also did practical conservation work and organized excursions and lectures to popularize the nature-conservation agenda. The actual designation of natural monuments fell to state authorities. In 1935, under National Socialism, the first German Nature Conservation Act was enacted. The legal meaning of the term *natural monument* was now restricted to smaller single objects, such as extraordinary trees; larger protected areas now were called *nature reserves.*

The concept of natural monuments was not uncontested among the promoters of nature conservation. Opposition came, notably, from the advocates of what in imperial Germany was called Heimatschutz (homeland protection), who represented a second wing of the emerging nature-conservation movement (Rollins 1997; Lekan 2004; Schmoll 2004). Dominated by artists and humanists, they called for the preservation of the aesthetic integrity of national and regional landscapes. Their understanding of Heimat was broader than the idea of a natural monument. Besides flora and fauna, it included landscape scenery, traditional village patterns, and folklore. Champions of Heimatschutz such as the Berlin music professor Ernst Rudorff and the poet Hermann Löns criticized the focus on natural monuments as too narrow and called for a comprehensive policy of landscape protection. The implementation of such a comprehensive approach of landscape protection remained rather limited.[6]

Nature conservation had a relatively strong lobby in Berlin and in the adjacent region of Brandenburg (Auster 2006; Stürmer 1991). In the nineteenth century, semi-academic naturalist associations that studied the regional flora and fauna had emerged; those associations subsequently became an important lobbying group for conservation. In addition, national or local associations that focused more directly on the care of natural monuments, on homeland protection, or on related goals of outdoor recreation were based in Berlin, or had local chapters there.[7] In 1922, the Volksbund Naturschutz (People's Alliance for Nature Conservation) was founded; until the 1970s, it would be Berlin's main forum for nature-conservation activism. Sponsored by major figures in the natural monuments administration,[8] it was devoted to a combination of natural-monument protection and homeland protection. During the Weimar period, the urban working class formed a more socialist-oriented wing of conservationism, which, until the rise of National Socialism, added further political weight to the nature-conservation lobby.[9] Nature conservation around Berlin also benefited from the existence of a university with a wide spectrum of botanical, zoological, and geological research. In 1926, the Naturschutz-ring Berlin-Brandenburg (Berlin-Brandenburg Nature Protection Trust) was founded; by 1929 it comprised about "30 associations and agencies" (Klose and Hilzheimer 1929).

The lobbying of conservation activists and local nature-conservation committees resulted in various measures to protect nature within or in the "close vicinity" of the city. In the years 1911–1932, the Berlin authorities designated sixteen nature reserves (Stürmer 1991: 77; Becker 1932.) The legal basis of these decisions were local police ordinances based on Prussian police law.[10]

Sentiments critical toward the city, industry, and technology, and also toward general features of modern life, such as uniformity or hectic time rhythms, abound in the discourse of early conservationism. Moreover, as Lekan (2004) has shown, nature conservation was inextricably entangled with matters of national identity and memory. In the wake of the romantic ideas of writers such as Wilhelm Heinrich Riehl, conservationists assumed an intimate relation between a nation's identity and its attachment to its vernacular physical landscape. Preserving nature, therefore, was as much about sustaining national or regional identity as it was about nature itself. Both Lekan (2004) and Schmoll (2004) have emphasized the similarities between Conventz's aspirations and the strivings for preservation of cultural heritage that accompanied the emergence of the German nation-state. Like museums and other cultural monuments, natural monuments were meant to preserve items of the past, and thereby to sustain the identity of the nation. This aspect is also evident in the use of the buzzword Heimat-natur (meaning nature of the homeland), around which much of the Berlin nature-conservation discourse revolved.

Nature-protection claims in the Berlin area connected these general considerations with another line of argumentation. In the same way as green planners had promoted the creation of parks, conservationists cherished nature as an antidote to the negative aspects of urban life. Accordingly, proximity to the city made the pieces of original landscape particularly valuable. These spaces allowed the urban dwellers to, at least periodically, experience the nature of their homeland in a most authentic form. Thus, during the conflicts about city extensions around the Grunewald, conservationists referred to the forest's "suitable location to the metropolis" (Potentié 1912: 3). In 1932 the city's commissioner for the care of natural monuments, Max Hilzheimer, argued that city dwellers could not afford to travel to escape the city's "desert of stone" and therefore depended on the "secret beauty" that was provided by "the surroundings of our city Berlin" (Hilzheimer in Becker 1932: 3).

Notwithstanding obvious parallels, there are also significant differences between the discourses of the two regimes. First, nature conservationists suggested that contact with "homeland nature" provided a more authentic experience than the artificially shaped nature of parks and greeneries. As Hilzheimer stated, urban greeneries lacked "the image of the free nature, where one can indulge at will, and in which trees and bushes grow without having been directed and put in place by human hand, where sources, creeks and rivers search their way according to their own laws" (1929: 7). Second, conservationists promoted other norms and values for using these spaces. They considered contact with nature a deep experience that allowed the visitors to feel the ties to their homeland from which they were deprived in the artificial environment of the city. Moreover, as the zoologist Friedrich Dahl stated in his guide to the animal life in the Grunewald, "the observer must have learned to see that what is interesting" (Dahl 1902: 1; see also Hilzheimer 1929: 7, 9). Besides claiming space for nature protection in the vicinity of the city, conservationists published guides and organized guided tours in which they popularized their specific way of perceiving and using these areas. Public lectures (increasingly illustrated with photographic images of extraordinary features), the regular expositions of the Berlin Museum for Natural History, and a great "nature-conservation exhibit" organized in 1931 were also used to create public awareness of homeland nature and its endangerment.[11] The target character of such campaigns was the Naturfreund (friend of nature), a category of person that pervaded conservationists' literature. A Naturfreund was a quiet visitor (usually a man, although women also were involved in such activities) who, alone or in company of a few fellows, quietly contemplated the beauty of nature. A Naturfreund would behave carefully and would refrain from any harmful interference, such as picking rare plant species.

As Regine Auster (2006) has argued, nature conservation in Berlin reflected the socio-political concerns of open-space planning and even had a significant left-leaning working-class wing. It is striking, however, that most of the politically influential conservationists promoted a rather exclusive, even elitist, model of experiencing nature. The Berlin variety of nature conservation reflected ideals of the educated middle class that conservationists and naturalists cultivated in their various associations and collective networks. In a rambling guide for the "friend of nature in Berlin," the paleo-botanist and geologist Robert Potonié (1922) explicitly distinguished his readership from the "ordinary tripper" (Ausflügler). In his description of one Berlin bog, Potonié noted with satisfaction that these trippers whom he supposed to be ignorant of the true character of that site (who presumably were of lower-class background) were kept out by a fence. He invited his reader to crawl under that fence and to carefully traverse the bog in order to contemplate the ferns, shrubs, and trees—some of them rare species—at the site. It was, he commented, "for us, those who want to get in closer contact with nature, rather than the ordinary tripper, that such sites are being preserved" (ibid.: 50).

In the 1920s, in line with Conventz's idea of natural monuments, conservationists' activities focused on sites that they considered particularly representative of the glacier-formed landscape of the Brandenburg region: fens, creeks, ponds, and a few sand dunes scattered around the metropolis. With urban development, many of these sites had been drained, had been built over, or had deteriorated as a result of indirect effects of intensified land use in their vicinity. Consequently, the bogs in the Grunewald were the first sites in the Berlin area to be designated nature reserves.[12] They were regarded as the most authentic pieces of the original natural landscape, even if they were located close to a big metropolis. Potonié (1922: 50) described them as the few places in the Berlin area where the "original nature of our homeland" (Urnatur unserer Heimat) was still visible.

The link with open-space planning was most apparent in conservationists' engagement in the public campaign for protection of the Grunewald that led to the purchase of the forest by the Zweckverband. From 1904 on, public protests mounted against projected city extensions and clearing measures and against the effect of urban water use on the forest's groundwater level (Wilson 2006). In popular publications and lectures, local biologists underscored the natural specificities of the Grunewald (Wahnschaffe, Graebner, and Hanstein 1907, 1912). In addition, they emphasized its value as a recreation area. Although their proposal to turn the whole Grunewald into a protected natural monument (see Potentié in Wahnschaffe, Graebner, and Hanstein 1912: 8) was never realized, the 1915 contract concerning the "permanent forest" prevented further encroachment into the area.

Remains of the traditional agricultural landscapes, cultivated forests, and even old gardens and parks were also turned into natural monuments if they displayed interesting features or were considered to be of special cultural significance. One example is the Pfaueninsel (Peacock Island), a small island in the Havel River near the Grunewald. Even single trees were designated as natural monuments if they were of an extraordinary size or shape or if they were tied to historical events or vernacular traditions (Schmoll 2004). By 1930, four trees in Berlin, all of them in suburbanized former suburbs, had received legal protection (Becker 1932: 65–71).

Birds were another focus of nature conservation (Schmoll 2004: 237–386). Nature conservationists cooperated with garden administrations to promote bird life in the city by providing nesting boxes, bird baths, trees, and bushes that would attract birds (Harrich 1931). The possibility of designating special bird reserves at the fringe of the city was discussed (Hilzheimer 1931). As Hilzheimer put it, however, birds in the city did not live under "natural conditions," so that their protection could only be justified by "aesthetic and mental [gemütlichen] reasons" or because they exterminated some harmful insects (ibid.: 262). In an interesting contrast to later urban ecology, Hilzheimer therefore denied that protecting bird life in the city had any potential relevance to scientific research.

After 1933, the German nature-conservation movement was largely supportive of the new National Socialist regime. For example, the Volksbund Naturschutz expelled Jewish members, and its representatives cooperated with the new government (Auster 2006: 165; Schütze 2005).[13] Indeed, the conservative and nationalist framings of nature as a basis of homeland identity seemed to match well with the "blood and soil" ideology of the National Socialists. Nazi ideologists reframed nature as a basis for the racial-biological strengthening of the people and considered lack of appreciation of nature a sign for the allegedly inferior state of other races (Lekan 2004: 153–203). Nature was thereby explicitly "racialized"—as Lekan put it—as the "living space" of a German Volksgemeinschaft (people's community).

As environmental historians of the period have shown, practical nature conservation was hardly able to benefit from these ideological leanings (Radkau and Uekötter 2003; Brüggemeier, Cioc, and Zeller 2005; Uekötter 2006). Under National Socialism, actual land-use policy was dominated by industrial and infrastructural projects that were extremely exploitative of nature. Moreover, the beginning of the war in 1939 implied a radical limitation of nature-conservation activities. However, the combined efforts of natural-monument protection and homeland protection achieved one important success in that period. In 1935, thanks to a combination of lobbying and personal support in the National Socialist government, Germany's first national nature-conservation act (Reichsnaturschutzgesetz) was issued. It extended the power of state authorities to designate nature reserves, natural monuments, and (a further protective category) landscape reserves. After the war, local authorities made considerable use of these legislative powers.

Reconstruction Planning after 1945: Organic Urbanism

After 1945, the discourse and politics of urban nature entered a new phase. The war had led to the destruction of vast parts of the former German capital. A contemporary survey estimated that about 28.5 square kilometers of the built-up area had been damaged in bombing raids or in the street fighting of the last days of the war (Fichtner 1977). Moreover, in order to compensate the shortages of fuel and food during and in the aftermath of the war, parks had been cleared and been used for growing vegetables.

In the ensuing debate among planning experts and the public on the reconstruction of the city, issues of "nature" and "green" were of central significance. The majority of planners viewed the destruction of the city as a unique and unexpected chance to get rid of the Mietskasernen and to realize a more openly structured Berlin. The architect Hans Scharoun (as cited in Bodenschatz 1987: 63) even spoke of a "mechanical loosening" (mechanische Auflockerung) that had paved the way for the creation of a "city landscape" (Stadtlandschaft). His colleague Max Taut (1946) hoped that a "new Berlin" would give more systematic attention to the city's "especially beautiful location in the landscape" and would integrate "water, forest and open space."[14] The reinterpretation of the destruction of the city into a "fortunate circumstance" (Allinger 1956: 6) became a stereotypical narrative and was repeated by many players in West Berlin's planning debate until the late 1950s.

City landscape and *loosening* were not merely buzzwords of the Berlin debate. As Durth and Gutschow (1988) have shown, those principles could already be found in the first plans for the reconstruction of bombed cities that were developed under the National Socialist regime. After the war, they represented the programmatic planning ideal that dominated postwar urbanism all over Germany. As Hans Bernhard Reichow put it in a trend-setting textbook (1948: 32), turning the city into a city landscape would make it possible to "combine the advantages of the big metropolis with those of a simple, close-to-nature countryside life." In the same vein as earlier green-space planners and nature conservationists, Reichow assumed that the landscaped city would allow residents to experience regular contact with nature. This would contribute to the "psychological strengthening" of the urban residents, and thereby prevent them from getting "uprooted" (ibid.: 28). For Reichow, the ideal city consisted of a hierarchical articulation of nested units of different size: cells, neighborhoods, districts (ibid.: 69, 70; see also Göderitz, Roland, and Hoffmann 1957: 24–25) in which an equally organic community life could thrive. The conservative garden designer Gustav Allinger (1950: 154) expressed hope that a city landscape would tie "the 'new city' tighter to its residents" and would turn it into "a living organism that is connected to nature."

The programmatic concept of the city landscape had many parallels with the old garden city ideal and with the modernist urbanism of the Athens Charter (signed in 1933 by the Congrès Internationaux d'Architecture Moderne), which also proposed vast open spaces. What Reichow and his contemporaries envisioned as a city landscape, however, did not have to do only only with the size and number of green spaces in the city. What distinguished this newly emerging regime from parallel undertakings was the preoccupation with the organic morphology of the future city, and its conceptualization as a means of engineering organic urban communities. Reichow portrayed the classical city as a mechanical entity that alienated its residents from nature. Against this he contrasted his ideal of an "organic urbanism."[15] That term pointed mainly to the morphology of the city as Reichow saw it from the bird's-eye perspective of the planner's table. Landscape features such as landform configuration, water, or vegetation were meant to guide the shape of urban form and to be integrated into a holistic city organism. Moreover, new green spaces were to be planted, to the greatest possible extent, with original vegetation. Johannes Göderitz, Rainer Roland, and Hubert Hoffmann (1957: 8) characterized their ideal as an "articulated and loosened city" (gegliederte und aufgelockerte Stadt). The units of the city were described as "limbs," "organs," or "cells" (ibid.: 25–27) that together would form the organism of the city. This organic approach was also evident in these authors' graphic representations of of their imagined ideal city. Streets were depicted in round forms, with arteries branching outwards (as in a leaf or an animal's body), and neighborhood units in organ-like shapes.

The Kollektivplan—the first master plan for the reconstruction of Berlin—was explicitly based on city-landscape principles. Designed in 1945 and 1946 by a planning team directed by Hans Scharoun, it envisioned a radically restructured city of dispersed modernist buildings, clustered into homogeneous neighborhoods, with large highways and vast green spaces.

Though the plan completely ignored the existing historical development, it fit the city neatly into the natural coordinates that were given by the Spree River and the geomorphology of the river's basin (Werner 1969; Bodenschatz 1987: 135–149).

The division of the city made the realization of a comprehensive master plan impossible. Moreover, as a schematic blueprint that projected an urbanistic ideal on the city, the plan had ignored many of the actual constraints that the reconstruction process was faced with. In fact, because water, electrical, and sewer lines had survived the war (Bodenschatz 1987: 136), reconstruction operated in a rather pragmatic and piecemeal manner, with the clearing of rubble and the rebuilding or renovation of existing properties. The 1947–48 Bonatz Plan and the 1950 Land-Use Plan (Flächen-nutzungsplan) of the West Berlin Senate refrained from making radical changes in the city's layout. The idea of the city landscape boiled down to a modest attempt at loosening the building development and a modernist separation of functions. Only the reconstruction of the destroyed Hansa-viertel for the 1957 Interbau construction exposition resulted in a coherent architectural ensemble that corresponded to the ideas of the city landscape (Bodenschatz 1987: 165–170; Wagner-Conzelmann 2007).

Although it did not follow the formal blueprint of a comprehensive city landscape, the shaping of green space was one of the priorities of urban reconstruction.[16] In March of 1949, after the division of the city, the West Berlin municipality launched its Green Emergency Program (Grünes Not-standsprogramm) for reconstructing green spaces and playing fields. The program was financed with the aid of Marshall Plan money (Migge 1953: 46). As in the early days of the Weimar Republic, most of the labor was carried out by unemployed people. Between 1950 and 1952, about 1,800 acres of green space were renovated in West Berlin. Planners were also concerned about the large amount of unhealthy dust that was produced by the ongoing work of cleaning up the city and the constant transporting of rubble. They expected that greeneries would filter dust particles out of the air around work sites (Fischer 1955). Others saw an immediate need to create attractive open spaces in order to lure children away from the ruins, which had become popular places to play (Witte 1952).

The ceremonial planting of a lime tree in the devastated Tiergarten by Mayor Ernst Reuter, which marked the start of the program, indicates the high symbolic value that was given to green space in the postwar years. Greening was seen as a way of restoring a sense of order in the damaged city. Moreover, with the intensification of the Cold War (notably the Russian blockade of the city during winter of 1948–49), it was also a demonstration of the political assertiveness of the West against the East (Witte 1952: 4). These motives were expressed in the motto Hilfe durch Grün (Help through Green) (Witte 1952), originally the title of an exhibition in the West German city of Essen and an urban gardening magazine. Whereas ruins were burdened with undesired memories of death and destruction, the growth of greenery was associated with the beginning of better times. Hans Bernhard Reichow, who had been involved in the reconstruction of parklands and rubble areas in Kiel, considered the construction of naturally shaped landscapes from the rubble of cities a means of "raising oneself above the psychosis of rubble [Trümmerpsychose] of narrow-minded people" (Reichow 1948). In this sense, greening resonated perfectly with the politics of memory of a society striving to separate itself from its Nazi past and preparing for economic growth.

Like their predecessors in the late nineteenth century, West Berlin's green planners saw greeneries as a remedy for the negative aspects of modern city life—as places where the "rushed city man" might find "health, comfort, and pleasure" (Rieck 1954: 102–103). In the 1950s, anticipating rapid automobilization, green planners advocated greeneries to reduce traffic noise (Dittmann 1958) and air pollution (Spindler 1958). All this was seen as particularly pressing in Berlin, where the loss of the countryside had deprived citizens of access to the surrounding countryside. After the blockade, the Cold War remained an important justification for green planning. In 1955 one observer used the phrase "Grosse grüne Freiheit Berlin" ("great green liberty Berlin") (anonymous 1955). In 1958, one planning official even claimed that green areas in West Berlin would strengthen "the power of resistance of this population at an advanced outpost" (Lohrer 1958: 122).

Although smaller in scale, the greening efforts shared with earlier city-landscape plans a preoccupation with organic landscaping. Just after the war, the botanist Kurt Hueck compiled a list of trees, bushes, and plants that would be the natural vegetation units of typical landscape forms in the Berlin area; that list was meant to serve as a template for the reconstruction of the city's greeneries (Hueck, no year). Hueck's map of the "original vegetation" in Berlin was included as a planning tool in the 1960 Berlin Planungsatlas (Akademie für Raumforschung und Landesplanung 1960: map 40). Willy Alverdes' reconstruction plans for the Tiergarten, formulated in late 1940s, were not intended simply to restore the prewar situation; they were intended to restore features of the alder swamp and the riparian forest that once had covered the area in which the park was located.[17]

Ideas of organic landscaping also influenced what was done with Berlin's rubble mounds (a major concern of Berlin's green planners in the 1950s and the early 1960s). Although a large amount of the rubble produced by the clearing process was reused for construction work, there remained a lot of material to be disposed of within the city and in its surroundings (Fichtner 1977). After plans to pour a layer of rubble over the Tiergarten were abandoned, several centralized dumping grounds were constructed, mostly on existing parklands and forests. In the ensuing years, those rubble mounds became the sites of ambitious cultivation experiments. Shaped as natural-looking hills and planted with dense vegetation, they were meant to enrich the landscape of the surrounding parks or forests.

In the 1950s, the city-landscape concept remained an important point of reference for green planners and architects who promoted the further expansion of green space throughout Berlin, notably by creating green-belts. In 1955, one observer welcomed the successful renovation of green spaces that had been completed so far but suggested that in the next phase priority be given to the "necessary dispersion of the inner city" (anonymous 1955: 28). The same observer envisioned a Berlin "that represents an organism connected to nature" (ibid.: 28). The land-care expert Kurt Hentzen relaunched a green-space proposal that was based on the idea of the urban landscape (Hentzen 1950, 1955). Hentzen had already developed this plan in 1936, under National Socialism, as a contribution to the debate about the renewal of the capital. Like Scharoun and Taut, Hentzen saw the destruction of the city as an opportunity to revise the mistakes of classical urbanism and to shape the city according to the landscape. He proposed linking the city's parks to a system of green spaces that would follow the lines given by rivers and fit into the geomorphologic outline of the Spree basin.

Green planning during the reconstruction period attempted to integrate parks with other green spaces such as playing fields, allotment gardens, and children's playgrounds. This led to the creation of green promenades along the city's natural waters and canals (Witte 1960: 218). In 1960, Berlin's director of green planning, Fritz Witte, urged that more attention be given to the value of the "connections and circumstances of the natural" (ibid.: 216). The creation of greenbelts was the main pillar of the Haupt-grünflachenplan 1960 (Principal Greenspace Plan 1960), which the West Berlin Senate developed under Witte's guidance. It provided for two orders of greenbelts. The first order were greenbelts, about 3 or 4 kilometers apart, that would structure the "amorphous development of the inner city area into areas . . . graspable for humans" (Witte 1960: 281). The areas were to be further divided by smaller green connections of a second order (ibid.: 281).

Existing development structures and private ownership of the respective areas stymied the realization of a coherent open-space system. With its Flächennutzungplan 1965 (Land-Use Plan 1965), officially endorsed by the Abgeordnetenhaus in 1970, the Senate officially renounced the plan of creating a coherent greenbelt system. Further greenbelts were only realized as part of the major city-extension projects of the 1960s (Falkenhagener Feld, the Markische Viertel, and the Britz-Buckow Rudow settlement, later called Gropiusstadt) (Schindler 1972: 484). For the Falkenhagener Feld, in Spandau, the garden architect Hermann Mattern designed a belt of public greeneries and allotment gardens that followed the valley of the Spekte, a small creek that stretched through the settlement area (ibid.). Whereas ideas of the city landscape still influenced such smaller-scale projects, the development of the general shape of the city was determined by other goals.

Recreation Space for a Walled City: Landscape Care

Although West Berlin was a profoundly urbanized enclave, considerable chunks of forests, agricultural areas, and other pieces of open landscape remained within its confines. Conservationists and planners underlined that they had become particularly valuable with the loss of the surrounding countryside. They did not see them only as helping to remedy the scarcity of green areas in the inner city. As the few parts of the natural landscape that visibly connected West Berlin with the surrounding countryside of Brandenburg, they were also of symbolic value for the identity of the walled city.

Thus, nature conservation was increasingly complemented by a concern for Landschaftspflege (landscape care) or Landespflege (land care), thereby evolving into a regime with a broader focus of concern. In 1949 the Berlin Senate had reappointed a voluntary Landesbeauftrager (land appointee) for nature conservation. (The position had been vacant during the war.) In 1955 it established the Naturschutzstelle (Nature Conservation Board), a successor to the Provincial Commission.[18] In 1956 the Naturschutzstelle became renamed the Landesstelle für Naturschutz- und Landschaftspflege (Land Board for Nature Conservation and Landscape Care). Under the guidance of Otto Ketelhut (a biologist by training), who was nominated full-time Landesbeauftrager in 1955, the Landesstelle developed the major outlines for nature and landscape protection in West Berlin.[19]

The term Landschaftspflege (meaning landscape care) had been coined in the 1930s and referred to a range of regulatory and technical practices that were meant to protect or maintain the shape of the rural landscape, mainly by integrating technical facilities into harmonic visual sceneries (Lekan 2004; Chaney 2008: 37; Runge 1998: 129–30). Under National Socialism, landscape care was supposed to complement industrial modernization and thereby to cater to a supposedly inherent German longing for a well-maintained organic land space (Blackbourn 2006: 251–310). One prominent example was the highway planning done under National Socialism to integrate highways with the surrounding landscape features (Zeller 2007). Another was the plan to reshape German-occupied territories in Eastern Europe to make them look like German landscapes (Blackbourn 2006). After the demise of National Socialism, proponents of landscape care kept their distance from the language of racial hygiene with which they had legitimized those earlier efforts, although conservative ideas about healthy landscapes and their positive effect on the mental and physical health of the people continued to inform landscape-care policies. As Chaney (2008: 114–147) notes, they also disentangled the term "living space" from its former geopolitical connotations and gave it a more domestic meaning focusing on the ecological health of the landscape—a meaning that also resonated with the later use of the term in urban species protection. Like organic urbanism, landscape care cherished the creation of organic visual sceneries. In addition to the organic morphology of the landscape, it also promoted the protection of ecological cycles and balances. In comparison with nature protection, with which it became institutionally aligned in the 1950s, landscape care justified its goals more directly, and in a less socially exclusive way, by appealing to the need to increase the amount of space available for recreation.

In line with such concerns, West Berlin protected more space. In the late 1950s, it possessed about 11 nature reserves and 30 landscape reserves (Ketelhut 1958: 63). Besides these reserves, about 633 smaller points of preservationist interest were protected as natural monuments (Ketelhut 1958: 61). The large majority of these were trees, but small fens and ponds, and even a colony of bees also fell under that category. More in line with the idea of landscape care was the increasing designation of landscape reserves. This category had been introduced with the Nature Conservation Act of 1935. It allowed authorities to put larger chunks of space under formal protection even if they did not fulfill the strict requirements for nature reserves. Preservationists' goals were catered for in these areas within the limits set by existing forms of land use, such as agriculture, forestry, or recreational activities. In the late 1950s, about 5,494 acres enjoyed the status of landscape reserves (ibid.).

The members of the Landesstelle also advised on the day-to-day decisions made by the Berlin administration that were likely to interfere with valuable aspects of nature and landscape. The Landesstelle made appeals against illegal construction activities, building permits, or other administrative acts that they deemed detrimental to nature and landscape. In line with the organic ideal of landscape care, it thereby paid ample attention to the scenery (Landschaftsbild) and the recreational value of the land. In the case of one suggested landscape reserve, it valued the "perspectives towards the appealing landscape of the Havel"[20] that made it a particularly valuable element of Berlin's "recreation green." Conversely, its members feared that "the landscape scenery would be significantly changed" if a project for a sand pit in another forest area went through. The Landesstelle also made proposals for the painting of technical items such as public conveniences, rescue centers,[21] or telephone switching units in order to fit them better into the surrounding landscape. The visual norms that guided these appeals and policies were based on an idealized image of landscape of the pre-industrial period. Whereas items that were connected to certain cultural traditions, such as castles or traces of agriculture, were seen as positive features, new building projects and industrial or traffic facilities were denounced as disfiguring the landscape.

Advocates of landscape care also called for the protection of areas that could serve as connecting pieces to other green spaces and thereby linked their own strivings to the contemporary goal of green planners to create integrated greenbelts.[22] Also in institutional terms, the aspirations of landscape protection of the 1950s converged partly with general attempts of green-space planning in Berlin. Thus, the General Greenspace Plan 1960 included all landscape reserves and nature reserves in its provisions.

Although landscape care legitimized its protective efforts with the recreation value of the respective landscape, there were many tensions between these two goals. On weekends many "trippers" flocked into the most popular parts of the green periphery and, according to the Landesstelle, caused many problems for the scenery and ecological balance of the landscape. The Landesstelle was particularly concerned about bathing and camping on the shores of the Havel[23] and parking in adjacent areas.[24] Similar problems were also discussed by landscape-care stakeholders in other parts of Germany, but in Berlin the problem was considered particularly dramatic because the enclosure of the city had resulted in a much higher spatial concentration of recreation activities.[25]

In his regular reports, Otto Ketelhut painted a positive picture of the achievements of nature protection and landscape care in West Berlin (Ketelhut 1958, 1961, 1957). This, however, should not distract from the contested status of these policies. Nature-conservation goals were mostly successful where they did not clash with powerful counter-interests or where they met the goals of other constituencies, such as by preserving recreation areas. For example, the Berlin waterworks, which had an interest in the unrestricted exploitation of groundwater in the area of the Grunewald, had initially opposed the attempts of the Landesstelle to designate the Grunewald as a landscape reserve.[26] Although the Nature Conservation Act had quite clear criteria for when a piece of nature had to be preserved, many protection proposals of the preservationists failed or were only realized after long political negotiations. In 1957, the local conservation activist Elisabeth Mecklenburg complained about the increasing pressure on the development of the land in West Berlin and about continual reduction of the size of the existing forest (Mbg 1957). In 1951, the Berlin botanist Ulrich Berger-Landefeld had called for protection of the Baumberge, a sand-dune area in the north of the city that was noteworthy both as an unusual landscape formation and as a sanctuary for rare species.[27] That initiative, however, clashed with the interests of the French military, which used the area as a training ground.[28]

Undoing Urban Nature: Infrastructure Planning and Extension Planning after 1965

In the mid 1960s, the ambitious postwar programs of urban greening lost their influence on West Berlin's land-use policy. For more than ten years, urban planning became preoccupied with the growth and territorial expansion of housing areas, industrial areas, and infrastructural technical facilities. This policy turn was markedly enunciated in the Senate's Land Use Plan 1965 (Flächennutzungsplan 1965, henceforth FNP 1965), a comprehensive land-use development scheme first presented in 1965 and, after various revisions, endorsed by the Abgeordnetenhaus in 1970.[29] Although the plan considered the entire city as its reference area, its specific provisions were confined to the West, where it provided for large housing and industrial projects on the outskirts. The plan also provided for new public transport facilities and a huge highway network. Other elements of the plan were the strict separation of urban functions within mono-functional areas and "ribbons" and urban renewal schemes for the inner city.

In these years, West Berlin followed the general trajectory of planning discourse in West Germany. The programmatic paradigm of city landscape or that of the loose and densely developed city became increasingly replaced by the goal of "urbanity by densification" (Krämer 2007). Equally important was the changing definition that local policy makers gave to the particular development needs of the walled city. Now they considered it of primary importance to provide ample building ground that would cater to the "needs of economic policy and social policy."[30] The Senate justified the FNP 1965 with the expectation that industry would need about 75 acres of new building space per year. Moreover, it was expected that the population of the Berlin area would reach about 5 million (without specification of the time period), and that they would need ample living spaces and an effective transport system.[31] Such urban extension would require huge public subsidies. According to Bodenschatz (1987: 186, 171), the dominant actors in the urban expansion and renewal projects were publicly financed housing companies, which cooperated closely with the local construction firms. The Berlin Help Act of 1962, which provided for tax relief on private investments of West German enterprises in construction projects in West Berlin, mobilized further sources of money (ibid., 171).

There was no space left in the program for the realization of the coherent green-belt system as it had been envisaged in the earlier Principal Green Space Plan. As the Senate explained in a draft plan in 1968, the expenses would have exceeded the recreation effect that would have been achieved by such a policy.[32] As a more pragmatic alternative, the FNP 1965 suggested merely merging adjacent green areas and making them more suitable for recreation. This pragmatic approach was also at the core of the 1972 Conception for the Creation resp. Use of Green and Recreation Facilities (Konzeption zur Schaffung bzw. Nutzung von Grün- und Erholungsflächen), in which the Senate advanced its ideas for future green planning. Instead of the creation of new green connections, this plan advocated the intensification or "activation" of existing green spaces by providing new leisure attractions (Szamatolski 1974: 457).

The plans for the development of urban infrastructure, notably the highway system, entailed significant cuts into the existing green space. An intended highway was to cut through the Volkspark at the Hasenheide and much of the greeneries that covered the sealed Luisenstädtischer Kanal in Kreuzberg (ibid.: 457).[33] Also, extension of the airport in Tegel was to require the cutting of trees in an adjacent park. As the Landesstelle maintained, this entailed a serious decline of the recreation quality of this park.[34] The Senate's plans also touched on the expense of the allotment garden areas in the city. Although the FNP augmented the area permanently allocated for garden plots (100 acres), it implied a reduction of the total amount of allotment gardens from about 1900 acres to only 940 (plus 100 acres of newly earmarked allotment space).[35]

The plan also included gigantic high-rise settlements on the outskirts. The Falkenhagener Feld in the West, the Märkische Viertel in the North, and the Berlin-Buckow-Rudow settlement (already under construction) would amount to about 35,000 new apartments (Bodenschatz 1987: 171).[36] Renewal plans for the six inner city areas were aimed at replacing the tenements with modern apartment housing. Large numbers of people were to move to the new housing areas on the outskirts. These housing projects were meant to offer an alternative to the densely built neighborhoods in the center. Although elements of the urban landscape ideal survived in these projects, planners and the public criticized their architectural gigantism (Rating et al. 1970).

In contrast to the urban-planning community, local gardening experts continued to lobby for greenbelts. "Where has the realization of the great idea of an uninterrupted green system gone?" the garden architect Helmut Bournot asked in 1970s. Berlin had a number of good parks, but they were "situated completely unrelated in the middle of the sea of buildings and flooded round by traffic" (Bournot 1970: 482). Norbert Schindler, who succeeded Fritz Witte as the director of the Berlin green space administration, defended the plans to create a green connection along the rivers against the explicit rejection by "promoters of 'urbanization via development densification'" (Schindler 1972: 490). More generally, the supposed "deficit of green," notably in the working-class districts in the inner city, became a focus of continuous debate among experts (Szamatolski 1974; Schmidt 1970; Heitmann and Muhs 1972; Bournot 1970). On the basis of a quantitative survey for the Senate, a working group headed by the Berlin professor of garden architecture Hermann Mattern claimed that there were huge differences in green-space provision between the luxury districts on the outskirts and socially less privileged inner districts such as Kreuzberg and Schöneberg (Heitmann and Muhs 1972: 12–14). Nature-conservation organizations and members of the Landesstelle were particularly concerned about the effect of construction projects on the landscape on the outskirts of the city.[37] They bemoaned the reduction of forest areas and traditional landscapes of agriculture.

Conclusion

Both nature conservation and green planning began as movements in which planners and social reformers attacked the dominant path of urban development, and thereby received broad support among the urban citizenry. In the early twentieth century, their concerns crystallized in four distinct urban nature regimes, each of which manifested itself in specific institutional infrastructures and policies. As a result, many spaces that otherwise might have been developed or might have fallen prey to intensive land use were kept free or were shaped as public greeneries. However, the growth policy of the late 1960s and the 1970s marked an important limitation to these strivings and reversed some of their results.

A prevailing theme in all these regimes was the deprivation of the city and its dwellers of nature. By creating parks and other urban greeneries, by protecting nature at the fringe of the city, or by shaping organic ensembles of buildings and open spaces, planners hoped to compensate for this alienation and the various effects that it was supposed to have on health, on moral attitudes, or the general quality of life. *Nature*, however, was not an unequivocal term in these debates. It was an ambivalent signifier that encoded different imaginations and values and articulated different urban utopias and pragmatic spatial policies. For nature conservationists, independent of their allegiances to monument protection or homeland protection, nature was identified with the countryside. This view had a temporal dimension too, as such spaces were supposed to represent earlier historical periods and therefore had to be defended against the dominant trends of modernity. From this perspective, "bringing nature to the city" was a logical contradiction, since green spaces implanted in the city would never approach the authenticity of non-urban and pre-urban spaces. To qualify as nature, vegetation or a landscape required some sort of intrinsic development that could not be enforced by humans. However, nature conservationists also advocated their own model of urban improvement—a model that entailed the protection of areas on the outskirts and their integration into a model of urban recreation. The regime of green planning relied on a similar identification of nature with the countryside. Parks were deliberately designed to evoke images of landscape sceneries that were in direct opposition to the actual city space in which they were located.

In the 1920s, green-space planning became more tied to socio-political goals that redefined the function of green spaces in terms of hygienic benefits, such as air, light, and recreation facilities. Although earlier visual codes of pristine nature continued to inform the design of parks, nature was now primarily identified with a specific form of physical culture. Adequate green spaces were thus considered valuable not primarily because they imitated the countryside but because they allowed citizens to engage in physical activities that were deemed healthier and more natural than ordinary city life. Thus the needs of the individual human body, as well as the needs of the urban population at large, became a measure of the value of open spaces in the city. Open-space planning, as had been prominently articulated in Berlin by Martin Wagner, thus became a kind of biopolitics—in the sense introduced by Foucault (1997)—of urban space. Biomedical norms of healthy physical exercise (or provision of fresh air) replaced the orientation to authentic nature. As the various struggles over land use document, there were no intrinsic limits to extending green spaces; there were only economic and political limits.

All these motives also reverberated in the utopia of the city landscape that dominated urbanistic imagery in the postwar decades. When experts on urban planning cherished the landform and the original vegetation as a basis for the orientation of urban development, they also embraced a pre-urban and non-urban state of nature as a model. Their attempt to engineer small neighborhood communities was oriented to the village and thus referred to the countryside as a model for the city. At the same time, they invoked the public-health considerations that open-space planners had raised in the early twentieth century. Pristine scenes and spaces that allowed citizens to engage in natural and healthy physical activities were important elements in urban planning. Planners' understanding of "nature" with respect to the city, however, put its main emphasis on the physical form: a city was more natural the closer its morphology came to the organic structure of the human body. It was mainly through this analogy that green spaces were considered to bring "nature" into the city. This could mean extending the city to the countryside in order to integrate it with its given features (as in various new housing estates at the fringe of Berlin) or it could mean transforming the open spaces within the city into greeneries.

Notwithstanding their different understandings, all these regimes framed nature as something like the opposite of the real existing city. Nature existed outside the real spaces of city—either in the spaces of the countryside, in hygienic norms, or in utopist concepts of organic cities. Thus, it had either to be preserved (as in conservationist discourse), to be implanted (as in the health-oriented discourse of open-space planners), or to be interlocked with the city (as in the design of organic landscapes). As we will see in subsequent chapters, these different motives lived on in debates about urban nature conservation in the 1970s and the 1980s.

Ecologists who advocated urban biotope protection sought a reconciliation of city and nature. At the same time, however, the aspirations of earlier green policies merged with, or existed in tension with, sets of meanings that emphasized different aspects of the city. Ecologists developed a notion of the city and nature that was more nuanced than those of their forerunners. Although not free from normative preferences for non-urban nature, they did not simply dismiss urban spaces as non-natural. The strict opposition of nature and the city around which earlier policies revolved was transformed into a more comprehensive view of different types of ecosystems or biotopes, each of which had at least some value or potential as nature. The emergence of this new regime is the focus of the next three chapters.

2 Ecology's Natures

In 1973 Herbert Sukopp published a programmatic article in which he pleaded for the recognition of the metropolis as an object of ecological research. In contrast to the widespread notion among ecologists that cities were always hostile to life, existing evidence on the urban flora and fauna had made it clear, according to Sukopp, that cities did indeed provide space for a considerable number of plant and animal species. Although in cities nature had been almost completely reshaped through intensive human land use, they should be conceived as variegated sets of ecosystems or biotopes that had their own characteristics. Rather than being a purely academic endeavor, Sukopp's program of an ecology of the city had a clear political mission: ecology was supposed to monitor and control the effects of human land use on urban nature and thereby create a basis for more rational planning of future cities.

Sukopp's call for an urban ecology was representative of the attempt of many ecologists at that period to link their expertise to the environmental problematique that concerned the public in industrialized countries (Halffman 2003; Söderqvist 1986; Cramer 1987; Bocking 1997; Kingsland 2005). At the same time, however, it was deeply rooted in a local research tradition that existed in Berlin. Since the nineteenth century, hikers, naturalists, field biologists, and other Naturfreunde had been active in the collection and exchange of knowledge about the local flora and fauna of the Berlin-Brandenburg region. After World War II, these activities focused increasingly on the territory of West Berlin, including the inner city. In the 1970s, studies of urban nature became a central focus of the Institute of Ecology at the Technische Universität Berlin, where Sukopp was a professor. Together and in close interaction with the politico-institutional constellations that I will describe in the subsequent two chapters, this ecological fieldwork was one the crucial pathways through which the urban biotope-protection regime emerged in Berlin.

Whereas many social and historical studies of scientific practice have focused on knowledge production in laboratories, hospitals, or museums, only recently has fieldwork been considered in a similar perspective. Whereas laboratories have been described as standardized ("placeless") spaces that are under the social and material control of scientists, it has been argued that the sites in which fieldworkers operate are much less socially exclusive and hence unruly (Henke 2000; Kohler 2002; Waterton 2002). A related theme concerns the strategies through which fieldworkers transform the particularities of the field into claims that are accountable as universal facts and that achieve the same credibility as data produced in placeless laboratories. Latour (1995) has described this as a cascade of translations through which specimens locally procured in the field become successively transformed into mobile and standardized inscriptions. Also, the historical studies of Robert Kohler (2002), Thomas Gieryn (2006), and Jeremy Vetter (2010) have traced material and discursive processes through which place-bound observations of fieldworkers were transformed into abstract, context-independent facts. Fieldworkers, however, do not always seek to disentangle their data from the context. In many cases they claim authority and credibility precisely from their intimacy with particular contexts, thereby opposing the abstraction ideals of laboratory science (Kohler 2002). As Kohler and Gieryn also have made clear, fieldworkers select spaces according to the specific observation possibilities that they offer (e.g., Chicago for urban studies), and thereby transform them into exemplary models for the generic phenomena that they study. As I will show here, the spaces of West Berlin had a similar function for the emerging urban ecology. Its destruction during World War II and its socio-geographic marginality made West Berlin an ideal model to tackle questions that were of general concern for ecology. In order to meet that goal, Sukopp and his co-workers translated their observations into generic models of the city that abstracted from the specific case on which they were based. How the practice of fieldwork and its knowledge products are implicated in the cultural trajectory of the concrete spaces that are their object has not been studied nearly as much.

In this chapter I will argue that the making of urban ecology not only mobilized Berlin's spaces for the construction of credible accounts that circulated within the orbit of academia; by recursively structuring the perception and to some extent intervening in their material order, they also redefined what Berlin and its spaces actually were. My focus will be on what I call ecology's *circuits of spatial observation*—that is, the reiterative and recursive process of practical engagement with and of production of data about the same space over longer periods of time. Such circuits exist when knowledge emerges from the observation practices in a space and, at the same time, feeds back into the way in which this space is perceived and treated, and how on this basis further observations of this space are carried out. I will describe how such circuits emerged from, and were maintained within, three configurations of fieldwork practices that coexisted in West Berlin: the spotting of species by a diverse network of observers; ecological surveys of exemplary sites that were conducted by individual or collective projects; and large-scale surveys which comprehensively mapped and zoned the city as a set of distinct nature-spatial units.[1] As I will argue, the observation results that fieldworkers produced in these circuits did not only locate biological entities, such as plant or animal species in pre-given space. They also inscribed meanings and attributes in that space, established or redefined spatial distinctions, and thereby carved out new spatial formations of nature. In other words, urban space was not only the passive background on which fieldworkers made their observations; it was also, at least partly, the product of their practices.

Although deeply embedded into the local circumstances of West Berlin, the project of urban ecology was not parochial. When Berlin ecologists researched their own city, they also reacted to developments elsewhere and appropriated some of their results and agendas. Conversely, they also contributed to such developments, feeding their findings and conceptual models back into the debates of an emerging academic community of urban ecology. At the beginning and at the end of the chapter, I will deal more systematically with this issue, focusing in the first case on the broader context of German ecology in which Sukopp began his work and in the second case on the relation between Sukopp's approach and other attempts to build an urban ecology.

Ecology in the Making

Many historians and sociologists have taken disciplines to be the institutional building blocks of academia. Accordingly, disciplines are collectives that are anchored in university departments, professional societies, and journals and that form internal markets for the exchange of knowledge among its members (Kohler 1982; Turner 2000; Rosenberg 1979; Lenoir 1997). Twentieth-century ecology, however, hardly corresponded to such an image of an organizationally integrated discipline. Ecology developed in a set of heterogeneous disciplinary, subdisciplinary, or transdisciplinary programs and in close conjunction with biology as well as with practice-oriented domains such as forestry, fishery management, agriculture, nature conservation, or environmental planning (Küppers, Lundgreen, and Weingart 1978; Schwarz and Jax 2011). This, however, should not be seen as a sign of weakness or a failure of discipline building; rather, it exemplifies the broad variety in which academic specialties can take form (Turner 2000; Frickel 2004). The absence of a coherent and stable organizational identity did not prevent the unfolding of a nexus of methodologically or thematically continuous research practices and discourses, even if these were conducted under the umbrella of shifting programmatic labels.[2]

Since being coined by the German zoologist Ernst Haeckel in 1866, the term *ecology* had referred broadly to the study of the relations between biological species and their environment, focusing either on single species (autecology) or on entire living communities or biotopes (synecology). In Germany until around 1900, *Ökologie* had often been used synonymously with *Biologie*, which by then was itself only a bridging label for quite distinct disciplines, among them botany, zoology, and physiology (Sucker 2002; Cittadino 1990). Moreover, as Nyhard (2009) has shown, around 1900 much work that later would have been labeled ecology evolved outside academia in museums of natural history, in zoos, and in schools. We might add the activities of semi-academic natural history societies, a tradition of civic science that has remained an important pillar of ecological fieldwork.

The first domain in which a substantial complex of ecological research developed was aquatic ecology. By the 1920s, limnology (the ecology of lakes) had taken shape in research stations, in state advisory commissions on fisheries, and in an international society largely dominated by German scholars (Schwarz 2003). The second domain was vegetation science. In continental Europe it developed into a distinct specialty called *plant sociology* (Küppers, Lundgreen, and Weingart 1978; Schwarz and Jax 2011; Nicolson 1989). It provided input for forestry, agriculture, and horticulture, and was therefore also covered by departments and professorships in these fields. In 1937 the newly created Zentralstelle für Vegetationskartierung (Central Agency for the Mapping of Vegetation) in Stolzenau became the organizational center around which an extended network of specialists assembled. Parallel to these endeavors, zoologists developed a systematic interest in animal ecology, a field that, however, gained much less institutional visibility.[3] Outside biology, ecology also became incorporated in geography, both as cultural ecology (an environmental deterministic approach to human culture) and as landscape ecology (Hard 2011).

In Europe (in contrast with the United States and Britain, where ecology established itself in disciplinary associations and journals),[4] these research programs developed relatively independently. Moreover, as biology established itself as a dominantly laboratory-based discipline, ecology suffered from a weakening of its academic status (Küppers, Lundgreen, and Weingart 1978: 82–83). As Wolfgang Tischler, who in 1963 received the first chair for ecology at a German university, later wrote in his autobiography, colleagues had even warned him that working in ecology would jeopardize his professional status in biology (Tischler 1992: viii).

In the late 1960s, the development of cybernetic ecosystem theories by American ecologists provided a common point of reference that helped to unite diverse strands of research (Kwa 1989; Kingsland 2005; Bocking 1997; Taylor 1988; Golley 1993). Faced with a broadening public concern about environmental degradation, ecologists also began to connect their research field with these emerging agendas and sought to give it institutional coherence.[5] This reorientation was stimulated by international initiatives such as the International Biological Program (1964–1974) and UNESCO's Man and the Biosphere program (since 1971), in which German researchers participated. In 1967 these activities had led to the foundation of INTECOL, an international scientific organization of ecologists. In 1969 the Gesellschaft für Ökologie (Society for Ecology) was established, which brought together ecologists from different disciplines, including technical areas that were considered relevant for combating environmental pollution (Küppers, Lundgreen, and Weingart 1978: 120). From the 1970s on, the West German government and the Deutsche Forschungsgemeinschaft (German Research Foundation) launched special funding programs for environmental science, which promoted ecological and other environment-related research (ibid.).

Besides the chair for landscape ecology created in Munich in 1972 and the Institute of Ecology created in 1973 in Berlin, ecology found recognition on the research agenda of many university chairs and departments, mostly in the domains of zoology and botany. A comparison of the issues of an inventory of West German teaching and research institutes shows an increase of organizational units that explicitly cover ecology from only 12 in 1964 to 17 in 1968 to 35 in 1973.[6] Although these are clear signs of an institutional consolidation, ecology's identity remained ambivalent, changing between an understanding as a biological sub-discipline and as a broader interdisciplinary field that blended into geography, psychology, or sociology. Moreover, the increasing prominence of *environmental studies* as a broader label in political programs and study programs (Küppers, Lundgreen, and Weingart 1978) created tensions with the notion of ecology as a distinct discipline.

Berlin was a venue for different strands of ecology throughout the entire twentieth century (Leps 1985). In West Berlin, ecological issues figured on the research agenda of the Zoological Institute of the Freie Universität, the Faculty for Agriculture of the Technological University, and a number of natural history societies. Since the 1960s, it was notably the work of Sukopp and a broadening network of collaborators that made this city a center of German ecology. Whereas their colleagues elsewhere devoted their fieldwork mainly to marine ecosystems, forests, rivers, etc., this group turned its own city into its most prominent study object. Thereby they became pioneers in a branch of research that hitherto had only occupied a marginal place on the agenda of ecology.

Taking Stock of Species Occurrences

In 1957, as a young botanist, Herbert Sukopp published an inventory of wild plant species that, since the end of the war, had been newly found in the area around Berlin, the so-called Mark Brandenburg (Sukopp 1957). The list appeared in the journal of the regional Botanische Verein (Botanical Association), of which Sukopp was a member and for which he acted as a journal editor. Further updates of this inventory followed in 1960, 1965, and 1967 (containing observations until 1961), under the common editorship of Sukopp and his colleague Hildemar Scholz (Scholz and Sukopp 1960, 1965, 1967). Drawing on the input of sundry observers, many of them amateurs, they reported the findings of species that were either rare or of concern for plant geography, and provided indications of their locations. Although the inventory covered the entirety of the Mark Brandenburg, it was striking how many of these new botanical findings were located within the confines of Berlin. To a considerable extent these were plant species discovered in the nature reserves or other sites on the outskirts of the city, which had long figured among the prominent sampling sites of local botanists. It was also the more central parts of the city, however, that increasingly attracted the interest of local botanists. The large bombed areas that had been left from the war had become covered by a constantly accruing vegetation—herbs, bushes, and later trees, many of them of foreign origin.

The exchange and collection of such observations of local species occurrences was one of the most basic circuits through which new ecological knowledge and urban natures emerged in postwar Berlin. A range of observers continuously roamed the landscapes within and around the city, tracking the locations of plant or animal species as well as even minor changes in their abundance and composition. Some of them, like Scholz and Sukopp, were scholarly biologists, who combined their professional or research interests with a commitment to the study of their local environment.[7] The majority, however, were nature lovers without any academic credentials in that field. They devoted their leisure time to the study of local wildlife or specific parts thereof, and many of them acquired remarkable observation skills and factual knowledge.

The historical roots of these inventory activities go back to the sixteenth century, when local natural history emerged as a new preoccupation in central Europe (Cooper 2007). It included the compilation of so-called floras—a term that designates not only the actual stock of plant species in an area but also the literary form in which the latter is described. Since the seventeenth century, Berlin botanists had compiled regional inventories of flora that focused on the Mark Brandenburg (Essholtz 1663; Gleditsch 1751; Schulz 1845; Ascherson 1954, 1864) or the proximity of the city (Willdenow 1787; Ascherson 1859). Corresponding inventories of animal species or specific groups thereof (faunas) were produced from the middle of the nineteenth century on (e.g., Schulz 1845). This period witnessed an increasing popular resonance of local natural history, mainly among the educated middle class (Phillipps 2003). When the Berlin professor Paul Ascherson published his flora of Berlin in 1859, he relied on the input of a number of fellow observers—most of them educated professionals, notably schoolteachers (Ascherson 1859: X–XI).

The incentive for such floras came partly from emerging fields of scholarship such as biogeography and vegetation science, which studied the distribution and spread of plant species. Local natural history, however, was not simply a service for academia. It was a form of collective connoisseurship through which its participants lived and expressed their commitment to the region and its places. Doing natural history was inextricably connected to outdoor recreation activities such as hiking or common excursions during which botanically or zoologically interesting locations were visited and new observations made. Conversely, reports of remarkable species occurrences functioned as directories for individual and collective excursions to these sites, so that naturalists could revisit and admire these findings. As fanciers of local nature, they also took part in strivings for nature conservation, contributing relevant findings and campaigning for the designation of monuments and reserves.

In the middle of the nineteenth century, local naturalists began to organize semi-academic associations for the purpose of coordinating their activities. The Botanischer Verein was founded in 1859 by regional botanists in the province of Brandenburg. After World War II, the association was reestablished in the Western sectors of Berlin, amounting to a membership of about 100 to 130 people (Sukopp 1957, 1967). For the study of fauna a number of smaller associations existed that devoted themselves to specific groups of species.[8] Besides these natural history societies, the Volksbund Naturschutz (People's League for Nature Protection) provided a further organizational network for the exchange and popularization of findings of the local flora and fauna.

Although natural history in West Berlin drew on a long research tradition, more specific factors determined its trajectory. For one, the political division of the city and the Cold War undermined the cooperative relations among botanists on both sides of the Iron Curtain. Since in 1952 the GDR had imposed restrictions on West Berliners' access to its territory, naturalists from this part of the city were no longer able to visit their former observation grounds in the surrounding former province of Brandenburg, and after the erection of the wall they were restricted solely to the territory of the Western sectors. Although the Botanischer Verein still had members in the GDR, and postal exchanges of knowledge with Eastern botanists never broke off entirely, West Berlin botanists now pursued their own observation work, mainly within the confines of their part of the city.[9] In the 1970s, détente between the two Germanys yielded a relaxation of the limitation of access to the East. Although in 1975 a West Berlin resident reported results of botanical excursions to vacant lots in Eastern Berlin, the wall remained the major boundary of naturalist fieldwork (Stricker 1975). As a result, spaces within or at the fringe of the city remained the main fields of West Berlin naturalists.

A second factor was the impact of the war on the biophysical structure of the city. As a result of its extensive destruction, the city had become extremely attractive to botanists. The rubble areas became cleared soon after the war, but remained undeveloped, to a large extent even until the 1980s. Although located in the midst of a metropolis, they were soon covered by dense vegetation, which contained many species that hitherto had been unknown in the area. As Hildemar Scholz wrote in his doctoral thesis on the "ruderal vegetation" of Berlin, the "unexpected spread of foreign species" in these destroyed parts of the city rendered previous work on the flora of the city completely insufficient (Scholz 1956: 1). And the inventory of the Mark Brandenburg that Sukopp and Scholz published in 1960 contained mainly findings from rubble areas, including the newly built rubble heap at the Teufelsee (Scholz and Sukopp 1960: 23).

A third factor was the increasing involvement of university-based ecology, notably through the person of Sukopp. During the 1960s and the 1970s, Sukopp himself was among one of the busiest observers of plant occurrences in West Berlin. Being both an active member of the Botanischer Verein and a professor at the Technische Universität, he stimulated the coordination of floristic observation in the city. In 1962 he published an announcement to local naturalists in which he bemoaned the piecemeal character of existing floristic records (Sukopp 1962). According to Sukopp, dramatic changes of the flora, which were partly due to the incursion of new plant species after World War II into the rubble areas, required a new and more systematic effort to compile a flora of Berlin.

According to the individual habits of the naturalists, species spotting happened in different forms, reaching from regular visits to confined spots, to more or less extensive exploration tours throughout the areas in Berlin. Individual site visits were often complemented by common excursions on which groups of naturalists visited sites of special interest, often under the guidance of a specialist. These excursions did not only serve for the exchange of knowledge; they were also collaborative forms of exploration in which species were recorded and observation skills were exchanged and practiced. Sometimes discoveries were made during the daily routine movements or in the direct living environment of an individual naturalist. For example, on his daily bike rides to his institute, the zoologist Fritz Peus noted that the devastated Tiergarten in the postwar years became populated by steppe bird species that hitherto had been unknown in Berlin (Peus 1952). Another naturalist observed new stands of plant species that settled in the rubble areas along his own street and reported them to Sukopp.[10]

On their exploration tours through Berlin, naturalists often focused on areas that other citizens rarely visited or which were private property. Before being leveled, sites with ruins were dangerous to enter. Later they were fenced off with billboards that naturalists had to pass in order to access the area. Naturalists also had to share these spaces with playing children and with homeless people, and often they aroused the suspicion of neighbors or passersby. When Scholz studied the rubble areas of Berlin, pedestrians often supposed that he was scrounging for scrap metal.[11]

The fieldwork practices also differed according to the particularities of the species observed. The findings were directly documented in a notebook, and, additionally, sometimes by picking herbarium species. Observers scheduled their work in accordance with the blossoming periods of plants or the daily activity peak of the species under study. Whereas plants could be revisited regularly by the same or different observers, the identity of animals that were sighted at different times or by different people was often impossible to determine (Löschau and Lenz 1967: 108). Besides the direct visual identification of animals, naturalists often had to rely on other means, such as listening to the sounds of birds or setting traps for insects. In his studies of Berlin vertebrate species, Victor Wendland analyzed the contents of owls' stomachs (Wendland 1971: 15).

Only after the information gathered by individual observers was pooled did a comprehensive picture of the local flora or fauna emerge. Observational data of particular sites thereby became integrated into a broadening circuit of observation that covered virtually the entirety of the Berlin territory. The systematic collection and documentation of broader amounts of material was mainly accomplished by a few individuals. For botanical research, it was Sukopp who had fulfilled this function since the 1960s. He maintained a central archive of floristic observations, which grew continuously and which, located at the Institute of Ecology, became the most comprehensive inventory of Berlin's flora. By running a herbarium of the Berlin flora at the Botanische Museum (Botanical Museum), Hildemar Scholz played a similarly important coordinating role. In faunistic studies this brokering role was played by Victor Wendland, a philologist by training, who devoted much of his life to the study of birds. His 1971 book on the vertebrates of West Berlin drew upon many observation records that he had received from fellow naturalists.

In describing the features of the local flora and fauna, naturalists drew on a complex set of observational categories and understandings. Inventories were typically arranged by species, drawing on established biological taxonomies of species. They indicated where various species were found, by whom, and sometimes also to what extent. In practice, the determination of species often required subtle visual skills. In particular, faunistic observers often specialized in the observation of specific groups of animals. Naturalists who were known for their special expertise were also consulted by their fellows when they needed help in determining species that they had found.

In their observation reports, naturalists deployed a range of locational devices through which they associated species with specific spaces and sites. At the most general level they attributed observed species to a certain geographical reference area, such as Brandenburg, Berlin, or (later) West Berlin. Single observation reports typically contained detailed indications of the sites at which the species were found within this area. These techniques of description differed rarely from those that Paul Ascherson (1859) had deployed in his flora. Owing to the increasing concern for the inner city, however, it was not only forests, ponds, or village borders that helped botanists to indicate plant occurrences. As for other urban dwellers, the street pattern and its representation on the city map provided ecologists with important landmarks. Indications of larger occurrences such as "numerous in the center" (Scholz and Sukopp 1960: 34) contrasted with very precise statements of place such as "area of the projected superstore hall at Beusselstraße" (ibid.: 27) or "front garden of the Humboldt-Universität" (ibid.: 31). In his notes on the rubble fields of the inner city, Sukopp even referred to single plots on the basis of the street number system.[12] Observers also drew on informal locational skills. For example, an indication such as "Wittenau, purple moor grass pasture in the South-East of the former pheasant-run close to the junction of the heather railroad and the freight railway" (Scholz and Sukopp 1960: 26) was understandable only if the reader already knew that the "heather railroad" was a suburban railway line, knew that a pheasant run had existed there, and was able to recognize or was already familiar with the existence of a purple moor-grass pasture in the neighborhood. Indeed, many parcels of land that were mentioned in the floristic registers seem to have been well known to local naturalists and regularly visited by them.

Another concern that structured the work of naturalists was the difference between the native and the exotic. In the late nineteenth century, research in so-called adventive floristics had developed subtle classifications that distinguished these plants according to criteria such as their time of introduction, their geographical origin, or the degree of naturalization (that is, the extent to which they had become a permanent feature of the new environment—see, e.g., Thellung 1918). Only plants that had existed in an area before the emergence of agriculture were considered truly indigenous. Plants that had arrived after 1492 (when Columbus arrived in America, and when worldwide trade and the exchange of seeds had intensified) were classified as neophytes.

Whereas gardeners and agriculturalists traditionally had welcomed and even promoted the importation of foreign species, in the early twentieth century nature conservationism and horticulture took an increasingly negative stance toward the incursion of non-native species (Gröning and Wolschke-Buhlmahn 1992; Coates 2006). Ecological concerns about the damaging effects of invasive species on existing ecosystems were thereby interwoven with a preoccupation with the preservation of culturally significant characteristics of the homeland flora. This trend can be observed in many countries in Europe and North America, and one might see this, at least partly, as a being tied to the maintenance of distinct national identities. As we have seen, Berlin naturalists also understood local nature as a feature of Heimat. This did not mean, however, that they necessarily preferred native plants to foreign ones. Indeed, newly arrived plants were among the discoveries that fascinated these naturalists most. Instead of condemning foreign plants, they tended to embrace them as an enrichment of their homeland nature. Notably, rubble heaps, walls, and burial grounds were the typical spaces where such plants abounded. For example, in 1896 one botanist reported having found new species in a ruderal area at a construction site (Behrendsen 1896). In the postwar years, it was in the rubble areas that newly arrived plants abounded. An example is the Jerusalem oak goosefoot (*Chenopodium botrys*), a plant of Southern origin that mushroomed in all rubble areas of the inner city (Scholz 1956; Sukopp 1971). A few decades later, Berlin ecologists even embraced these new species as objects of urban nature conservation.

Naturalists assumed that species were not distributed randomly. Whether a site was dry or wet, the kind of soil, the exposure to the sun—all these were features that were assumed to be associated with specific kinds of plant or animal species. For example, they characterized rubble areas by the combination of plant species that could be typically found there. Likewise, birdwatchers developed a subtle sensitivity for the kinds of generic spaces in the city that were associated with certain bird species. Naturalists were able to make these judgments more or less intuitively on the basis of the shared experience of fieldworkers. They read the city and its spaces through the lens of its ecological function for various animal and plant species. Scholarly biologists also employed systematic mapping methods to establish the relations between plants and sites. For example, in his thesis (1956) Scholz mapped the distribution of *Chenododium botrys* to probe the spatial conditions that determined its presence.

Another characteristic of naturalists' views on local nature was an emphasis on its dynamic character. By drawing on their personal acquaintance with a site, as well as observation reports of fellow naturalists, they were able to discern how species became rare or even fully disappeared from their area. In 1966, when Sukopp compared the number of plant species that were recorded after 1945 with the findings that Ascherson had reported in 1859, he arrived at the troubling conclusion that 124 species had disappeared. The loss of species was thus three times the gain from newly immigrated species, which Sukopp (1966) calculated at 42.

What kind of space was performed in the context of these practices and the circuits of observation that they sustained? In the most general sense the existence of a record of its flora and fauna that focused on the territory of Berlin inscribed these entities into this space. As much as the species became semantically associated with Berlin, the city, at least for the members of the ecological observation culture, became another space. In a more dynamic sense it underwent changes like the decline, loss, advent or settlement of some species. The city, however, was not constructed as a homogeneous unit in these practices. Through the itineraries of field-workers, and through the locational specification of their observations on lists and maps, spots in the city were constituted as *occurrences* of specific species or of combinations thereof. Such occurrences were more or less precisely located in geometrical forms, and they rarely had clearly defined boundaries, but taken together they formed a new spatial pattern of nature in the city.

Practices of the Ecological Survey

Although species spotting continued to sustain extended circuits of obser-vation, it became complemented in the 1960s by another configuration of fieldwork practices. These were surveys in which one or more researchers systematically explored the flora, fauna, and ecological conditions of one or more exemplary sites. The goal of such surveys was not simply to gather information about distinct occurrences of species, but to yield in-depth knowledge of the ecology of the site. They typically resulted in a set of liter-ary, numerical, and cartographic representations of an observation area, which were published in the form of journal articles or monographs. Besides a structural description of the actual state of the biotope, they com-prised a historical analysis and at least, a rudimentary functional analysis of the dynamics of the component parts.

The observation circuit of site surveys ordered space differently than the spotting of species occurrences. The latter evolved along paths and consti-tuted occurrences as distinct spots as the fundamental spatial entities of urban nature. Surveys, in contrast, employed three spatializing strategies that were absent in species-spotting practices. The first was a strategy of spatial demarcation that singled out specific observation areas and deter-mined their boundaries. Such demarcation practices worked both on the ground (by orientating research on physical landmarks, plotting areas, and so on) as well as on the map. The resulting areas could be of different shapes. Often they were parasitic to pre-existing administrative boundaries.

These clearly circumscribed spatial envelopes functioned as workplaces and units of documentation. The second strategy was an inventory of these areas that sustained circuits of observation and integrated newly acquired data (often from different observers) with data from former surveys or other spotting practices. The third strategy was differentiation. Through juxtapositions with other spaces, qualitative or quantitative differences were constructed that defined the identity of any of these spaces. Observational spaces and the generic spaces that they were supposed to represent were thereby grouped under ecological space categories such as vegetation formations or biotope types and by the amount of species found on them.

The Conservation Surveys

The roots of ecological site surveying in Berlin go back to the early twentieth century. Local amateur naturalists and scholarly biologists have often scrutinized areas that they considered particularly interesting, often to establish or monitor the effects of conservation measures (Chappius 1930–31, 1931; Wahnschaffe, Graebner, and Hanstein 1907, 1912). Such preservation issues motivated a series of surveys of West Berlin nature-conservation areas that Sukopp began around 1960. In contrast to the piecemeal inventory of individual species that had characterized earlier surveys of nature reserves, Sukopp aimed at a "scientific" approach that was based on advanced concepts and research methods of scholarly ecology. In this regard, his studies dovetailed with a general trend in German ecology to expand its expertise to practical nature and landscape conservation. The location of the Berlin nature reserves in close proximity to the city, however, gave Sukopp's surveys a special significance. They became landmark studies of the influence of the urban environment on the conditions of insulated biotopes, a theme that remained a constant thread in Sukopp's future research.

The starting point was a two-year inquiry into the bogs in West Berlin that Sukopp carried out for his doctoral thesis (1958). These "oligotrophic bogs" were particularly interesting to Sukopp because they had once covered about 8.7 percent of the surface of Brandenburg (ibid.: 42). As was noted in chapter 1, they were widely appreciated as the most original pieces of nature in Berlin and some of them belonged to the "classic destination(s) of Berlin's botanists" (ibid.: 81). However, as Sukopp established, many species in those areas had been lost or retreated, and the original vegetation formations had changed their character. Sukopp attributed these changes to intensification of land use, lowering of groundwater, air pollution, or the incursion of too many people. As he put it, these nature reserves were

Figure 2.1
A bog in Berlin, circa 1980. (photo: W. Tigges)

"small and even the smallest enclaves within the sphere of human influence" (ibid.: 148) and therefore were extremely vulnerable to such
influences. He also claimed that active measures of maintenance and restoration were needed to keep the bogs in shape. Drawing on his findings,
Sukopp wrote a memorandum urging political authorities to launch such
measures (1958).

Further studies in the 1960s and the 1970s extended Sukopp's approach
to all nature reserves in West Berlin (Sukopp and Köster 1970; Sukopp 1961,
1962, 1960; Sukopp and Straus 1968; Sukopp 1970; Sukopp and Böcker
1975, 1971; Sukopp 1973; Sukopp, Brahe, and Seidling 1986; Sukopp
and Auhagen 1979). They were collective endeavors in which Sukopp
cooperated with local amateur specialists. In the 1970s, the survey of Berlin
nature reserves evolved into a formal research program of the newly
founded Institute of Ecology. This program, known as Wissenschaftliche
Grundlagenuntersuchungen Berliner Landschafts- und Naturschutzgebiete, was directed by Sukopp and received logistic support from the Berlin
administration.

Besides nature reserves, Sukopp also studied other areas in Berlin which he considered endangered by human encroachments. Most significant were his studies on the waters, which suffered from pollution and the effects of various recreation activities (Sukopp 1969; Sukopp, Markstein, and Trepl 1975; Sukopp 1963). Sukopp was particularly concerned about the constant retreat of the reeds on the banks of the Havel. He blamed the waves produced by increasing boat traffic, notably leisure motorboats, for causing this damage. According to Sukopp, the bathers who flocked to the riverbanks on warm summer days caused further damage. It was due to his successful lobbying that in 1969 a special act was issued by the Berlin parliament that provided for restrictions on boating and bathing.[13]

For his doctoral thesis, Sukopp repeatedly visited the ten bogs under study. For the studies of the Havel, a boat was used to explore the reeds. Fieldwork also included picking herbarium species or the placing of traps for animals. By the deployment of these various means, Sukopp and his co-workers engaged with the materiality of the sites they studied, transforming them occasionally or permanently into places that were productive of ecological knowledge. The rather marginal location of the sites and their legal status as nature or landscape reserves gave them much leeway to operate freely and to appropriate them for their own epistemic purposes.

Sukopp also made frequent use of the technique of plant-sociological sampling. In contrast to the term *flora*, which designated the simple stock of plant species that happened to grow in a reference area, the term *vegetation* referred to the way in which these plants cluster in larger entities that are physiognomically coherent, the so-called plant communities. It has been a basic tenet of the continental European tradition of plant sociology (Nicolson 1989) that a fixed number of such plant communities exist, and that they depend on the environmental conditions of their sites. Such plant communities are neatly distinguished by a fine-grained taxonomy of classes and sub-units. To this end, a coherent part of the vegetation would be chosen, and within a rectangular sampling spot all plants and their relative abundance would be recorded. This sampling technique had to be performed by a skilled observer, who was not only to work on single species determination but also to visually identify and demarcate suitable sampling spots.[14] Results of such spot surveys figured prominently in all the ecological monographs on Berlin's nature areas, both in the form of numerical composition tables and in the form of maps on which hachures or colors divided the area into distinct vegetation zones. In the 1960s, permanent sampling spots were installed in some nature reserves so that the development of the vegetation could be monitored.

The categories botanists had previously used to describe the vegetation were assumed to represent naturally developing plant communites and, therefore, deliberately disregarded the influence of humans. Many of the vegetation types, however, that Sukopp and his colleagues found in the bogs and in other nature reserves deviated systematically from the models of plant sociology. Sukopp (1959: 155) introduced the term *Degradations-gesellschaften* (meaning degradation communities) to account for such variations. As he put it, such communities emerged when human encroachment had distorted the natural trajectory of vegetation communities and thereby pushed their development onto a different track.

Ecological site surveys also encompassed systematic exploration of the non-biotic conditions provided by the sites. Researchers took samples of soil (Sukopp 1959: 46). Sukopp also relied on additional sources, such as recent and historical temperature and precipitation records, results of geological drillings that had been carried out next to one bog, and records of the groundwater level that had been collected by the local waterworks. To some extent, his fieldwork crossed over to the practices of an environmental historian. He collected old maps, topographical descriptions, and administrative documents that contained information about the history of the sites.

Although nature reserves and landscape reserves had long been seen as valuable pieces of the Berlin landscape, they achieved a new meaning through ecologists' research activities. Besides underlining the ecological features that made the area outstanding, the surveys also framed them as highly vulnerable pieces of the urbanized space. Owing not only to their location but also to their ecological interconnections, they had become part of the man-made environment of the city.

Surveying Bombed Lots and Other Wastelands

Sukopp's and his colleagues' site surveys also looked at the various wastelands that existed in the city, notably in bombed areas but also at some abandoned railway facilities. Rubble areas had been attractive observation grounds for botanists. Some tended to focus on areas that they visited regularly, thereby slowly moving from the recording of species to rudimentary site surveys. This is well documented in the field notes that Sukopp took on his regular visits to the rubble areas of the Schöneberg, Luisenstadt, Tiergarten, and Kreuzberg districts in the early 1960s.[15] Sukopp organized his notes according to the street names and plots, which he inspected repeatedly, thereby producing an ongoing record of their vegetation development. The first botanical publications on Berlin rubble areas, published

Figure 2.2
Wastelands in a bombed area of Berlin, circa 1960. (photo: Herbert Sukopp)

Figure 2.3
Herbert Sukopp taking stock of ruderal flora in a bombed lot, circa 1960. (photo:
Alexander Kohler)

in the late 1950s and the early 1960s, also reveal that their authors had carried out plant-sociological sample surveys (Düll and Werner 1955–56; Kohler and Sukopp 1964).

The primary goal of these surveys was to characterize the vegetation and to predict in what direction it would further develop—its so-called succession.[16] Since the 1950s, botanists who studied rubble areas in bombed cities in Germany had noted that the specific combination of species did not comply with existing plant-sociological categories (see Lachmund 2003). They considered them as typical for a special vegetation type: ruderal vegetation (a term based on the Latin word for rubble, *rudus*). As Scholz had stated in his thesis in 1956, it was characterized by the abundance of neophytes, often of Southern origin, that benefited from the environmental conditions of the city, especially the warmer climate. In 1955, Düll and Werner, two biologists at East Berlin's Humboldt University, established a succession scheme for sites with different characteristics within the rubble (1955–56). In the course of time, when the rubble areas became greener, more systematic patterns of the ruderal vegetation became visible. In the early 1960s, when many rubble areas had become covered by dense groves, Kohler and Sukopp were able to distinguish two dominant vegetation types: at sunny and dry sites the *Robinia pseudo-acacia* woods, and at cooler and shady sites the *Sambucus nigra-acer negundo-woods*. Kohler and Sukopp assumed that the process of succession was still going on and expected that it eventually would lead to some kind of forest. They expected, however, that, owing to the specific circumstances of the city, the result of the succession would differ significantly from the pine-oak societies that originally had covered the region. They maintained that it was not possible to know what this final state would look like (1964: 396). In the following years, however, Sukopp and his co-workers developed a detailed scheme of the series of plant communities through which the succession developed at these rubble sites (Sukopp 1971). In this context, Sukopp also established the category of the *Chenopodium botrys* association, a plant community that existed exclusively in Berlin. These examples show how the need to adapt to pre-existing local conditions of Berlin resulted in new ways of understanding and classifying the vegetation as an object of ecological knowledge.

Plant-sociological surveys were carried out in many different spots of the rubble, at small sampling squares often no bigger than a few square meters. In some cases, ecologists selected long-term sample areas that they visited regularly to produce records of the vegetation composition over time. In the mid 1960s, Sukopp and his co-workers Wolfram Kunick and Ursula Hennig carried out a vegetation survey of an entire block in the district of Kreuzberg (Sukopp 1971). The area contained a plant-sociological sampling square that Sukopp inventoried repeatedly between 1964 and 1967.

Most outstanding was the research trajectory of a small triangular site next to the Lützowplatz in the district of Tiergarten, later also called Dörn-bergdreieck. Sukopp had visited the site for the first time in 1961. Back then it differed little from the large zones of bombed areas that spread along the Landwehrkanal in the Berlin district of Tiergarten. Sukopp repeatedly revis-ited the site during the following years. In the 1970s, the space developed into one of Berlin ecology's primary research venues. Sukopp and other col-leagues from the Institute of Ecology regularly inspected the site. In addi-tion to floristic and vegetation studies that had been carried out by Sukopp since his first visit to the site, research was now also extended to the soil by Marlies Runge, who wrote a doctoral thesis on Berlin's soil in the early 1970s, as well as by the institute's professor of pedology (Runge 1975; Institut für Ökologie 1983; Blume and Runge 1978). In 1978 and 1979, the Institute of Ecology also used the site as a teaching ground for its practical courses in urban ecology. In preparing the courses, and during the courses themselves, a large amount of data on different ecological features was produced, including flora, fauna as well as the non-biotopic factors such as soil, climate, and hydrological conditions. This long-lasting observation circuit and the fact that the site had remained undeveloped in the midst of the urbanizing environment made it into a kind of model system for Berlin ecologists. Here they were able to follow the interplay between the environ-mental conditions of a large metropolis and the composition of the flora and the succession of the vegetation over decades. Although the results from the area were never published in a separate monograph, Sukopp kept an expanding file in his archive at the Institute of Ecology. Results were also reported in various publications by Sukopp and other Berlin ecologists. In the 1980s, when a hotel project was launched, Sukopp even claimed that it had become the world's most extensively documented urban ecosystem and therefore should be preserved. (See chapter 5.)

Berlin ecologists' surveys of ruderal areas did not remain restricted to bombed plots. They also sought to trace the succession of ruderal vegetation in rubble heaps (before they were shaped by gardeners), landfills, railway areas, and industrial areas. In the 1980s, two abandoned railway stations—the Potsdamer Güterbahnhof and Personenbahnhof (also called Gleisdrei-eck) and the Tempelhofer Rangierbahnhof (also known as Südgelände)—received attention as survey grounds. Although geographically located in

West Berlin, they had been part of the railway system that was run by the GDR railway organization before they were transferred to the West. (See chapter 4.) Ecologists and nature conservationists alike were impressed by the dense vegetation that had accrued in these areas since the 1950s. The surveys that ecologists conducted there, partly produced in commission of the Berlin administration, underlined their considerable species diversity (Kowarik 1986; Kowarik and Langer 1994). They also afforded valuable insights into plant-sociological patterns and the succession path of the urban ruderal vegetation.

Although the basic methods were the same, fieldwork in wastelands differed in many respects from that conducted in the nature reserves. Ecologists had to share these spaces with all kinds of other visitors, whose activities potentially interfered with their research. Wastelands were used for walking dogs and as places for children to play. For some years a snack bar was located at the fringe of the Lützowplatz, and from time to time circus animals grazed there. At night the area often turned into a place for homeless people to sleep as well as a meeting place for prostitutes and their clients.[17] Other problems faced the researchers who studied the railway areas of the Reichsbahn. Ulrich Asmus conducted his first survey of the Gleisdreieck before it was officially returned to the West. This meant that he had to pass loopholes in the fences to enter the area, and he had to contend with security personnel and watchdogs.[18]

The main problem that threatened this line of research was the increasing pressure for redevelopment. As early as 1964, Sukopp and Kohler wrote, with some regret, that many succession areas had to give way to the ongoing reconstruction activities (Kohler and Sukopp 1964). In the early 1980s, the IBA, an international construction exhibition, entailed a further step of redevelopment, notably in Kreuzberg. Protests by Sukopp and other ecologists could not prevent the edge of the Dörnbergdreieck from being redeveloped with postmodern residential buildings. (See chapter 4.) In 1986, the construction of a hotel led to a complete loss of this well-established ecological observation area. Even if they were not redeveloped, wastelands suffered from decisions by the administration, which regarded them as useless, ignoring ecologists' needs. For example, in 1978 the district administration had removed—or "destroyed," as ecologists put it—some parts of the vegetation at the Dörnbergdreieck in order to make the site less attractive as a venue for prostitution (Auhagen 1978).

Most people tended to see wastelands as eyesores. Although located in the city, they were rather marginal spaces, used only for shadowy activities. With their surveys, however, ecologists reframed these sites as a new environmental entity—as *urban ruderal areas* or *urban ruderal vegetation*. Among their distinct features were relatively high species diversity and the occurrence of site-specific vegetation formations. Ecologists cited the local character of the soil and the climate of the city to explain the specific composition of plants that settled in these areas. Moreover, increasing railway traffic was mentioned as a reason for the spread of foreign seeds to Berlin. For ecologists, wastelands were the prototype of an urban ecosystem: their marginality within the urban environment allowed nature to thrive almost unimpeded, and yet it was a nature that was largely determined by the human-made environmental conditions of the city (Kowarik 1991).

Surveying Horticultural Spaces

Urban greeneries such as parks and cemeteries always had been among the favorite observation grounds of ornithologists. Amateur botanists had also compiled descriptions of the planted and wildlife flora of these spaces (Arndt 1941; Czepluch 1966; Anders 1979). In the 1970s, Berlin ecologists carried out more comprehensive surveys of greeneries. For example, when Wolfram Kunick studied larger sample areas in various parts of the inner city, he also included some areas in parks, the largest being the southern half of the Tiergarten (Kunick 1974). Two further surveys of the Tiergarten (Berger-Landefeld and Sukopp 1962; Sukopp et al. 1979) were carried out by the Institute of Ecology to evaluate the effects of an envisaged highway and tunnel project. Later, parks figured in the so-called biotope mapping survey on Berlin that was carried out by a working group under Sukopp's direction (Arbeitsgruppe Artenschutzprogramm Berlin 1984; Martens and Scharfenberg 1982–83). Ecologists thereby also covered other planted greeneries, such as lawns along roads, private gardens, or allotment gardens.

Again, the ecological surveys comprised inventories of plant species and analyses of vegetation samples as well as observations of fauna. As the fieldworkers established, originally planted species had spread widely in the parks and many wild species had added to the plant coverage. Whereas gardeners tended to consider the latter as weeds and tried to minimize their presence, ecologists considered them as the authentic result of the natural conditions in that biotope. Also, when planted species showed a tendency to spread on their own, this was a clear sign for ecologists that they had become successfully integrated into the park's ecosystem. For this reason, many studies of parks included them into their floristic and vegetation inventories (Kunick 1974; Berger-Landefeld and Sukopp 1962: 108; Sukopp et al. 1979: 30; Martens and Scharfenberg 1982–83). As Kunick noted in his study on the Tiergarten, it was in fact often impossible to determine if plants had grown spontaneously or if they represented "a very successful planting from the first period after the (parks J.L.) reconstruction" (Kunick 1974: 108). Kunick faced similar problems with the inspection of private lawns. He often witnessed species that had grown wild from bird seeds, and that then were sometimes tolerated by some garden owners whereas others immediately tried to erase them (ibid.: 53–54). This shows how the boundary between the natural and the culturally shaped became increasingly fuzzy but was continuously reestablished by the contextual judgment of the researcher.

Many private gardens could be viewed only from the street so that the back yards systematically escaped the sight of the observer (Lenz 1971: 47).[19] Residents tended to view these activities with suspicion or even reacted hostilely when encountering fieldworkers around their property. Kunick remembers that he was sometimes mistaken for the local rat catcher.[20] When an amateur botanist inspected private lawns in the inner city, he received "depressing and threatening comments" from the residents who watched him from balconies and open windows (Stricker 1975: 6). Fieldwork on gardens therefore included negotiations with landowners or tenants. In some cases, residents were also valuable sources of information. One fieldworker told me that he especially appreciated his contacts with Turkish immigrants, who often came from rural areas and had a rather intimate knowledge of the soil and plants in their courtyard gardens.[21]

When studying parks and gardens, fieldworkers were also attentive to ways in which people interacted with the flora and fauna. Victor Wendland (1971: 32) witnessed the common practice of pedestrians feeding squirrels when he studied animal life in Berlin parks. In his studies of garden vegetation, Ulrich Asmus learned how urban lawns differed according to the way in which they were kept by their owners. Whereas some owners kept their lawns systematically free of any weeds, others were much more tolerant, so that the lawns produced vegetations that came close to agricultural meadows or ruderal sites. Such observations helped ecologists to explain the differences that existed in flora and fauna throughout the city, and on this basis to make suggestions for more wildlife-friendly ways of maintaining parks and gardens.

Among the parks, the Große Tiergarten was the most impressive for the ecologists. In 1970 and 1971, Wolfram Kunick established 448 species of fern and flowering plants in the park (Kunick 1974). This was also reconfirmed by the list of species that Herbert Sukopp and his co-workers put together in the summer and fall of 1977 (although they found a slightly smaller number of species) (Sukopp et al. 1979: 30). This distinguished the Tiergarten markedly from all other Berlin parks, which had considerably lower numbers of species. Faunistic surveys came to similar conclusions. Between 1963 and 1977, the amateur ornithologist Klaus Anders made about 1,092 inspection tours of the Tiergarten, thereby developing a dense record of the park's avifauna (Anders 1979: 9). As he established, with about 76 species, the Tiergarten was "by far the richest park in Berlin with respect to bird species" (ibid.). Data that other observers collected on bats and other small mammals in the Tiergarten (cited in Sukopp, Anders et al. 1979) further underlined the park's outstanding diversity of species. Even soil samples were collected in the park that revealed a considerable spectrum of soil life (ibid.: 54–59). Sukopp and his research team attributed the Tiergarten's species diversity both to its size (according to the dominant ecological theory at that time, species diversity correlated positively with the size of a biotope) and to the variety of differently shaped sites within it (ibid.: xi). As the researchers maintained, the Tiergarten was exceptional in this respect. Although situated in the city, its ecological features made it more similar to a park landscape on the outskirts than to "an urban biotope in its proper meaning" (ibid.: 45). In order to shed light on the environmental conditions on which this species diversity thrived, the Tiergarten survey also included the exploration of other factors, such as climate, soil, air pollution, and the hydrological regime of the ponds.

Ecologists attributed the smaller number of species in other parks mainly to their smaller size. Furthermore, they claimed that these ecosystems suffered from severe "distortion" caused by the large number of park visitors or gardening activities. This was the case with the Volkspark Rehberge, which a team under Sukopp's direction explored between 1978 and 1979 (Sukopp 1984). Although that park was relatively large, it contained only a small number of species. Most astonishingly for the ecologists, the species of the area's original forest were completely gone (ibid.: 52). According to the surveyors, even the Tiergarten showed some serious effects of environmental disturbance. They blamed the presence of people and loose dogs for the declining density of some bird species (Anders 1979: 15; Sukopp et al. 1979: 49). They also diagnosed poor air quality (mainly due to the gas emission of cars) and, periodically, poor water quality in the ponds (ibid.: xi). Moreover, the soil along the roads showed relatively high levels of lead and de-icing salt (ibid.: xi).

Among the planted species, ecologists gave specific attention to trees. They commented positively on gardens with native trees or fruit trees, which approximated the natural vegetation or agricultural landscape that once prevailed in the area and which also provided attractive living spaces for birds and other animal species (Drescher and Stöhr 1980). Even the greeneries of the modern settlement Märkische Viertel, which a few years earlier had been criticized by conservationists for their damaging effects on the preexisting environment, proved to have developed interesting new ecological qualities. In summer 1977, students from the Technische Universität inventoried the birds of the settlement (Breitenreuter et al. 1978). Not only did their inventory reveal that the area had attracted a significant number of birds (which it also attributed to the habitats that birds found in the roofs and walls of the buildings); it also showed that the attractiveness of the settlement for birds differed considerably among various kinds of green spaces. The authors of the study took this as a clue to how ecological "enrichment" of such housing estates was to be achieved (ibid.).

Parks and gardens are areas that planners and citizens had always considered as nature in the city. For ecologists, however, it was not primarily the visual features or the recreational function of these areas that associated them with nature. Like the bogs or the ruderal areas, they constituted a distinct urban biotope that displayed its own floristic, faunistic, and physical characteristics. More than just a descriptive category, this notion of a biotope turned into a norm that defined how the respective spaces should be handled. What ecologists valued in park and garden biotopes were certain indicators of their environmental quality, such as species diversity, the presence of declining or rare species, or more generally, species and morphological features that they considered site-specific for the ecological conditions of the park biotope. This brought ecologists into opposition with dominant gardening practices. They bemoaned that ornamental plantings prevented the growth of a site-specific flora. Similar criticism was raised against garden owners, who sometimes aggressively combated the wildlife plants on their lawns (Kunick 1974, 1978). In other words, what from the perspective of gardeners was just the shaping of a better park or garden (according to the dominant aesthetic principles) was reframed by ecologists as a disturbance of the requirements of the ecosystem.[22] On these grounds, some ecologists embraced alternative models of gardening, such as the "wild gardens" of the Dutch horticulturalist Louis Le Roi. (See chapter 6.) Even more so than with other biotopes in the city, it is clear that such norms did not so much point back to any preordained natural state. Even the current Tiergarten, notwithstanding its valued ecological features, was the product of its re-creation in the postwar period, and it reflected the aesthetic ideal of its designer. Looking back at that period, the ornithologist Wendland even wrote about the reconstruction of the Tiergarten as the "destruction" of the biotope of certain bird species that had settled there after the war (1971: 17). It was through the observation circuits of ecological fieldwork and the resulting representations that horticultural spaces had achieved a new public meaning, which itself became the basis for new norms concerning how such spaces should be valued and how they should be maintained.

The Ecological Ordering of Urban Space

Ecologists did not only collect observations on single species or exemplary sites in the city. Beginning in the late 1970s, they also aimed at a comprehensive structuring of the Berlin territory according to ecological criteria and thereby redirected the former circuits of observation. This third configuration of fieldwork practices resulted in spatial classifications, statistical indicators, and cartographic renderings that together represented Berlin as a variegated complex of flora, fauna, and living spaces. It was not only the ordering of this specific place, however, that was at stake here. By formalizing local observations as context-independent re-presentations, they also helped to establish *the city* (or *the urban ecosystem*) as a generic object of ecological knowledge.

Ecological landscape research has relied largely on the so-called structuring of natural space (naturräumliche Gliederung), which distinguished landscape units according to the dominant formations of the physical landscape (Meynen and Schmithüsen 1953–1962). In geobotany and biogeography, classification schemes and maps had been used that divided geographical spaces into distinct vegetation zones. In addition to serving scientific purposes, such spatial orderings functioned as normative criteria for practical measures of landscape care, including the creation of organic landscapes in cities. According to these classification schemes, and on the resulting maps, the Berlin area consisted mainly of glacially shaped plains and valleys. In terms of its original natural vegetation, it belonged largely to the pinewoods (in the river valley) and oak-hornbeam woods (in the elevated plains) that once covered the entire region of the Mark (Akademie für Raumforschung und Landesplanung 1960: map 40).

Such spatial categorizations and maps also figured in the discourse of Berlin ecologists when they discussed the natural preconditions from which the flora and vegetation had developed. The Berlin ecologists maintained, however, that these natural entities were only of limited value for the structuring of the urban environment (Zacharias 1972: 55; Kunick 1973: 13, 14; Sukopp 1973). Accordingly, the physical structures of the city and the various impacts of urban life on the original flora had resulted in a highly anthropogenic landscape in which the units of the structuring of natural space were rarely visible. For their own structuring of the city, they therefore took a different tack. For them human impact was the dominant factor that shaped the environment of the city, and hence the most appropriate criterion according to which its spatial units should be divided and mapped.

Since the 1960s, human influence had been the pivotal theme of Sukopp's ecological thought (Sukopp 1968, 1969). Like many contemporary vegetation scientists, Sukopp assumed that most of the existing vegetation in Europe was anthropogenicly molded. Important ways in which humans were supposed to have had such an impact were agricultural and forestry practices such as clearing, moving, cultivation, fertilization, and, more recently, the use of herbicides. Instead of treating nature and culture as two fundamentally opposed categories, Sukopp suggested a nature-culture continuum on which vegetation types or landscape formations could be located according to the extent of human influence or *hemerobia*. Following the Finnish vegetation scientist Jaakko Jalas (1955), Sukopp distinguished four different degrees of hemerobia and used the relative number of invading and lost native species as quantitative indicators of these degrees.

It was not until the late 1960s, however, that Sukopp began to explicitly address the city as the generic object of his research. Whereas formerly he had used evidence from Berlin to refine the concept of human influence, he now used it as a basis for an ecological conception of the city. At the same time, he gave the word *influence* a broader meaning, encompassing not only the effects of human interventions on the vegetation but also their effects on the fauna and the non-biotic components of the urbanized environment. In 1968, Sukopp articulated this view in a schematic cross-section of the "modifications of the biosphere in a metropolis" in which he depicted the multiple effects and changing degrees of urban development from the outskirts to the center (Sukopp 1973).

Although still recognizable as a representation of Berlin, Sukopp's cross-section was reduced to its basic abstract form, thereby becoming a formalized model of the supposedly universal features of any metropolis. In a condensed visual form it displayed the basic tenet of Sukopp's ecology, namely that the city was an environment in which the huge impact of man had transformed almost all pre-existing natural conditions. Having been reprinted in almost all major publications by Sukopp and other Berlin urban ecologists, the cross-section has become an emblematic expression not only of the city as a generic object but also of the Berlin style of urban ecology.

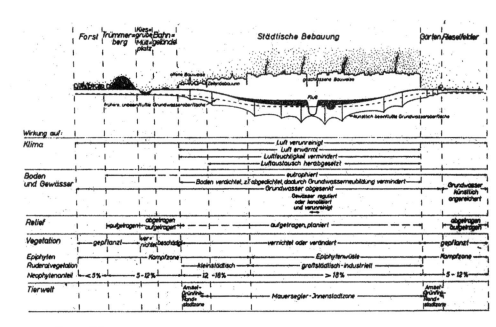

Figure 2.4
A cross-section of Berlin showing the different zones of human influence on the biosphere. The denominations at the top translate to forest, rubble heap, gravel-pit/landfill, railway area, urban development (open, terrace, closed), river, gardens, and sewage farm. Although highly schematized, the city is still recognizable as Berlin. (source: Sukopp 1973)

In 1968, Sukopp presented the scheme for the first time in his habilitation lecture. Both the scheme and an elaborated version of the lecture were then published in 1973 in a programmatic essay (based on a lecture that he had given in the same year in Vienna) in which Sukopp outlined the goals of an ecology of the city (Sukopp 1973). As Sukopp posited, human influence was pervasive in urban environments, yet it did not work homogeneously over the whole city: "In spite of the fact that some features are shared by the entire space of the city, it is not possible to conceive of the city as *a single* location. In contrast, it is a living space, that resembles a mosaic, composed of many different small locations." (ibid.: 93) He therefore considered the classification and characterization of the components of this mosaic as central tasks of urban ecology. Sukopp had already maintained in an earlier publication that some plant species were systematically tied to specific land-use patterns (Sukopp 1969). His examples had been plant species that abounded systematically at bathing places or along canal slopes. In addition, Sukopp assumed that urban development also had more diffuse effects that decreased gradually from the center to the outskirts, as he also suggested in his scheme of the city.

The ecological structuring of the entire vegetation of West Berlin was realized by Wolfram Kunick in his doctoral thesis (Kunick 1974). The thesis was supervised by Sukopp, and together with a parallel thesis on the Berlin soils (Runge 1974) was financially supported by the Deutsche Forschungsgemeinschaft. Kunick spent the entire vegetation periods of 1970 and 1971 collecting floristic data in Berlin. Since a comprehensive survey that would have covered the West Berlin area seemed unrealistic, he concentrated on seventeen sample areas. Material that Sukopp had collected earlier provided further evidence from another sampling site (Kunick 1974: 73). At about one square kilometer, these sampling areas were much larger than the small sampling spots on which plant-sociological inventories had focused. They were spread widely over the city, and most of them comprised a variegated spectrum of city-specific locations. Kunick selected and delineated the sampling areas in such a way that his survey would yield highly standardized and comparable observations. Kunick thereby followed physiognomic features of the land, such as land-use changes or the block structure of developed areas (Kunick 1982: 5).

Originally Kunick had planned to draw partly on plant-sociological methods to analyze his sample areas. As he explained later, the patterns of vegetation formations turned out to vary on such a small scale, that they were of little help for structuring the city into larger spatial units (ibid.: 2). Kunick therefore restricted his fieldwork exclusively to an inventory of species. On the basis of qualitative and quantitative comparisons of species occurrences, he established a strong correlation between the species composition and the structure of the city. Following the model of Sukopp's scheme, he divided the structuring of the city into four concentric "development zones" from the densely developed area in the center to the external periphery, each presenting its distinct composition of species. The result of his analysis was a floristic map that represented the extent and the boundaries of these four zones in the Berlin territory (Kunick 1982, figure 5). In a second step, Kunick traced the change of the floristic composition according to the dominant forms of land use in West Berlin—what Sukopp had described as a mosaic of living spaces. Kunick differentiated the land use into sixteen categories, among them building structures (such as open, terraced, and closed), various kinds of urban greeneries, industrial areas, construction sites, and different types of forests and agricultural land. These structures were of a much smaller scale than the zones and were scattered widely throughout the entire city, many of them through all four zones.

An important medium through which this structuring of the city materialized was the map. Previously, ecologists had mainly used two types of maps: "dot maps" of individual species and plant-sociological maps that represented only specific sites. A dot map divided a space into neatly distinguished observation areas, usually according to a geometrical grid scheme, and indicated by a dot where certain species were observed. Maps of the entire city were of a different character. Kunick's maps were directly derived from the official 1970 outline map of the Senate for Housing and Construction. In that map the differences in color and shape of the signature were used to distinguish five types of development: closed development, terrace and row development, open development, villages, and industrial and service utilities.[23] Kunick relied on the structures shown in this map to distinguish four main development zones whose floristic composition he described on the basis of the sample areas located within them. To fine-tune the delineation of the areas of common floristic composition, he drew on the distribution areas of dot maps of plant species that he considered representative for each of these areas.

Development zones and land-use types remained important units for structuring urban space also in other works of Berlin ecologists. In the programmatic essays on urban ecology that Sukopp later co-authored, the textual descriptions and tables were structured according to land-use types. A few years later, Sukopp and the ornithologists Hinrich Elvers and Hermann Matthes established a close relationship between Kunick's floristic zones and the distribution of bird populations in the city (Sukopp, Elvers, and Matthes 1982). The comprehensive Biotope Mapping Survey of West Berlin that a working group of Sukopp carried out between 1979 and 1983 distinguished 57 "biotope types" that were directly derived from dominant land-use patterns. (See chapter 3.) In the 1980s, the mapping and structuring of the Berlin environment was further pursued in the context of an interdisciplinary research project of the Institute of Ecology. Some of these maps were included in the Environmental Atlas of Berlin that the administration published from 1985 on. The human impact on the natural environment remained the main focus here, and it was cartographically established for different aspects of the environment. The mosaic structure of the environment that Kunick had described in terms of land-use structures was thereby further differentiated according to more specific criteria of the related ecological disciplines (pedology, climatology). One result of this work was the Ökochoren-karte (map of ecochores), which divided the city into 69 spatial units characterized by dominant vegetation formations, soil, climate, and specific degree of hemerobia. The map, produced for the 1986 Environmental Atlas, was also included as an appendix in *Stadtökologie: Das Beispiel Berlin*, a seminal book by Sukopp and his colleagues that was published in 1990.[24]

By classifying and mapping the city, ecologists substantiated a generic understanding of the city as an ecological phenomenon. It was through categories such as floristic zones, biotope types, and ecochores and their various subdivisions, that they made sense of the various ways cities were supposed to affect various dimensions of the living and non-living environment. At the same time, they also gave a new meaning to the specific city in which they carried out their research. The well-established structuring of urban space into development types and districts was translated into a jigsaw of biotopes, each of which had its own ecological characteristics, suffered from different kinds of problems, and needed specific kinds of care. As we will see in the next chapter, this new spatial ordering took on broader political meaning in the ecological planning strategies of the early 1980s.

Instituting Urban Ecology

Studies of urban wildlife as they were conducted in Berlin not only led to a remarkable accumulation of data; they also contributed to the establishment of the Institute of Ecology at the Technische Universität. Before the 1970s, the Faculty of Agriculture had already been engaged in some ecological research. In 1973, however, the Institute of Ecology was established. It would soon develop into an important center of ecological research.

The establishment of the Institute of Ecology was one result of a broader reform of the university's organizational structures and educational programs that involved dissolving the Faculty of Agriculture. The three departments (Abteilungen) of the new institute—Applied Botany, Pedology, and Horticulture—had hitherto been separate institutes within the Faculty of Agriculture. In 1974 a department of Ecosystem Research and Vegetation Science was added, with Sukopp as director. Three further departments were created in 1976, focusing on bioclimatology, limnology, and regional pedology (the latter split from the existing pedology department). A professorship and a department of human ecology never materialized (Institut für Ökologie 1976: 2). It is also remarkable that no distinct department of animal ecology was created. It was mainly through limited cooperation with the Freie Universität, where a zoological institute existed, that also faunistic perspectives were fed into the work of TUB ecologists.

The trajectory of Sukopp's urban ecology was connected to the Institute of Ecology in at least four important ways.

First, the institute was the immediate context through which Sukopp established his academic career.[25] Soon after receiving a doctorate from the Freie Universität in 1958, he became an assistant to the botanist Ulrich Berger-Landefeld on the Faculty of Agriculture at the TU. Berger-Landefeld held the chair at the Institute of Applied Botany, which was meant to provide insights that were operational for the fields of agriculture and landscape architecture. Berger-Landefeld represented an experimental ecological approach that had little bearing on Sukopp's own work, which was based almost exclusively on his fieldwork on Berlin.[26] Berger-Landefeld died in 1967 and was succeeded by Reinhard Bornkamm, an experimental plant ecologist from Göttingen. Sukopp passed his habilitation and achieved the status of a Privatdozent (private lecturer). This allowed him to supervise dissertation projects, to acquire research funds, and to play a more significant role in the politics of the institute. When in 1969 a second professorship was created at the institute, the position was given to Sukopp. In 1972, Sukopp's professorship remained part of the Department for Botany, as the Institute for Applied Botany was called after its integration into the Institute of Ecology. In 1973–74, a call from the University of Hohenheim allowed Sukopp to renegotiate the terms of his appointment.[27] The Technische Universität had a strong interest in keeping Sukopp in Berlin and obliged his demands to create a new department under his directorship. The denomination on Vegetation Science and Ecosystem Research was proposed by Sukopp and reflected the research agenda that he had developed in the previous years. New assistant professorships that were created for the department, and a regular research budget provided Sukopp with a significant increase in his research capacity. When Sukopp retired, in 1995, he was succeeded by his former doctoral student Ingo Kowarik.[28]

Second, the institute emerged from and facilitated an interdisciplinary network of researchers who cooperated in the ecological study of Berlin. Such cooperation had started with the Project Urban Ecosystems that Sukopp had launched with Hans-Peter Blume, a pedologist, who had come to the Faculty of Agriculture in 1970.[29] Around 1970, Bornkamm also participated in cooperative research with Sukopp and Blume, providing experimental background insights into the spread of the neophyte species *Chenopodium botrys* (Sukopp et al. 1971). Another partner in the collective research was Manfred Horbert, a climatologist, who in 1976 was appointed as professor and chairman of the new Department of Bio-Climatology. Both Blume and Horbert cooperated with Sukopp in writing expert reports for the Berlin administration and in the ecological mapping of Berlin. Although Bornkamm was not involved in the inventory of the Berlin ecosystems, the plant-physiological and vegetation development studies that he conducted, often outside Berlin, yielded conceptual insights for urban ecology. He also maintained a sustained interest in roof vegetation, which was of practical relevance for urban biotope-promotion measures (Bornkamm and Auhagen 1977; Jacobshagen, Bornkamm, and Heinze 1977). Through the interdisciplinary Arbeitsgruppe Ökologie und Umweltforschung (Ecology and Environmental Studies Working Group) that founders of the institute initiated in 1972, they also maintained contact with researchers from other Berlin institutions (including the Freie Universität) and with amateurs (Sukopp 1977). The institute thereby became not only an intellectual habitat that united researchers with a common research agenda but also a "center of calculation" (Latour 1987) through which the observation circuits of ecological fieldwork were integrated and coordinated.

Cooperation among different branches of ecology had been the explicit mission of the Institute of Ecology since its founding. In a policy statement that Sukopp and Blume drafted in 1971, they referred to the cooperation that had already developed as a model for the future institute.[30] The need for cooperation beyond traditional disciplinary boundaries remained the basic justification for the existence of the institute and for the addition of new departments throughout the 1970s. Members of the institute invoked the newly evolved ecosystems theory, which distinguished different systemic components yet also maintained their integration into a functional whole, as a foundational metaphor for the operation of the institute (Sukopp and Schneider 1979; Institut für Ökologie 1983: 4; 1974: 1). The jurisdiction of the departments was thereby identified either with specific components (climate, soil, waters, plants) or, as with Sukopp's department, with the entirety of the system (Institut für Ökologie 1974: 1). A teaching system with many interdisciplinary courses, which were directly connected to the research agenda of the staff, was a further factor that facilitated cooperation among the institute's members. This does not mean, however, that the institute was homogeneous. Whereas Sukopp devoted nearly all of his research to Berlin, others were only partly or temporarily involved in this kind of research or followed completely different lines of research.

It was no accident, however, that the most intensive cooperation that evolved in the institute was in urban ecology. As with naturalists spotting regional species, the urbanized landscape within the confines of West Berlin was the immediate territory to explore. Although individual institute members also studied remote fieldwork sites, this was especially relevant for larger cooperative projects that involved different people, including doctoral students, and that were often carried out over longer periods of time, or that had only a low budget available. In fact, much of the institute's research output was generated in student projects and qualification theses, which were carried out in the city. Moreover, Sukopp's earlier research on the Berlin vegetation and his broader conceptualization of urban ecosystems provided a focal point for students of other ecosystem levels. Indeed, Sukopp himself actively maintained such contacts. Finally, via Sukopp's reputation as a voluntary Landesbeauftragter (see chapter 4), the Berlin Senate became an important demander of ecological knowledge. This meant that ecologists had to increasingly cater to practical concerns of the Senate which were limited to the territory under its jurisdiction.

In this context, it is also relevant that the institute was itself part of the university' Specialist Division Landscape Development (Fachbereich Landschaftsentwicklung), the organizational umbrella for different environmental planning institutes. As it was understood by the founders of the division, ecology was supposed to guide "activities of landscape planning and landscape shaping" (Technische Universität Berlin 1972: 48) and to contribute to the teaching and research in this area. Actual research cooperation, however, remained largely restricted within the boundaries of the institute. Although Sukopp and his colleagues emphasized the practical implications of their research for planning and also contributed to local planning processes, they pursued these goals largely independent from the other institutes of the division.

A third way the Institute of Ecology contributed to the consolidation of urban ecology was through the social reproduction of a distinct community of scholars. Not least through its interdisciplinary teaching of landscape planning, it was able to draw large numbers of students into its lecture halls and field seminars. They participated as tutors in the program, worked on dissertation projects, and, later, often pursued research careers in Berlin and elsewhere. As one interviewee who had studied there recalled, it was particularly popular among students to opt for "basic sciences" instead of the more aesthetically or administratively oriented strands of expertise that were taught at the other institutes.[31] Notably, Sukopp was very successful in attracting doctoral students. By 1995 he had supervised 130 undergraduate theses and 32 doctoral theses (Kowarik, Starfinger, and Trepl 1995: 18). Many of the students thereby focused on aspects of Berlin's ecosystems or even served as co-workers in Sukopp's projects. The bulk of these students were graduates in landscape planning, the program in which the institute performed its main teaching function. A good example is Sukopp's later follower Kowarik, who took a degree in landscape planning before pursuing a career as a vegetation ecologist. Also among Sukopp's doctoral students were a few biologists who had graduated from the Freie Universität. Students experienced urban ecology as a timely issue and later became themselves important agents, not only in promoting ecological aspects in the practical planning field, but also in forging a local research community.

Finally, the Institute of Ecology functioned as a platform for its members' vision of urban ecology and helped to establish Berlin as an internationally recognized center of urban ecological research. Since the 1960s, Sukopp had maintained close relations with botanists from East European countries— relations he maintained while a professor at the institute. In 1980 the institute hosted the Second European Ecological Symposium. Organized by a group of British ecologists and Reinhard Bornkamm, it brought a large number of ecologists from various (mostly European) countries to Berlin. Besides presenting their results, Berlin ecologists organized excursions to the most significant sites of their fieldwork. A team of ecologists had prepared a book-long excursion guide that drew on decades of ecological research in the city (Sukopp 1982). Not only the Berlin researchers themselves but also the object of their studies—the flora and fauna of Berlin—thereby achieved broad visibility in the worldwide ecological community.

In 1980, the institute explicitly advanced a collective self-understanding as a center of urban ecology. Such status was, for example, claimed in the ten-year report that the institute published in 1983 (Institut für Ökologie 1983: 4). This identification of the institute with the urban topic was also underlined by the choice of the title illustration, a drawing of a "railroad landscape" with a dense cover of typical neophyte trees and bushes that almost hid the contours of an adjacent (possibly abandoned) factory. On other occasions, too, Sukopp promoted a narrative that featured the development of Berlin as a center of urban ecology, invoking its enclosed character (Sukopp 1990: 244; Sukopp and Schneider 1979).

Urban Ecology in Berlin and Beyond

Sukopp's ecology took shape in local fieldwork practices in Berlin, but also formed part of an increasing interest in cities among European and American ecologists in that period. I will use the remainder of this chapter to sketch some of the continuities and differences between Sukopp's work and approaches to urban ecology that were promoted elsewhere.

Before the 1960s, the term *urban ecology* had been mainly identified with human ecology as it was represented by the Chicago school of sociology, and also by the work of some biologists.[32] As Kingsland (2005) and Cittadino (1993) have described, since the 1920s ideas for an ecology of human-dominated environments were also sporadically articulated by biologists in the United States. In the 1950s, when the director of the Missouri Botanical Garden, urged ecologists to move from the jungles to the cities, he wanted them to observe human behavior and solve human life problems (Kingsland 2005: 240–241). An example from Germany is the so-called Kommunalbiologie that Hermann Peters, a biologist working for the city administration of Stuttgart, proposed (Peters 1956). Peters pleaded for a kind of social hygiene that had clear eugenic undertones, and for a holistic understanding of cities and their human inhabitants. Although Peters mentioned a few studies of plant and animal species (notably vermin) in cities, plants and animals were of only marginal concern in his project. Conceptually demanding models of cities as biotopes of non-human species, such as that proposed by the Vienna zoologist Wilhelm Kühnelt (1955), remained rare exceptions.

As with ecology in general, it was the rise of ecosystem theory that gave a new twist to the development of urban ecology. The term *ecosystem* had been coined by the British biologist Arthur Tansley in 1935, but was significantly elaborated in the 1950s and the 1960s by North American ecologists such as Raymond Lindeman, Stanley Auerbach, Howard Odum, and Eugene Odum (Kwa 1989; Kingsland 2005; Bocking 1997; Taylor 1988; Golley 1993). In their hands it became a cybernetic model that represented organisms (including human beings) and their physical environments in terms of self-regulating flows of materials and energy. These theorists believed that the proper functioning of the ecosystem relied on the maintenance of a stable state of equilibrium. Quantitative measurements of energy consumption and material flows and the representation of the results in schematic diagrams became the hallmarks of the new research practice that they introduced into their discipline. Although this approach had first been developed mainly for the purpose of studying the biological and health effects of radiation (an agenda that clearly reflected the Cold War preoccupation with nuclear weapons), it was subsequently extended to the study of all kinds of environmental problems.

In the late 1960s, American system ecologists claimed that cities could be understood as special types of ecosystems (Odum 1971: 11). Such claims were not much more than conclusions by analogy from the knowledge that ecologists had generated about other ecosystems.[33] Heinz Ellenberg, director of the German division of the International Biological Program, pointed to a lack of substantial research that would allow a classification of what he called "urban-industrial ecosystems" (Ellenberg 1973: V, 23, 237–8). The most substantial project on a city to be carried out within the framework of the International Biological Program was a system-ecological analysis of Brussels by a group directed by the botanist Paul Duvigneaux (Duvigneaud 1974; Havelange, Duvigneaud, and Denaeyer-De Smet 1975). Cities figured more prominently in the Man and the Biosphere Program, which started in 1971 and which even included a special subprogram on "Metropolitan Areas as Ecosystems." The topic materialized in various international conferences and in pilot projects in Hong Kong, Frankfurt, and other cities.[34] Sukopp variously referred to these activities as a broader context of his work, and also contributed results to some of the conferences that were held in this program. Also in North America, many system-ecology-based studies of cities, or models of urban ecosystems were developed (e.g., Sterns and Montag 1974).[35] In these ecological studies, cities were seen as critical spots where the environmental crisis took shape and where it needed to be addressed. In line with the dominant preoccupation of system ecology, attention was given to the circulation of materials and energy through the urban environment—the so-called metabolism of the city (Wolman 1965; Duvigneaud 1974; Odum 1971). As these authors maintained, cities differed from natural ecosystems in that they drew their energy input not directly from sunlight but from the consumption of resources that were brought into the city from other ecosystems (e.g., fossil fuels and food).[36] This "parasitic" relation to other ecosystems was seen as one of the fundamental problems of the modern city.[37] This approach is well captured in the schematic representation of Brussels that Duvigneaud included in many of his publications, and is in a way, a system-ecological equivalent to the ecological zoning of the city that Sukopp had modeled on Berlin.

    Another concern of urban ecologists was the quality of the city as a "habitat" for human beings. A biological view thereby tied in with understandings developed in the tradition of sociological urban ecology. In their study on Brussels, the Duvigneaud group included studies of the spatial distribution of certain groups of inhabitants in the city (Jouret in Havelange, Duvigneaud, and Denaeyer-De Smet 1975). Duvigneaud also talked about the "pathologies" of the city, such as pollution, civilization diseases, and crime (Havelange, Duvigneaud, and Denaeyer-De Smet 1975: 6). He

also gave various recommendations for "improvement," such as densification, greening, and embellishment (ibid.: 31). But, like other proponents
of a system-theory approach, Duvigneaud never tied his research systematically to the goal of protecting wildlife species in the city, as Sukopp did in
Berlin.

It was clearly in reaction to these international developments that
Sukopp and his colleagues began to use the term *urban ecology* to characterize their work. Until the early 1970s, Sukopp had considered his studies
on Berlin mainly as geo-botanics, vegetation science, or nature conservation.[38] Only beginning with his landmark article in 1973 did he label
his work systematically as *ecology*, occasionally prefixing that word with
urban.[39] In 1978, when Sukopp, Blume, and others urged more ecological
studies of cities, they mentioned that more and more people lived in cities
and that population density was a main feature of cities. This resonated
with similar statements by Duvigneaud and by the Man and the Biosphere
program. Whereas Sukopp hitherto had seen practical implications of his
work in nature protection, now, in line with the Man and the Biosphere
program, concern for the quality of urban life began to figure more prominently in his work. Plant species were now also considered as indicator
species whose mapping should help in evaluating the quality of the urban
environment (e.g., air pollution). With the work done for the Species Protection Program, conservation goals were extended more systematically to the
inner city. (See chapter 3.) Whereas other promoters of ecological research
on cities followed the precepts of ecosystem science more closely (see
Wächter 2003: 113–117), Sukopp appropriated these ideas only selectively.
He used terms such as *urban ecology* and *urban ecosystems* in his writings,
and, as we have already seen, he invoked them to legitimize the interdisciplinary organization of the Berlin institute. The substance of his own
approach, however, remained soundly based on the observational fieldwork that had developed in Berlin since the 1950s (such as floristic inventorying and the classification of spatial units). Whereas, for example, much
of the Duvigneaud group's research on Brussels was concerned with the
quantitative estimation of the city's energy balance and the material input
and output of materials of the Brussels agglomeration, there is no equiva-
lence of this in Sukopp's work. More quantitative system-oriented studies
had been undertaken by Berger-Landefeld and by Bornkamm (1987).
Although those studies were carried out at a field station located in the
city, they did not have the city as an object, and they were developed relatively independently of Sukopp's urban ecology agenda.[40] The difference
between Sukopp's and Duvigneaud's approaches becomes apparent in their

graphical descriptions of the urban ecosystem. Whereas Sukopp's scheme (based on Berlin) featured the flora, fauna, and other ecological factors throughout the different zones of the city, Duvigneaud developed a scheme (based on Brussels) that was dominated by arrows and boxes intended to represent the throughput of material and energy in a blackboxed urban ecosystem (Duvigneaud 1974; Havelange, Duvigneaud, and Denaeyer-De Smet 1975: 46).[41]

Another important difference was that Berlin ecologists gave the city as a human ecological phenomenon relatively little attention. An analogy between the growth of the city as a human agglomeration and the dynamics of the concept of plant-sociological succession that Wolfram Kunick drew in the introduction to his book (1974: 10) was purely illustrative and had no counterpart in Kunick's research practice. Although Sukopp and his colleagues repeatedly emphasized ecological research's potential for improving the living conditions of city dwellers, he did not see human beings as integral biological components of the urban ecosystem. In the 1960s, Sukopp, drawing on work by the ecologist Karl Friedrichs, had characterized man as "pitted against nature" (Sukopp 1968: 68), although he admitted that man was, at the same time, "rooted" in nature. For the practice of vegetation research, this meant that man should be considered as a "super-organic factor" (ibid.) and not as an element of the "living community" under study. Although Sukopp never made this explicit again, it remained a basic understanding of his work on human influence. Humans appeared generally as the cause of environmental change, but not as an integral part of the ecosystem. No attempt was made by any of the Berlin ecologists to include the sociological segregation patterns or other aspects of the social ecology of the city in empirical research or in a theoretical conception of the urban ecosystem, as the Duvigneaud group in Brussels did. Sukopp also remained distant from more radical versions of human ecology that sought to move beyond entrenched differences between humans and the ecosystem (e.g., McDonnel and Picket 1993). This probably would have changed if a chair in human ecology had been established at the institute. It is more plausible, however, to see the failure to establish such a chair as a reflection of the relatively low significance that human ecology had for the research practices of Sukopp and the other ecologists at the institute.[42]

Although Sukopp related his research rhetorically to the system-theoretical approach (for example in the denomination of his department), this did not have much impact on the way in which he actually conducted his research.[43] Field observations of vegetation remained the basis from which he approached the urban ecosystem, and for which he sought cooperation with other ecological disciplines. In this respect Sukopp's position was in line with a more widespread skepticism among West German biologists about an excessively far-reaching quantification of ecology.[44] Also in international debates, it was the detail of its floristic analysis and its explanations of floristic composition by environmental conditions of the city that were seen as hallmarks of the Berlin approach (Goode 1989: 861; Adams 2005: 140). For those who promoted a broader understanding of ecosystems, this was a severe limitation of Sukopp's approach. For example, the Canadian ecologist Anne Whyte (1985: 17) cited the Berlin studies as an example of ecological work that had shied away from developing an "integrated approach" to the city.

To be sure, Sukopp and his co-workers were not the only ones who based their ecological view on fieldwork in the city. In the 1970s, Sukopp's work had thrived within a network of intellectual exchange that, besides West German vegetation scientists, comprised scholars in Eastern Europe, the Netherlands, and Finland with whom Sukopp shared data and similar concerns for the human influence on vegetation.[45] Local observers had compiled data similar to those compiled by the Berlin naturalists for many other cities; examples in West Germany include the work on Münster by Rüdiger Wittig, Dagmar Diesing, and Michael Gödde (1985), the plant-sociological surveys by the so-called Cassels school of open-space planning (Kienast 1978; Hülbusch et al. 1979), and the urban surveys by Wolfram Pflug (1980) and the Department of Landscape Ecology at the Technische Universität München (Duhme 1983).[46] It is also notable that a sustained strand of urban conservation studies in Britain in the 1970s relied on botanical and zoological fieldwork.[47] Kingsland (2005: 292–256) also describes developments in U.S. urban ecology in the 1990s that moved beyond the black-box vision of the urban ecosystem that had underlay earlier approaches of system ecology.

There was a combination of features, however, that set Berlin's urban ecology apart from related research activities in other cities. First, there was the relatively strong institutionalization of urban ecology in a complex social network of research that had its central node in an academic university institute but also comprised organizations of amateur naturalist and landscape planning professionals. Second, there was the extent to which botanical methods and categories structured the conception of the city (notably at the expense of animal geography and more structural landscape categories as they were used by landscape ecologists). This was due to Sukopp's individual career, but it also reflected the organizational design of the Institut für Ökologie, which had originated partly from a botanical institute and which never provided much space for zoology. Third, an enormous amount of data had been collected, over many years, by a complex and well-integrated network of observers. As we have seen, this was a combined effect of a long-standing tradition of naturalist field-work and the erection of the wall around West Berlin. Not only did this enormous amount of data enable Sukopp and his colleagues to trace developments over long time periods in much more density than in other cities; it also enabled them to achieve an extraordinarily subtle differentiation of urban space into various types and subtypes of urban biotopes. Finally, Berlin's ecology stood out for the mutually constitutive interaction of its academic research with conservation planning in the city—a point that will be elaborated in the following chapters. Although all these features could also be found in other research programs, their specific combination in one city gave Berlin's urban ecology a distinctive character.

Conclusion

In this chapter I have traced the various sets of practices through which ecologists in Berlin accumulated environmental knowledge about their city. Although basic items of ecological knowledge were generated in distinct moments of fieldwork—often individual site visits—these were continuously interwoven through the exchange of knowledge between different fieldworkers and other consumers of ecological knowledge. At the same time, the knowledge that was accumulated in these networks became a basis for revisiting fieldwork sites or for selecting new sites. It was through such temporally and spatially extended circuits of observation that ecological knowledge of the city was accumulated. In the beginning this happened almost exclusively in the extended networks of species spotters and their naturalist associations, then through academic surveys of sites, or the entire city, organized by Sukopp and his colleagues. Eventually, the Institute of Ecology became a broker of field knowledge, recursively accumulating observations from within the city and connecting these observations with ecological researchers in other cities and towns. All these modes built on one another, incorporating their former results and directing further lines of research. Amateurs did not lose their relevance with the evolution of new fieldwork modes, even if the social center of gravity of this research moved from the naturalist associations to the Institute of Ecology.

Like other scientific knowledge, ecological data cannot simply be treated as a transparent window to the world. As I have tried to make clear in this chapter, ecologists and their amateur partners did not simply produce descriptive representations of preexisting spaces. The new environmental knowledge, and the practices through which it was produced, *performed* these spaces. They configured them as new natural entities, and they endowed them with new meanings and identities. As we have seen, the natures that conservationists had always been concerned about, such as the fens and lawns in the surroundings of the city, were also a primary focus of this work. Through the mapping of their long-term development, they became redefined as highly vulnerable and influenced by the intensive land use that dominated the urban landscape around them.

What was most significant about the approach taken by Sukopp and his co-workers was the extent to which it covered even the built-up areas of the city. A patch of ground next to the market became a habitat for a foreign species. An ugly bombed plot between housing areas became a ruderal habitat with an interesting succession pattern. Urban parks such as the Tiergarten turned from pleasant recreation spaces for human visitors to biotopes, and their embellishment by gardeners hence became a threat to its potential of biodiversity.

In the 1960s, such orderings of space were still restricted to a relatively limited community of practice. Neither ordinary citizens nor policy makers took much notice of how naturalists and the ecologists read the surface of their city. Soon, however, the spatial categories and commitments of the ecologists affected wider spheres of Berlin's public life. In parallel with much of the research work described in this chapter, Sukopp and his colleagues began to occupy a new role in the public-policy process of their city. Precepts of their discipline that had evolved over decades of hard fieldwork thereby became translated into problem claims and political technologies around which new public constituencies assembled. It is this consolidation of the nature regime in the domain of public policy and planning to which I will turn in the following chapter.

3 The Emergence of a Policy: Ecologists and the Species Protection Program

In December of 1978, the Abgeordnetenhaus approved a Nature Conservation Act, which obliged the Senate to develop a Landscape Program (Landschaftsprogramm) that would include a Species Protection Program (Artenschutzprogramm).[1] In contrast to the piecemeal approach of traditional nature conservation, the Landscape Program was meant to stipulate development goals that would help to improve "nature and landscape" throughout the entire territory. Whereas other parts of the program covered the "natural household" and the recreation function of the landscape, the Species Protection Program aimed specifically at the protection and promotion of wild plant and animal species. After nearly ten years of preparatory work, both programs were ultimately enacted in 1988.[2]

It was through the development of the Species Protection Program that ecological knowledge practices coalesced with politico-administrative practices and processes to form an institutionally embedded local nature regime in Berlin. Sukopp had largely conceived the very idea of a Species Protection Program. He also directed a working group that carried out the scientific inventory (biotope mapping) on which the program was based. In 1984 it published an expert report which contained a draft version of the program. The Species Protection Program, however, was not just a simple application of a ready-made body of ecological knowledge to a political problem.[3] In this chapter I will show how the program evolved in a process of negotiation and alignment that mediated between competing rationales and practices of ecology and the administrative system.

First, ecological knowledge making for the Species Protection Program merged with what Hannigan (2006) has called "environmental claims making"—that is, the discursive activities through which environmental conditions get framed as problems that deserve public attention and systematic policy responses. Throughout its developmental phases, the program was connected to such claims-making activities: first they paved the way for the acknowledgment of the problem of biotope loss on West Berlin's public agenda, then they were involved in the negotiation process through which the actual program took shape, and finally—as the subsequent chapters will show—the program itself became a resource that actors mobilized to bolster claims in specific land-use conflicts in the city. Although ecologists such as Sukopp were central actors in this process, it would be misleading to view this claims making as a unidirectional process guided by a few promoters, as Hannigan's original model had suggested. I understand claims making here rather as the overall dynamics of mutual enforcement and alignment through which claims were formed by ecologists and appropriated by their intended or factual audiences, including the merging of such claims into larger packages or the splitting of claims into new claims that developed along different trajectories. Besides the rhetorical appeal of claims and the effects of some favorite institutional circumstances as they are listed by Hannigan (ibid.: 77–78), this includes what Latour (1988) has described as the "enrollment" of actors through the "translation" of their interests. This means that a claim is presented in such a way that it resonates with claims already acknowledged by the audience so that the audience will accept the suggested solutions. As Latour put it in one of his later publications (1999: 179), this means that a link is created between two actors "that did not exist before and that to some degree modifies the original two." Once they entered the political arena, the conservation claims formulated by ecologists became aligned with established problem claims of green policy, and the two sets of claims merged in a new planning discourse.

Second, the Species Protection Program became mobilized as a resource to transform urban nature into spatial entities that were governable within the orbit of the organizational procedures of the Berlin administration. As James Scott (1998) has argued, modern state institutions employ simplified and schematic representations that disentangle phenomena from their concrete context and thereby turn them into objects amenable to organizationally homogeneous policies. Likewise, scholars working from a governmentality perspective have emphasized the role of formalized knowledge systems (e.g., accounting) in creating "fields of visibility" in which political programs can focus and act (Oels 2005; Luke 1995; Agrawal 2005; Dean 1999; Murdoch 1997). Ecological knowledge played a similar role in the shaping of the biotope as an object of policy-relevant calculation, control, and intervention. In contrast to Scott and governmentality scholars, however, I do not assume here that certain formal characteristics such as abstraction or formalization were themselves already sufficient to make ecological knowledge operative as a technology of governing. As I will show, categories such as the biotope and the various sub-categories that were connected to it had to be actively aligned to the categories and procedures through which the communities of policy making and administration operated. They had to become *boundary objects* (Star and Griesemer 1989; Bowker and Star 1999: 296–298)—concrete or abstract objects that are plastic enough to suit the information needs of different communities of practice yet, at the same time, stable enough to maintain a basis for the coordination of action across communities. Much of the work on the Species Protection Program was about the transformation of techniques, categories, and spaces of ecological fieldwork into boundary objects that resonated with the interpretive and practical needs both of ecology and of the administration.

In this chapter I will trace these processes of negotiation and articulation. I will show that the claim for species protection emerged in the context of a broader reform debate in the Federal Republic in which traditional nature conservation was replaced by a comprehensive approach to landscape planning. I will then turn to the working group of ecologists that, under Sukopp's direction, conducted a survey and developed an ambitious proposal for the program. The work of the group materialized in a number of knowledge products that circulated as boundary objects between ecology and administration: a set of "Red Lists" (Rote Listen) of endangered species, a planning-oriented biotope survey, and proposals for specific planning provisions. In the last three sections of this chapter I will trace the administrative and political negotiations that followed the publication of the ecologists' proposal and show how the final version of the program took shape. Although the ambitious goals of the original working group were constantly watered down in this process, the urban biotope became established as the target of an institutionalized regulatory structure.

Articulating a Problem Claim: Landscape Planning and Species Protection

The way to the Species Protection Program had been paved by a long-standing campaign in West Germany for the introduction of landscape planning as a complement to the classical strategies of nature conservation. Since the early 1960s, leading representatives of landscape care and nature conservation had criticized the approach of existing nature-conservation legislation.[4] Accordingly, only new regulatory competences would allow authorities to effectively counter the ongoing deterioration of nature and landscape. Systematic landscape planning schemes were meant to provide for restrictions on land use as well as active measures of landscape development, and to integrate all these provisions into a spatially comprehensive approach. Together with the impact regulation that will be discussed in chapter 6, the introduction of landscape planning was one of the main innovations of the new West German Nature Conservation Legislation (Bundesnaturschutzgesetz) that, in 1976, replaced the Nature Conservation Act of 1935.[5] On this basis, the Länder drafted their own Nature Conservation Acts, in which they substantiated this planning instrument within their own jurisdictions.

Although this legal reform was motivated by apprehension about the deteriorating state of the environment, its advocates framed this problem claim in different ways. In 1961, landscape-care experts who had gathered at a conference on the south German island of Mainau had made a "healthy landscape" the main object of their programmatic Green Charta (Grüne Charta von der Insel Mainau 1961). In contrast to earlier ideas of landscape protection, the concept of a healthy landscape was understood in one broad sweep to include the landscapes of recreation, housing, industry, and agriculture.[6] The Deutscher Rat für Landschaftspflege (German Council for Landscape Care, a semi-official think tank established after the Mainau conference), the Bundesanstalt für Naturschutz and Landschaftspflege (Federal Agency for Nature Conservation and Landscape Care) and the Arbeitsgemeinschaft Deutscher Beauftragter für Naturschutz und Landschaftspflege (Working Group of German Appointees for Nature Conservation and Landscape Care) were the main forums through which these issues were kept on the agenda.

The 1970s witnessed a differentiation of species protection as a distinct problem claim within this reform discourse. A catalyzing role was played by the Bundesanstalt für Naturschutz and Landschaftspflege, which invited ecologists and nature-conservation experts to the First Species Protection Seminar in the Federal Republic of Germany in 1971, and the follow-up conference The Changes of Flora and Fauna in 1975. After the Federal Nature Conservation Law had been passed, a number of workshops at the Land level and the 1980 Deutscher Naturschutztag (German Nature Conservation Day—the annual meeting of West German nature-conservation organizations and institutions) were devoted to the topic. Although landscape-care experts dominated the broader debate on landscape deterioration, the issue of species protection was mainly promoted by ecologists. One important sponsor was the ornithologist Wolfgang Erz. As the director of the department of nature conservation at the Bundesanstalt für Naturschutz and Landschaftspflege (from 1968) and as the proxy of the federal German nature-conservation appointee Bernard Grizmek (1970–1972), he acted as a coordinator of the debate on Species Protection Programs and developed many of the guiding principles that such programs should follow. Herbert Sukopp also played a central role in this debate. As an academic ecologist and as an expert advisor to the Land of West Berlin, he participated regularly in these discussions, and—drawing on his data and experiences from Berlin—exerted a decisive influence on them.

For the proponents of species-protection programs, environmental degradation presented itself in the form of a dramatic decline in the number of plant and animal species, both in distinct geographical areas and worldwide. Whereas earlier nature conservationists had been concerned about the loss of individual species that they considered characteristic of a regional landscape, this discourse refashioned the problem in terms of quantitative estimations of the overall decline and endangerment of species. This narrative revolved around the conviction that species diversity—or, more broadly, biological diversity—was the ultimate criterion for the ecological quality of the environment.[7] Surveys in various parts of the world and international conferences featured the loss or decline of species as a pressing global environmental problem.[8] Starting in 1960, the International Union for Conservation of Nature and Natural Resources regularly published a *Red Data Book of Endangered Species*, first focusing exclusively on mammals but later also including other animals and flora (MacCormick 1989: 39–41).[9] In Germany, such data existed in the early 1970s mainly on bird species.[10] Since 1966 Sukopp had presented evidence on the ferns and flowering plants in Berlin, and for Germany he estimated a loss of at least ten species (Sukopp 1966, 1972). In 1971, the German section of the International Council for Bird Protection put together a Red List of endangered bird species in West Germany (Deutsche Sektion des Internationalen Rats für Vogelschutz 1971). One direct implication of this framing of species loss was the permanent need for data. As evidence of the loss of species was often uncertain or absent, the compilation of Red Lists for species other than birds was seen as one of the main pillars of a future Species Protection Program (Sukopp 1972: 69).

Another feature of the new species-protection discourse was the framing of the problem in a broad ecological perspective that identified species protection with the protection of "living spaces" or biotopes. Accordingly, the general decline of species was mainly due to the loss or degradation of the biotopes in which these species lived and on whose ecosystemic conditions they depended. This implied that direct measures of species protection such as restrictions on picking plants or hunting animals (although this remained a big issue for some bird species) were considered of limited value. Species protection was instead to be applied throughout virtually the entirety of the territory. It had to be ensured that the areas in which species lived, or could live, were kept, or made as suitable as possible, for the settlement of a large number of species. Since the mid 1950s, this view had already successively been adopted by the International Union for the Protection of Nature, leading to the launch of an international program for wetland protection in 1961 (MacCormick 1989: 40).[11] With similar arguments, promoters of biotope protection in West Germany called for the extension of protected areas such as nature and landscape reserves to secure the ecologically most relevant species. In contrast to the earlier initiatives by the IUPN, this did not concern only specific areas; in order to guarantee an effective protection of species, conservation experts advanced an application of biotope-protection measures throughout virtually the entire territory. This meant that the development of land use should be directed in such a way as to promote a high number of species. In addition to the prevention of technical encroachments, this implied the development of less intensive forms of forestry and agriculture and the integration of species protection in recreation planning. A species-protection program, it was now thought, should be a systematic means of coordinating such measures within a certain reference area.

Conservation experts asserted that their problem claims were not just based on a sentimental or ethical apprehension of plants or animals (Erz 1972: 9). By postulating an intricate interdependence that tied species to their ecosystems, they emphasized the ecological role that even single species played for the functioning of an ecosystem (Heydemann 1985). In line with the diversity-stability hypothesis that had come to dominate ecological thinking in the 1960s, Erz (1972: 9) and Sukopp (1972: 68) posited that a broad diversity of plant and animals species was required to maintain the stability and balance of ecosystems.[12] Thus, while a functioning household of nature was necessary to maintain the diversity of species, the reverse was equally true. As Erz suggested (1972: 8), species-protection measures were of crucial importance for the preservation of the overall household of nature, a link that, in his view, had been largely neglected in the dominant discourse of landscape care.[13]

In the 1970s and the 1980s, claims about species protection helped biological ecology to establish itself as a mission-oriented field of expertise and to establish ownership of a newly acknowledged environmental problem. More specifically, it allowed the discipline to align issues of nature conservation with the broader take on landscape planning that evolved in these years. It had been typical for advocates of landscape planning to present their own approach as a modern and science-based alternative to the supposedly passive character of traditional nature conservation. By reframing nature conservation in terms of species diversity and by associating the problem with a planning-oriented ecological science, advocates of species protection were able to present an alternative form of nature conservation that appeared to be as modern and as planning-oriented as landscape planning was. It allowed ecologists and nature conservationists to pursue their own agenda within the wider framework of nature-conservation reform that had been promoted by the landscape planning movement.[14]

Another aspect that has to be taken into account here is the wider belief in science-based planning that dominated the political debate of the late 1960s and the 1970s (Metzler 2005; Nolte and Gosewinkel 2008). As Kai-Uwe Hünemörder (2004) and Jens-Ivo Engels (2006) have argued, this belief was also an important trait of the Environmental Program (Umweltprogramm) through which the new Social-Democratic/Liberal federal government (formed under chancellor Willy Brandt in 1969) sought to address the newly emerging problem of environmental protection in the early 1970s. With the call for species-protection programs and landscape planning schemes, conservationists met and benefited from this broader political preoccupation with expert-based planning.

It was largely as a result of Sukopp's direct involvement that species protection became a salient component of the Berlin Nature Conservation Act. Although the federal legislation had allowed Länder to develop special regulations on species protection, it did not yet include provisions for species-protection programs. When the ruling Social-Democratic/Liberal coalition presented a draft regulation to the Abgeordentenhaus in 1977, it only sketched the basic principles of the Landscape Program, and mentioned the Species Protection Program as one of its goals.[15] The final law detailed specific demands for the program. They had been proposed by the responsible parliamentary committee, which itself had based its opinion explicitly on publications by Erz, Sukopp, and the ornithologist Gerhard Thielke.[16] One of Sukopp's main successes was the provision that the development of sound "scientific foundations" was provided for as a formal pillar of the Species Protection Program.[17] This regulation paved the way for the biotope-mapping survey that Sukopp carried out on behalf of the administration, through which he established urban ecology as a major voice within the West Berlin planning debate of the 1980s.

The Berlin regulation was especially strong from a legal viewpoint as it explicitly provided for a Species Protection Program as a distinct element of its Landscape Program (§ 28). Moreover, whereas in the other Länder comparable programs only had a consultative status, the Berlin Landscape Program was internally binding for future decision making by all administrative bodies. On the basis of the program, the districts would later develop landscape plans (Landschaftspläne) and green ordering plans (Grünordnungspläne) to specify the policy on a smaller scale. Also, land-use plans and building plans—the basic instruments of urban planning—had to be made consistent with the Landscape Program. Exceptions were permissible only in "justified singular cases" as they were defined by the development planning clause of the Federal Building Code (Bundesbaugesetz) (Muhs 1979: 452). As we will see later in this chapter, however, this original intention later gave way to a weaker interpretation of these legal provisions.

In December of 1979, the Senate which was formed by the Social Democratic Party (SPD) and the liberal Free Democratic Party (FDP) set the first formal steps for the Landscape Program. In 1981, in the wake of a political corruption scandal that led to the replacement of Mayor Dietrich Stobbe, elections were held that resulted in a minority Senate under the Christian Democratic (CDU) mayor Richard von Weizsäcker. After 1983, it continued its work in a coalition with the FDP.[18] It was under the terms of three different CDU/FDP Senates (as of 1984 under Weizsäcker's successor Eberhard Diepgen) that the Landscape Program was eventually finalized. During most of this time, the lead management fell into the domain of the Senate Department for Urban Development and Environmental Protection. Newly formed in 1981, this Senate department was in charge of comprehensive urban and structural planning, including the burgeoning field of environmental policy. It thereby complemented the work of the Senate Department for Construction and Housing that dealt with housing and infrastructural policy. The task was fulfilled by its division of nature conservation, green planning, and landscape planning.

In order to prepare the Landscape Program, the Senate commissioned a number of ecological research projects. Specifically for the Species Protection Program, smaller preparatory studies were conducted that focused on exemplary areas or groups of species, and, as of 1981, the extensive survey that led to the draft version of the Species Protection Program.[19] The Senate also commissioned studies of the landscape scenery and recreation areas that were carried out by landscape and urban planning consultants.[20] In the fall 1981, the Senate decided together with the Umweltbundesamt

(Environmental Protection Agency) to develop an Environmental Atlas. Besides the survey that was financed by the Senate, the Umweltbundesamt financed a whole range of other maps that featured different aspects of the nature household, including the ecochore map that was discussed in the preceding chapter.[21]

Although the Berlin Nature Conservation Act provided for a deadline of 5 years, it took the Senate and the Abgeordnetenhaus until 1988 to endorse the eventual Landscape Program. This delay was mainly due to the gap between the complex research design and the financial support provided by the Senate. A second reason was the shortage of staff on the side of the administration. In the first years, only one and a half appointments were available for the preparation of the landscape program. This changed with the establishment of the Senate Department for Urban Development and Environmental Protection, where under the direction of Klaus Ermer a special working group for the Landscape Program was established.[22] A third reason was political quarrels about the content of the program. The decision of the Senate in 1984 to harmonize the Landscape Program with the new Land-Use Plan 1984 (Flächennutzungsplan, henceforth abbreviated to FNP 84) entailed painstaking negotiations among the responsible sections of the administration that took about three years.

Urban Ecology and the Urbanization of Species Protection

Although the debate in Berlin followed the principles set out in the broader discussion in West Germany, it featured specifically the urban environment as its object area. The problem claim of species loss converged with the research agenda of urban ecology and Sukopp's re-conceptualization of the city as a mosaic of biotopes. Whereas in his earlier efforts on the behalf of nature protection Sukopp had been mainly concerned with the protection of the remains of nature at the fringe of the city, he now also considered the inner city as a target for nature-protection measures. "City and nature are not opposites," he wrote in 1979. "A large number of plants and animals are able to live in direct proximity to the urban dweller. It is the aim of nature conservation in the city, to protect these organisms for cultural, sociological, hygienic and ecological reasons." (Sukopp 1979: 31) Such conservation was necessary, according to Sukopp, because many of these species were endangered and much of the potential living space for plants had been lost to intensifying urban development. To counter this trend, Sukopp advocated designating nature reserves and integrating species protection into the dominant forms of land use (Sukopp, Kunick, and Schneider 1979; Auhagen and Sukopp 1983; Sukopp 1979).[23] In 1978 Sukopp promoted this view at a hearing of a special committee of the Abgeordnetenhaus that reworked a draft version for the Berlin Nature Conservation Act.[24] As Sukopp argued, nature conservation in the city did not intend to "retransform the city into a paradise of nature."[25] He argued, however, that various organisms shared the environment of the city with humans. Sukopp proposed that areas close to nature which existed outside the developed part of the city should be made nature reserves. At the same time, Sukopp also pointed to the value of the built-up parts of the city. As Sukopp elaborated, not only private and public gardens performed a positive ecological function within the city, but equally, the green strips in the middle and at the edges of streets. And he claimed that the many ruderal areas that existed on abandoned lots in the city, "the open spaces which one shyly hides behind walls of advertisements" as he put it, performed a positive ecological function.[26]

Sukopp assumed that nature in the metropolis could only be preserved by "an active development which had to be based on ecological and scientific knowledge."[27] This was the reason why he considered a "scientific-ecological survey of the developed and undeveloped part of the landscape and its ecosystems," or as he also called it, "biotope mapping,"[28] as being indispensable. Only the results of such a survey would allow sound judgments of land-use claims, the prognosis of the impact of encroachments or a systematic choice of areas to be made nature reserves.

During the process of preparing the Species Protection Program, Sukopp and his co-workers presented further rationales to back their claim for urban species protection. As Sukopp and his co-author Axel Auhagen maintained in a 1983 article, biological diversity had a number of different functions worldwide, at a national level, and in the inner urban area (Auhagen and Sukopp 1983). As regards to the city, Auhagen and Sukopp assumed that species contributed generally to the functioning of urban ecosystems. Second, they noted the role of urban species as research objects for basic research in biology and ecology. Here they referred explicitly to their own field of urban ecology that required ecosystems in the inner area of cities that were intact and rich of species. Finally, they maintained that a large number of species and ecosystems had an important function of "recreation and homeland protection" (ibid.: 10–11; see also Sukopp 1979). They also claimed that the species protection would strengthen people's sense of place (Heimatgefühl) and responsible social behavior (Auhagen and Sukopp 1983: 11)—an argument that, notwithstanding Sukopp's otherwise reform oriented rhetoric, was very much in line with earlier ideas of nature conservation. In continuity with the former discourses of conservation and green-space planning, Sukopp and his colleagues also referred to the specific situation of walled West Berlin, which in their eyes made all these rationales even more significant than in other cities (Arbeitsgruppe Artenschutzprogramm Berlin 1984, 1: 25; Auhagen and Sukopp 1983: 10, 11).

In parallel with his specific work on the Berlin Species Protection Program, Sukopp also promoted urban biotope protection at the federal level. Around 1980, some West German cities launched "biotope mapping" surveys that would prepare species-protection programs for their area. A key role in this context was played by the Arbeitsgemeinschaft Biotopkartierung in besiedelten Bereich (Working Group Biotope Mapping in Settled Areas), which was set up in 1978 by representatives of various Länder (Sukopp, Kunick, and Schneider 1979). That group (which until 1996 was directed by Sukopp) became a general forum through which the issue of urban nature protection was kept on the agenda and transmitted to different city administrations. Whereas other members of the working groups (notably the representatives of Bavaria) advocated an approach to biotope mapping that was based on a broad landscape-care approach, Sukopp drew strictly on his research approach in Berlin, pursuing standards that would transform urban species protection into a form of applied urban ecology.[29] The survey results that Sukopp's author group later published in Berlin were thereby also seen as a pioneering attempt to develop a standard methodology for planning-oriented urban ecological surveys.[30]

As we have seen, nature conservationists had justified the protection of natural landscapes and the creation of green spaces with the alleged hostile features of urban life. Sukopp's views on urban species protection retained this old view of nature as a panacea against the ills of the city. What was at issue here, however, was no longer the compensation of the unnatural city by the nature outside the city. Urban ecology instead aimed at the transformation of the city itself to a set of biologically diverse environments. With this idea, the program moved beyond the agendas of traditional nature conservation as well as classical green planning. For example, it broke with the fundamental anti-urbanism of the traditional nature-conservation movement. Rather than treating city and nature as a pair of opposites, Sukopp emphasized the existence of nature within the city. Nature for him was as much part of the city as the city was natural. This was most significant in Sukopp's view that cities developed their own characteristic form of "nature": its spontaneous vegetation and related fauna. At the same time, the program retained a critique of all those aspects of urban development that would minimize the living spaces of such flora and fauna. Much of the work of the Species Protection Program consisted of establishing and maintaining the boundary between the shaping of valuable urban natures and the destruction of nature by the city.

It has not been self-evident for urban ecology to couch its practical mission in terms of species protection. Although Sukopp had always been an ardent advocate of nature-conservation measures, before the late 1970s even he had not yet considered the inner city as an object for conservation. Duvigneaud, in his work on Brussels, never showed a comparable interest in protecting urban species as a goal in itself. Why then did this focus on nature conservation become so central for Sukopp? First of all, Sukopp had always operated in both arenas. He was a nature conservationist, and as an expert advisor he was deeply involved in the advisory community that promoted species protection. Through advancing urban biotope protection, Sukopp merged these two areas of engagement into a common political concern. More important was the context of legislative reform in which the Species Protection Program developed. The debate on a new nature-conservation law offered Sukopp an opportunity structure to extend the domain of ecological expertise to the city. By linking urban issues to nature conservation—to "translate" them, as Latour would say—urban ecology was able to achieve an important institutional underpinning to position itself in a field that had been occupied by the urban planning profession. It was not in the construction and development law that ecologists found a route to practical relevance in the city, but via the Nature Conservation Act. Sukopp therefore always underlined that the Nature Conservation Act had to be applied to the entirety of the territory (e.g., Auhagen and Sukopp 1983).

The Construction of Endangered Species: Red Lists for West Berlin

A first step through which the Species Protection Program materialized was the compilation of Red Lists of Endangered Species. This was meant to provide criteria for the evaluation and protection of biotopes as well as for the design of special action programs which aimed at the reintroduction or promotion of species. As a spokesman of the federal Arbeitsgruppe gefährdete Farn- und Blütenpflanzen (Working Group Endangered Fern- and Flowering Plants), Sukopp had already edited the first list for the FRG on this group of species (Sukopp 1974). Sukopp invited specialists on specific groups of species—most of whom were amateur naturalists—to contribute to the development of the Red Lists of Endangered Plants and

Animals in Berlin (West).[31] In June of 1980, at a conference with local government officials, the group presented its results (Sukopp and Elvers 1982). As the organizers posited, among a total of 1,269 plant species that had been registered in Berlin, 137 (10.8 percent) were "extinct" (or had "disappeared") and 487 (38 percent) were "endangered" (ibid. p. II). Likewise, 210 (12.5 percent) of 1,685 animal species were "extinct" or had "disappeared," and a further 698 (38.9 percent) were "endangered." The group explained these results by the degradation of biotopes in the urbanized environment. The group, however, also had some good news: a few species that were endangered because their original biotopes had disappeared had found alternative biotopes, notably in urban wastelands (Auhagen and Sukopp 1982: 16). The basis of these Red Lists was the extensive knowledge that local naturalists and ecologists had accumulated about West Berlin's wildlife. At the same time, the lists encoded a number of methodological conventions which were negotiated at the national as well as the local level. First, the establishment of lost and endangered species was based on a comparison of contemporary records with the flora and fauna of about 100 to 150 years ago. As Sukopp and his colleague Hinrich Elvers explained, this was due to the fact that precise records from earlier times were not available. In the case of Berlin, Ascherson's 1859 book *Flora der Provinz Brandenburg, der Altmark und des Herzogthums Magdeburg* had provided the backdrop for such estimations. This meant, however, that this historical state also functioned as a normative yardstick for the diagnosis of the contemporary quality of flora and fauna (Sukopp and Elvers 1982: 2), as opposed to some earlier, probably even pre-historic situation.[32] This reference standard fit well with the historical origins of the German nature-conservation and landscape-care movements, which were more concerned with preserving a pre-industrial and pre-urban landscape than with some sort of prehistoric wilderness.

Second, the Red List implied assumptions about critical thresholds beyond which a group of species were to be considered rare or endangered. Whereas, for example, Sukopp (1970) considered plant species as "rare" when they were to be found at fewer than eleven locations in Berlin, the amateur naturalist Walter Stricker (1977) based his own list of rare plants on a threshold of only five stands. Since ecologists rarely possessed precise data about the quantitative size of the populations under consideration, they based their judgments on data on the spatial distribution of a species and its changes over time (Sukopp and Elvers 1982: 2). The Berlin group used the number of its separate populations within the observation area as an indicator for the degree to which a species was endangered.[33]

Another issue that was discussed in the group concerned the status of so-called neophytes. Former Red Lists had referred only to indigenous and archaeophythic plants—that is, plants that had already been established in the flora before 1500. And neophytes had been explicitly excluded from the federal Red List that Sukopp had co-authored (Sukopp 1974; Korneck and Sukopp 1988). This choice was motivated by a largely negative evaluation of neophytes among nature conservationists. Such species are often seen as intruders into the indigenous flora that caused environmental degeneration, falsified the character of the indigenous flora, or were generally of less ecological value (Gröning and Wolschke-Buhlmahn 1992; Eser 1999). However, as Auhagen and Sukopp claimed in 1982, neophytes were so abundant in the flora of a big city (about one-fourth of the area of closed development) that they should be included on the Red Lists (Auhagen and Sukopp 1982: 5). On this basis, the Berlin group estimated that 48.1 percent of ferns and flowering plants were either lost or endangered (otherwise it would have amounted to 54.3 percent) (ibid.). Although they expected a further rise in the overall number of neophytes in the city flora, some neophyte species had themselves become so rare that ecologists considered them endangered. In the second version of the Red List for Berlin, which appeared in 1991, Kowarik defended this approach against critics. Among other things, he maintained that this was a necessary consequence of a value-neutral stance in the compilation of Red Lists; accordingly, they were just meant to register the disappearance of settled plants, independently of how one judged their ecological value (Kowarik 1991).

The decision to include neophytes on a Red List was significant for the positive evaluation of the man-made nature of cities that had evolved in the biological fieldwork culture of West Berlin. Rather than seeing them as falsifications of homeland nature or as signs of environmental degradation, as is the case in much contemporary conservation discourse, the Berlin group presented them as plants that were more adapted to the urban conditions. It claimed that, at least in urban biotopes, neophytes should be accepted and, if necessary, protected. This did not mean that the distinction among these categories completely lost its meaning. This was underlined by the separate estimation of the quantitative amount of species endangerment. Moreover, that neophytes were judged positively with respect to the inner city did not mean that they were also welcome in the more rural biotopes in the outskirts. What concerned the ecologists was not so much whether the species were alien as the extent to which they contributed to or reduced species diversity within a specific ecosystem.[34]

The Red Lists were compiled specifically for the species-protection program, but as boundary objects they quickly achieved a much broader significance in the day-to-day life of the Berlin administration. Nothing could better underline ecologists' and conservationists' claims, or their criticisms of certain urban projects, than the presence of "Red List species" at a certain location. They turned complex findings of biological fieldwork into simple indicators that could be easily communicated to broader audiences. The systematic marking of species as "lost" or "at risk," together with the metaphorical association of the alarming color red, made the lists particularly suitable for creating a public sense of urgency. At the same time, the lists appeared as an objective record instead of a mere normative judgment. As Erhard Mahler, the director of the responsible section in the Senate's environmental department, stated in 1991, "by their pure existence, that is by the 'factual power,' the Red Lists had achieved a significance that goes far beyond the weight of a simple statement of opinion of consultants or experts" (1991: 14).

Conservation Planning and the Biotope-Mapping Project

Berlin ecologists were convinced that the decline in species could be halted only if the biotopes on which these species depended were effectively protected or restored. This was why, besides the Red Lists of species, an inventory and ecological evaluation of the existing biotopes in Berlin was a second pillar on which the Species Protection Program rested. The Berlin Nature Conservation Act only generally provided for such an inventory (§ 28, 2, 1). Its specific goals and methodological design remained a matter of negotiation among the ecologists and the Senate. The landscape planner Klaus Ermer, who coordinated the work in the Senate, acted as Sukopp's direct counterpart. As a former student at the Technische Universität, Ermer had already been involved in drafting a first research program the Landscape Program (Ermer, Kellermann, and Schneider 1979/80). In his new position at the Senate, he played a major role in the mobilization of organizational and financial support for the survey. At the same time, he also sought to influence the trajectory and the content of the survey in order to make sure that the results would meet the specific needs of the planning administration as much as was possible.

The Quest for Data: From Comprehensiveness to Sampling

Originally the survey was meant to provide a comprehensive inventory of the land-use structure, as well as the flora, vegetation formations, and fauna of its distinct biotopes (Sukopp, Kunick, and Schneider 1979). Such a comprehensive approach was only realized in the district of Kreuzberg, where two pilot projects were carried out in 1978 (Kunick 1979; Asmus, Martens, and Scharfenberg 1983).[35] The scarcity of financial resources and staff forced the working group to revise its methodology continuously and to search for a less laborious approach. In 1979 the landscape-planning consultant Christian Schneider (Schneider and Ökologie & Planung 1979: 17) suggested restricting fieldwork to representative sample areas covering only about one-fourth of the territory under consideration. Schneider also suggested drawing systematically upon existing material on Berlin's biotopes to reduce the actual amount of fieldwork.[36]

Around 1982, even this turned out to be too ambitious to be achieved within the given time period. The budget that Ermer had at his disposal for external expert reports expanded only slowly between 1979 and 1983.[37] New data had been produced solely in the context of expert reports, carried out by the Institute of Ecology between 1979 and 1981, that focused on a few especially significant areas.[38] At the same time, Ermer's department was under time pressure to identify priority areas for nature conservation and to defend them against alternative development interests. The Christian-Democratic government, which came new into office in 1981, had decided to synchronize the stipulation of the Landscape Program with the new land-use plan (FNP 84) and to finalize both in 1984. In addition, it developed Guidelines for Urban Development (Leitlinien zur Stadtentwicklung), which were meant to define general development goals for the city. It was critically important for the Senate department to receive survey results that could be related to the territorial scale on which these programs operated and would help it to bolster spatial claims against other political priorities. In March of 1981, Ermer therefore urged the survey group to accelerate its work so that it would be able to present at least some provisional results by May of the same year.[39]

In order to meet this challenge, Axel Auhagen, a full-time management assistant at Sukopp's advisory office, proposed a further revision of the methodology. Although he still envisioned a more intensive period of fieldwork in 1983, he gave priority for a "quick mapping" (Schnellkartierung) of the land-use structure, which was to be carried out in the same year.[40] Instead of detailed studies of the flora and fauna of single areas, it aimed at the cartographic inventory of the land-use structure of the territory. This was in line with Kunick's and Sukopp's earlier claim that, in a city, human land use was the dominant factor determining the composition of the flora and fauna. The group relied on official documents of land use in Berlin and on aerial photographs. Rudimentary field observations helped them to further differentiate the land-use categories according to the ecological features (Institut für Ökologie 1982).

Although originally meant to be only an intermediate result, land-use mapping paved the way for what the authors of the report later described as their "reduced methodology" (Arbeitsgruppe Artenschutzprogramm Berlin 1984, I: 78). To avoid time-consuming fieldwork, they took the idea of extrapolation from sample areas one step further. Biotopes in the city were now classified under larger categories: the so-called biotope types, which were roughly consistent with land-use categories, and which were assumed to each represent ecologically homogeneous conditions. Instead of inventorying these areas, the group only analyzed data from existing studies and extrapolated their results to the entire category. Although the group finally arrived at nearly the same number of biotope types (ibid., I: 57), some land-use categories had been further differentiated or coalesced in order to make them consistent with the data on species occurrences (ibid., I: 83).

Owing to the time pressure and the change of method, the species-protection survey produced much less new evidence on Berlin's flora and fauna than had originally been intended. It was largely a secondary analysis of existing data on about 600 sample areas (Arbeitsgruppe Artenschutzprogramm Berlin 1984, I: 80). In principle, the survey group considered all data that had been produced since 1945. Additional information was achieved through interviews with "experts on the terrain" (geländekundige Fachleute) (ibid., I. 110) and through a questionnaire that had been sent to nature-conservation organizations, public agencies, and individuals (ibid., I: 79). Whereas the group had originally planned to apply one standard method of data collection, it was now faced with a heterogeneous body of data. In order to maintain at least some minimal consistency, the group decided to use only data from inspection areas that "could clearly be delineated and localized" (ibid., I: 78). Still, it remained difficult, and in many cases impossible, to relate all this information to the areas that were indicated on the biotope map (ibid., I: 108). On the other hand, information about the biotopes was uneven—for example, some industrial areas had only recently begun to attract the systematic interest of ecologists (ibid., I: 187), and some military areas were not accessible to the survey's participants. Even the scale on which data about biotopes was gathered and represented on the map was determined by the financial and time pressures.[41]

Another implication of this methodological change was that the survey lacked a systematic analysis of the correlation of floristic and faunistic data, on the one hand, and land-use respective biotope categories, on the other. The "reduced methodology" implied that the biotope types were deduced from existing land-use categories instead of being inductively constructed on the basis of new fieldwork data. Although the authors of the report justified this by the hypothesis that in the city the actual land use was the main determinant of the distribution of plant and animal species (ibid., I: 44), they made it clear that this had never been confirmed at the level of single biotope types. That would have required a broader database and the use of quantitative correlation analyses as they had been used in Kunick's thesis. Owing to the circumstances under which the survey was produced, however, the supposed influence of urban land use had changed from a research topic to a resource from which knowledge claims were derived.

This trajectory shows that the biotope-mapping survey had become increasingly determined by the time pressure and the practical demands of the urban planning institutions by which it was commissioned. It rarely complied with Sukopp's intention to make it a pioneering experiment in planning-oriented ecological research. In the final report, Sukopp made no bones about his disappointment with the outcome of the survey. As he declared in the preface to the report, the survey only roughly met the quality standards required for a Species Protection Program (ibid., I: 3). For Sukopp it was clear that further research was needed, both to better understand the urban ecosystem and to effectively guide practical nature conservation. It is significant, however, that the alleged shortcomings of the Berlin survey—however serious they might have been from an academic perspective—had no visible impact on the legitimacy or the operability of its results or on the planning measures that were based on it. Here we can see how the inclusion of ecological observation work into an administrative context also changed the standards on the basis of which knowledge claims came to be justified and stabilized.

The Biotope Type as a Mapping Unit

The notion of the biotope type as a classification device for spatial units had already been introduced in earlier biotope mapping surveys outside of settled areas and was also used in a series of surveys of Bavarian cities that were carried out in parallel with the Berlin project (Brunner et al. 1979;

Kaule, Schaller, and Schober 1979; Sukopp, Trautmann, and Schaller 1979). These surveys focused on a few sample areas that had been selected beforehand as potential priority areas. The classification of these biotopes was based on vegetation structures or other features of the physical landscape. Although the Berlin survey included similar biotope-type categories, notably for the nature reserves and the more rural areas at the fringe of the city, it differed from these earlier surveys as it structured the entirety of urban space into 57 distinct biotope types. Urban development types such as different forms of open or flat-block building, industrial areas or urban infrastructures such as harbors, railways tracks, streets, filter beds, landfills, and construction sites, thereby became re-framed as distinct biotope types (Arbeitsgruppe Artenschutzprogramm Berlin 1984, I: 96–98). Often these categories were further subdivided according to typical differences in their vegetation features. In a similar way, open spaces such as parks, cemeteries, agricultural areas, waters, and forests were neatly differentiated according to specific land-use criteria and landscape-ecological features.[42]

The biotope types were used as a basis for various maps (including an unpublished working map and the program map), and they structured the textual presentation of evidence on the Berlin flora and fauna. The survey group represented the Berlin territory as a set of mappable tokens of these classificatory units, the actual biotopes. As the group defined it, the word *biotopes* "means circumscribed areas of a biotope type on the biotope-type map" (ibid., I: 194). Since biotopes of the same type were supposed to be ecologically homogeneous; they were also considered comparable in their ecological significance and conservation needs, so the same "assignments of action" (ibid., I: I75) were designated for them. Biotope types were also the level at which an ecological evaluation of biotopes operated, and for the most critical among them a Red List of Endangered Biotope Types was compiled. The biotope types were also the territorial building blocks of larger "development zones" for which the group defined different measures of biotope protection and development.

The first version of the biotope-type map was a montage of several sheets of the official map of Berlin on which the biotope areas were delineated with blue ink. In the final report, a smaller-scale version of this map was reproduced under the title "value of the biotopes." The coloring of certain areas in red and pink indicated the results of an ecological evaluation that had been carried out on a limited number of biotopes. Another version of the map displayed the city as a mosaic of differently colored areas, which signified different sets of biotope types. The explicit theme of this map was the practical conservation and development measures that

Figure 3.1
A halftone reproduction of a map of different biotope types in Berlin from the 1984
Artenschutzprogramm of the Senatsverwaltung Berlin für Stadtentwicklung und
Umweltschutz. In the original, colors refer to biotope types that require the same
kinds of practical measures. (courtest of Senat Berlin)

were proposed for each biotope type. The other thematic plans that were produced for the Species Protection Program were basically rearrangements of the units of the biotope map for more focused and selective purposes.

In contrast to maps of homogeneous biotope-type areas, dot maps of species or detailed vegetation maps, such as were usually produced in academic work of Berlin ecologists, did not play a central role in the survey. Originally, Sukopp and Kunick intended to produce such maps in the Berlin survey (Sukopp, Kunick, and Schneider 1979), and ecologists continued to proclaim that they were particularly valuable for the preparation of species-protection programs (Schulte et al. 1986). The value of dot maps lay mainly in their precision and in the complexity of information that they afforded in their display of specific spaces (Kunick 1984). For the creation of planning information, however, such complexity would have been more of an obstacle. Actually, planning assignments and the documents in which they were stipulated focused not on specific spots, but on larger parcels of space. The structuring of the city into biotope types linked biological knowledge to the institutionalized classification practices of urban planning and thereby facilitated communication between academic ecologists and administration personnel. Biotopes were defined on the same scale and by similar criteria as the spatial units of urban planning schemes. Evaluations and practical recommendations that focused on these areas could thus be directly transcribed into thematic maps or planning schemes. Moreover, as the authors noted in their report, biotope types were preferred to conventional forms of biological mapping—notably the detailed mapping of so-called plant communities—because they could easily be identified by planning officials who did not possess any academic biological training (Arbeitsgruppe Artenschutzprogramm Berlin 1984, I: 33).

Although the biotope map integrated a vast amount of environmental information about urban wildlife and environmental conditions, it fit very well with the interpretative habits of its recipients in the planning administration. The taxonomic and cartographic structuring of the city was thus not simply a continuation of earlier ecological research. It resulted from an approximation of ecological knowledge and the circuits of observation through which it was produced to the representation needs and practices of the planning administration. In a similar way as the Red Lists, the mapping units could therefore function as conceptual boundary objects that linked ecology with the planner's world.

The Value of Biotopes

One of the pillars of ecologists' authority in planning matters was their claim not only to produce empirical insights into urban nature but also to be able to evaluate urban spaces according to allegedly objective and scientifically valid standards. Standardized evaluation procedures have come to be widely used in landscape planning since the 1970s. A famous example was the so-called Vielfältigkeitsfaktor (diversity factor), which the landscape planner Hans Kiemstedt (1969) developed to give quantitative expression to the "recreation effect" of an area. The problem of evaluation had also been discussed by conservation biologists since the late 1970s.[43] Sukopp (1970, 1971) also had called for an "objective evaluation" to establish the scientific-ecological significance of nature reserves. By comparing the results for each reserves with respect to a set of general ecological characteristics, he had ranged them into three levels of conservational value (1970, table 1).

In the biotope survey the evaluation focused on the whole spectrum of biotope types and biotopes rather than only on nature reserves. It was the basic function of the evaluation to provide a rational fix of potential conflicts between conservational aspirations on the one hand, and competing land-use claims on the other. The authors thereby referred to a principle that Wolfgang Erz, some years before, had illustrated in the form of a pyramid (Erz 1978; Arbeitsgruppe Artenschutzprogramm Berlin 1984, I: 125). Accordingly, the intensity with which an area could be claimed for nature-conservation purposes diminished with respect to the intensity of established forms of land use as well as the overall size of the area. Thus, areas in which land use was absent and which represented only small patches, i.e., the top of the pyramid, should be turned into strict nature reserves. At areas with the most intensive land use and large size, i.e., the bottom of the pyramid, nature conservation should be confined to small "accompanying measures" that did not interfere with the dominant form of land use. In the middle range of this spectrum, Erz suggested giving relative priority to nature conservation or, when land use became more dominant, stipulating selective restrictions on this land use. The outcome of the evaluation procedure helped to place biotopes in this pyramid.

The actual evaluation procedure consisted of various separate selection processes that built on one another in a cascade-like manner (Arbeitsgruppe Artenschutzprogramm Berlin 1984: 123).[44] The criteria that guided these procedures corresponded to the wider rationale of the Species Protection Program. That program focused on species' occurrences within the respective biotopes or biotope types, which were estimated in both terms of biological (or "biotic") diversity and rarity. The first term referred to the number of species in one biotope as well as structural elements such as plant communities or habitats/ecosystems. Although diversity was generally viewed as a positive feature of a biotope, the consistent application of this criterion faced some practical dilemmas. Sukopp (1970: 138) had already discussed this when he evaluated the Berlin nature reserves. At that time, he had maintained that estimations of species diversity would require precise data on the number of species in the respective areas that were often not available. Moreover, he viewed biological diversity as dependent on the size of a biotope so that comparisons should ideally focus only on biotopes of the same size. This was barely the case with the biotopes that had been inventoried in Berlin. As Sukopp had concluded, biological diversity could therefore only be "one, albeit important, criterion among others" (1970: 138; see also Sukopp 1971: 186). Furthermore, the survey team maintained that an excessive number of species could also be a sign of ecological disturbance. In evaluating individual biotopes it therefore referred to an average range of species diversity in the respective biotope type as a yardstick, instead of the absolute number of species.

In addition to the overall diversity, the group referred to the presence of individual species (or, to a lesser extent, plant communities) that were rare in the biotope or the biotope type under consideration (Sukopp 1971, 1970; Arbeitsgruppe Artenschutzprogramm Berlin 1984, I: 143). This criterion was relatively easy to operationalize through sightings of Red List species. Sukopp (1971: 188) had already suggested that the occurrence of rare species could also be taken as an indirect indicator of species diversity, since both conditions typically went together in the same biotope.

As a third criterion, the survey introduced the "biotic potential" of a biotope type. It was specifically introduced to evaluate the "heavily sealed" biotope types of the inner city. These included smaller open spaces adjacent to buildings such as courtyards or gardens, which were characterized by a high "hostility to life" (Arbeitsgruppe Artenschutzprogramm Berlin 1984, I: 117). Depending on the degree of sealing and the intensity of horticultural maintenance, however, they were supposed to have a potential function as a biotope for some desired, albeit not specially threatened, species.

Finally, biotopes also ranked higher on the evaluation scale if they presented a very low grade of hemerobia (Sukopp 1970) and therefore were considered representative of the original landscape (Arbeitsgruppe Artenschutzprogramm Berlin 1984, I: 116).

With the evaluation of biotopes, the survey moved far beyond the knowledge-making practices that had so far dominated ecological field-work. Their role was not to create new empirical knowledge claims about urban ecology, but rather to translate the latter into land-use claims that could be justified and defended in administrative-political conflicts. Ecologists thereby opted for a method that operated with standardized criteria that could be quantified, added, and weighed. As they declared, the results were often only estimations and conclusions, which were not scientifically valid. This does not mean that they considered them false or unjustified, but only that they considered them less demanding and, purely formal criteria of objectivity. The achievement of the procedures that they developed in the survey was seen in the higher transparency of the underlying evaluation criteria (ibid.: 111). In this respect they followed a common trend in modern politics to create transparency and accountability by creating standardized forms (Porter 1995). This made criteria accessible to the relevant publics in the policy process even if these were not able to recapitulate the empirical details that had led to these results. At the same time, however, this procedure also had a concealing aspect. Although dealing with issues that have an intrinsic normative quality—what kind of nature do we want to preserve?—the procedure ended up with the attribution of values to the biotopes that appeared as their quasi-objective traits.

The Experts' Proposal

Sukopp's working group not only produced a survey and an evaluation of the existing state of Berlin's flora and fauna but also made a proposal on how these insights should be put into the planning practice. In close cooperation with Ermer's department, the group developed four thematic plans, which represented the first draft of the future Species Protection Program (Arbeitsgruppe Artenschutzprogramm Berlin 1984, III). The plans were drawn up in the form of maps and textual explanations and provided for different types of measures to protect or enhance the ecological quality of the Berlin biotopes. In addition, the report of the working group also suggested "urgency measures" for extremely endangered species, the so-called Species Support Programs (Artenhilfsprogramme).

At the core of the four schemes was an operative knowledge that combined widely approved conceptions of conservation science, local field observations of Berlin ecologists, and the practical skills of landscape planners. This knowledge took shape through a relatively open process of collective reasoning in which general conceptions and technical skills were adapted to the specific requirements of urban planning. The translation of conservation claims into plans required that they be packaged into compact assignments related to clearly circumscribed planning areas. Thus, all the plans had biotope types as their basic territorial building blocks. Through the concept of the "development-zones," the group synthesized these assignments further, so that their scale matched the urban zoning through which the FNP 84 operated.

There was hardly any area in the city for which the group did not assign some practical advice. However, it was a guiding principle of the Species Protection Program that its measures should respect the existing patterns of land use (Arbeitsgruppe Artenschutzprogramm Berlin 1984, I: 5). The classical measure of nature conservation—the designation of reserves and monuments—was therefore restricted to only a few biotopes, which scored at the very top of the evaluation scale. Many of the proposed reserves represented biotope types on which nature conservation had always focused: fens, ponds, wetlands, etc. which were typically located at the geographical margins of the city. Landscape reserves were also proposed in some green spaces, such as the Tiergarten. The most innovative aspect of the Species Protection Program was its provision for reserves on urban wasteland. Sites such as the central railway track triangle and the adjacent former freight station, as well as the Südgelände in the district of Schöneberg, scored comparatively high in the evaluation procedure (ibid., I: 162).

A salient goal of the proposed program was maintaining or even developing the living conditions for plants and animals. For example, some meadows were to be mowed regularly, or certain plant species were to be removed. Other technical interventions aimed at more far-reaching modifications of biotopes in order to enhance their value or even transform them into a more valuable biotope type; these latter interventions were called development and thereby semantically differentiated from the mere maintenance of an existing biotope type (ibid., I: 140–141). Following earlier suggestions by Sukopp, the authors of the report proposed that water be brought back to dried-up bogs and wetlands. In built-up areas they called for environmentally friendly gardening practices (no cutting of hedges, no use of herbicides, tolerance of spontaneous vegetation, no removal of fallen leaves), the transformation of lawns into meadows, and the systematic greening of houses' facades and roofs. For each of the twelve "biotope-development zones," broader development goals were formulated which integrated the measures at biotope level into a coherent large-scale policy (ibid., II: 549–717).

The attempt to technically enhance the quality of Berlin's wildlife places the Species Protection Program within a range of recent policies of "nature development" (Keulhartz 1999; Helford 1999; Belt 1993) or nature restoration (Helford 1999; Gobster and Hull 2000; Gross and Hoffmann-Riem 2005). In contrast to these other attempts, however, the Berlin program was not based on a single overarching historical reference (such as the pre-colonization savannah of the Chicago area or the Dutch marches in the direct aftermath of the glacial age.) Its development goals were instead specified on the basis of the classification of urban nature into a plurality of biotope types and the standards by which those biotopes were evaluated. Historical references—which went no further back than to the state of the landscape before the industrialization of agriculture and urbanization— were thus significant only with regard to "close-to-nature biotopes" or those that were meant to be transformed into such biotopes. One case was the Tegeler Fließ, a relict of the glacial wetland that had suffered severely under intensified land use and recreation, for which the program suggested re-naturalization measures (Arbeitsgruppe Artenschutzprogramm Berlin 1984, II: 671–674). Other development measures aimed at enhancing bio-diversity or creating retreat areas for species within humanly shaped areas. Examples are the proposal to promote "site-adequate and region specific park trees" in housing areas (ibid., II: 589). The program also suggested to transform ornamental greeneries into wasteland biotopes (ibid., II: 576), or to renewed sprinkling of the abandoned sewage irrigation fields (ibid., II: 670). Whereas the first proposal was clearly inspired by the model of pre-industrial agricultural landscape, the latter two proposals sought to promote biotopes, that were profoundly industrial and urban, yet with regard to their species characteristics were considered particularly valuable.

As has already been noted, the proposed plan was also largely inspired by the concept of biotope networking. A coherent network should allow a smooth exchange of animals and seeds among the individual biotopes, and thereby sustain the city's biodiversity. An equal distribution of biotopes throughout the city, the existence of smaller intermediary biotopes (Tritt-steine), and large connecting linear structures were considered prerequisites of a viable network of urban biotopes (ibid., I: 161–163). For example, the program proposed a large greenbelt from the central Tiergarten to the wasteland areas around the Potsdamer Platz to the Südgelände and further down to the southern fringe of the city. The contours of such a future network in West Berlin were mapped out in the form of specific "biotope development zones" in the fourth plan of the program (ibid., I: 60–61).

Instead of simply applying a ready-made conservation conception to Berlin, the planners had to actively adapt the concept of networking to the local circumstances. As they admitted in their report, the conditions for the realization of such a system were rather limited in the city (ibid., I: 162). The lack of detailed ecological insights into the actual exchange of seeds in the city implied that its design had to rely mainly on estimations. Moreover, the persistent disturbance by dominant land-use practices made it rather difficult for a biotope network to function in the optimal way. Finally, the restrictions given by other, principally accepted forms of land use, constrained the possible selection of elements for such a network. At the same time it was an advantage of the concept of biotope networking that it allowed ecologists to build effectively upon understandings and material structures that had been deeply entrenched in the history of modern urbanism. Urban-technical infrastructural systems such as canals, power facilities, roads, and railway tracks had already been built and conceived of as network-like structures. In their report, ecologists reworked the meanings of these networks in bio-ecological terms: greenbelts, roads with their accompanying green strips, railroads, canals—all of these became reframed as potentially valuable linking elements of a biotope network.

Although in principle it encompassed the entirety of the space, this was not a program that radically challenged the structure of the city. First, measures were basically kept within the confines of biotope types which were defined on the basis of existing land-use forms. This foreclosed any attempt for more radical changes to the type or spatial distribution of land use throughout the city. Second, the evaluation of biotope types was explicitly balanced against other land-use claims so as to coordinate them in the most complementary way. The definition of the "present land use" was thus of eminent political importance, since it prefigured the space within which species-protection measures could be legitimately imagined within the program. In line with the idea of modernist city planning, it was based on the assumption that each area has one dominant function. Since people actually tend to use these areas in multiple, overlapping, and often contradicting ways (and also in socially illegitimate ways, such as for loitering or prostitution), singling out a dominant function necessarily gave priority to one function over the other.

Finalizing the Program

The deadline for the official submission of the Species Protection Program expired in January of 1984. The draft proposal was the only way in which the program had materialized in that year. Together with the results of the biotope-mapping survey it formed part of the "Grundlagen des Artenschutzprogramms," which were published in two volumes and a set of maps by the Technische Universität's division of landscape development (Arbeitsgruppe Artenschutzprogramm Berlin 1984, I–III). The Senate issued a brochure in which it presented the draft proposal of the Landscape Program to the wider public (Senator für Stadtentwicklung und Umweltschutz 1984). Besides the four plans of the Species Protection Program, it included a further draft plan that focused on scenery and recreation. Other aspects of the nature household were covered by a couple of thematic maps that had been produced for the Environmental Atlas but which had not yet been developed into a distinct planning scheme (ibid.: 46–47). Together with a parallel draft proposal for a new FNP 84, these plans were also shown during a public exhibition by the Senate in June.[45] Although in itself not officially endorsed as a program, the draft proposal received broad public attention and became an important landmark for the public discussion on nature protection and green planning in Berlin. In the following years political actors referred variously to the stipulations of the draft program when they made claims for nature protection in the city.

There was no direct avenue, however, from the draft proposal to the official Landscape Program that was submitted by the Senate in 1988. First, substantive parts of the Landscape Protection Program were still rudimentary. They had to be elaborated and integrated with the Species Protection Program into an internally coherent package of planning schemes. Second, and more importantly, the Senate insisted on the need to harmonize the content and procedures of the Landscape Program with that of the new FNP 84 (Senator für Stadtentwicklung und Umweltschutz 1988: 4).[46] Since the provisions of the draft programs that had been presented in 1984 included many contradictory provisions, this required a controversial process of fine tuning in which some provisions of the original proposal were sacrificed. For the Senate this linking was also a formal excuse for the delay of the Landscape Program, which had been due in January of the same year.

After Sukopp's research commission had ended, the technical work on the program became concentrated within the unit of landscape planning at the Senate Department for Urban Development and Environmental Protection.[47] The unit played an important role in defending land-use claims of the original draft against caveats of other actors within the administration. The revised draft of the Species Protection Program that the Senate presented to the citizens and public stakeholders in 1986 and the final document that was submitted to the Abgeordnetenhaus in 1988 were the results of the cumbersome negotiation among these administrative units.

In his role as a Landesbeauftragter, Sukopp tried to bolster the claims of the initial report against potential counter-claims. The Landscape Program and its relation to the FNP 84 figured variously on the agenda of the Council which critically accompanied the process.[48] The Council also did everything to defend areas that were earmarked in the draft of the Landscape Program against alternative development interests. In addition to the justifications for a species-protection program that he and his co-workers had presented in the preceding years, Sukopp mobilized economic arguments. Sukopp and the Kassel economist Ulrich Hampicke cooperated on a cost-benefit analysis of the proposals of the species-protection program. In April of 1986, Sukopp and Hampicke gave a press conference at which they presented the results of the study. The message was clear: Nature conservation was "not as expensive as many people think" (Berliner Zeitung 1988). For only 4 million D-Marks—which, as Sukopp and Hampicke underlined, was only 0.01 percent of the GDP of the city—the measures could be fully realized. Sukopp and Hampicke even argued that this could be done in a way that was financially neutral for the public household.[49] As the press coverage reveals, Sukopp and Hampicke used this event to directly support the Landscape Program against the critique of other parts of the administration that wanted to designate more space for economic development. (See, for example, Spandauer Volksblatt 1986.)

Parliamentary debate and citizen and stakeholder participation were further arenas of negotiation. In the Abgeordnetenhaus, as had previously happened with the Nature Conservation Act, consent existed regarding the general need for the Landscape Program throughout all factions. The belated elaboration of the program, which according to the law was to be launched in January of 1984, was criticized in parliamentary debate. In October of 1985, an official parliamentary question (Große Anfrage) and a related motion by the opposition Social Democrats voiced critical concerns about the state of landscape planning in Berlin, including the Species Protection Program.[50] In the ensuing parliamentary debate, delegates from the SPD and the newly elected Alternative Liste für Demokratie und Umweltschutz (Alternative List for Democracy and Environmental Protection—the West Berlin equivalent of the West German Green Party) called attention to various examples of ongoing environmental deterioration, which, they argued, could have been avoided if the Landscape

Program had already been in force. In their eyes the species-protection survey provided ample evidence that the alarm bells should start ringing, and they demanded that practical conclusions be drawn from that material. "The Senate," one SPD representative complained, "has produced much printed paper on these problems. The as-is state is undoubtedly rather well described, but a description of the state is not enough."[51] In May of 1983, parliamentary motions by the Alternative Liste (henceforth AL) had already aimed at launching immediate action programs which would precede the Landscape Program and which were very much in accord with the immediate action proposals made in the survey report.[52] The survey report had not yet been published by this time, but the AL could draw directly on Kunick's research and the report on the "central area" to support its claims. The AL renewed these motions in June of 1984, but did not find the support of the majority in the Abgeordnetenhaus.[53]

According to the Nature Conservation Act, the proposal for the Landscape Program had to be publicly presented in order to invite opinions by citizens and public stakeholders. The coordinated proposals for the Landscape Program and the FNP 84 were presented in May and June of 1987 in a central exhibition for the whole city and in ten smaller exhibitions, one in each district. After further re-elaboration of the plans, a final presentation took place in March and April of 1988 (Senator für Stadtentwicklung und Umweltschutz 1988). During this procedure the Senate received 1,326 demands for changes in details in the Landscape Program. Most came from individual citizens, others from citizens groups, conservation organizations, and so-called "agencies representing public interest."[54] As one might expect, nature conservationists and environmental citizens groups called for more and more rigid nature-conservation measures than those stipulated in the proposal. Other objections were raised by landowners and by clubs of equestrians, yachtsmen, and marksmen. Although only a few of these opinions resulted in changes in the plan, the Senate was obliged to consider them all and to provide justifications when rejecting them.[55]

Owing to the procedural connection that had been established between the two programs, the debate on the Landscape Program tied in with the fierce conflict that was simultaneously being fought over the FNP 84. Whereas the FNP 1965 (not endorsed until 1970) had foreseen a radical expansion of urban development and infrastructure, the Senate had explicitly revised these goals. Based on a much more modest prognosis of the future population increase, the proposed FNP 84 earmarked less space for development, and thereby left considerable leeway for protection goals of the species-protection program. The Senate explicitly rejected the guiding paradigms of the 1960s, with their preference for asphalt and concrete. In contrast to the former plan, it tried to create space within the city by densification instead of extending the developed area.[56] Yet it still estimated that the population would grow and foresaw various areas as reserves for spatially expansive development projects. Most of those spaces were currently being used as allotment gardens, agricultural areas, and other kinds of open space. These claims were supported by a coalition of local business and real-estate representatives.[57] The FNP 84 therefore faced much opposition from nature conservationists, citizen activists and the parliamentary opposition who called for a "Green FNP."[58] In 1986, with similar intentions, the Sachverständigenbeirat für Nature und Landschaftspflege (Advisory Council for Nature Conservation and Landscape Care), the former Landesstelle, asked that an Environmental Impact Assessment be carried out for the FNP 84.[59] Within the Abgeordnetenhaus, critics also bemoaned that the Senate had begun to enforce the legislative procedure of the FNP 84 in 1987 relatively independent of the Landscape Program, although the proposal for the latter had already been finalized. This meant that, instead of being based on the Landscape Program, the FNP 84 eventually restricted the Landscape Program's regulative scope. According to the critics, however, the provisions of the Landscape Program had originally been intended not to be compromised for the FNP 84, but, in contrast, to guide the latter's construction.[60] The opposition also criticized that the Senate had submitted the FNP 84 before the Landscape Program, which, in their view, ran counter the original intention of making the latter a "basis and green yardstick of the Land-Use Plan (FNP 84, J.L.)."[61] The Senate and the parliamentary majority, on the other hand, argued that the economic considerations that motivated these provisions were of equal value for Berlin as considerations of nature conservation and recreation planning. In parliamentary debate they sought to present the FNP 84 as a balanced compromise that acknowledged both of these conflicting goals. Its representative explicitly distanced himself from the "space guzzling and blessedness of concrete" (Flächenfraß und Betonseligkeit) of the past and emphasized that the new proposal was considerably greener than the earlier FNP.[62] At the same time, the Senate and the ruling parties insisted on the formal priority of the provisions of FNP over those of the Landscape Program.

The draft version of the Species Protection Program had been significantly reshuffled during this process. These changes reflected the compromises that had been made in the coordination of the Landscape Program and the FNP. Some nature or landscape reserves that had been proposed in the initial plan were diminished or completely omitted. Agricultural areas could be preserved only to a much smaller extent than originally intended in the draft. Citizens' participation also resulted in a few changes in the program proposal, mostly by earmarking protected areas for smaller biotopes, overlooked in the survey, that interested citizens had claimed were worthy. In two cases designations of areas as network biotopes were removed from the plan, and in one case a pond that had been designated as a natural monument was put in a less restrictive legal category (Senator für Stadtentwicklung 1988: 52–53).

The elaboration of the plans by the Senate's working group also resulted in a change in the planning format (Senator für Stadtentwicklung und Umweltschutz 1988: 4). The final version of the Species Protection Program consisted of only one single plan, which was displayed on one map. Together with three other plans—focusing on the household of nature, on recreation, and on landscape scenery—it constituted the entire Landscape Program. The provisions that in the first draft had been presented in four different plans were thereby only represented at the level of the "biotope development zones." Biotopes of special interest were emphasized by additional colors or hachures. Special cartographic symbols were used for laying down areas for nature-conservation and landscape-conservation measures, which hitherto had been represented on a separate map. The nomenclature of the development zones, the biotopes to be developed, and the cartographic boundaries were also slightly reshuffled during the elaboration process. Most importantly, the integration of the program into one single map entailed a considerable loss of specificity. Whereas the former (sub-) Program for Maintenance and Development and Action included very detailed proposals at the level of biotope types throughout the same city, the final Species Protection Program only stipulated maintenance actions at the level of zones. The program thereby followed a tendency that was already visible during the work of the working group: Relatively specific information that was based on fieldwork had been translated into larger categories of space that fit with those of other planning schemes. In contrast to the fine-grained pattern of the biotope types, the development zones were largely homogeneous with the scale of the planning zones of the FNP. In this way also the formal organization of the Landscape Program was harmonized with the FNP.

Another result of the policy process was a weakening of the formal power of the Landscape Program. The FNP 84 always had priority over the

Landscape Program. In contrast to the 1984 decision to harmonize the two programs, this allowed the Senate to accept more far-reaching stipulations, since, in the process of implementation, they would be overruled by the FNP anyway.[63]

Although the final version of the Species Protection Program that was enacted by the Berlin parliament in 1988 was much more modest and less specific in its spatial provisions than the 1984 proposal, it was still one of the most far-reaching programs of ecological planning that had ever been approved for a German metropolis. It still contained crucial provisions of the draft program, such as the network of biotope, protected wastelands in the city that had a very innovative character. As a new programmatic principle, both the Landscape Program and the FNP embraced the preservation of the Grüne Mitte (Green Center) (Landschaftsprogramm—Artenschutzprogramm 1988: 66–67). The term Grüne Mitte referred to the Tiergarten, the adjacent Spree Valley, and a number of surrounding wastelands areas. They were considered important species reservoirs in the inner city, as well as a recreation area and an aesthetic landmark for the city's future development.

In the long run, however, the development of the city did not fulfill the ambitions of this plan. In terms of its administrative status the plan was a rather weak instrument. In contrast to conventional urban land-use plans, it was not legally binding but had only a consultative function for the administration. Furthermore, in order to become effective, many of its provisions had to be transposed into Bereichsentwicklungspläne (Area Development Plans, through which the Senate coordinated planning in larger areas of the city) as well as into specific landscape plans of the districts, the smaller administrative districts of West Berlin. Since the intentions of the plan often conflicted with the adverse interests of administration units as well as the districts, this proved to be a rather long-lasting process in which the content of the plan became further watered down. Eventually, a major setback for the Species Protection Program was caused by the fall of the wall in 1989. Both the unification with East Berlin and the subsequent assignment of many central biotope areas to new development plans for the renewed German capital required a major revision of the original Species Protection Program.[64] In June of 2004, the Senate issued a new version of the Species Protection Program that also covered the former East Berlin. In accordance with the new political realities of capital planning, many of the former provisions for nature preservation on the former territory of West Berlin were omitted from this program (Senator für Stadtentwicklung und Umweltschutz 1994).

Conclusion

The Species Protection Program represented a major shift in the history of urban planning in Berlin. In contrast to the urban growth policy that had dominated West Berlin in the 1960s and the 1970s, it sought to integrate aspects of biotope protection and species protection into the shaping and management of urban space. Although such goals continued to compete with growth-oriented interests, and although they largely respected the status quo of urban land use, the program gave a political meaning to the urban nature that local ecologists had put on the agenda. The problem claims that guided this program differed from former open-space policies, which had framed their strategies in terms of public health, recreation, or horticultural aesthetics. By literally putting them on the map and into other officially endorsed documents, the program acknowledged wildlife species, their biotopes, and the ecosystems of which they formed a part as new entities that belonged to urban space. The mosaic of urban zones of land-use on which planners focused their analyses and policy recommendations thus became redefined as a mosaic of living spaces which all deserved a specific form of care. Even if this did not mean that they were protected to the extent that its most ardent advocates would have liked, they became objects of public debate, causes of political frictions, and arenas in which struggles were fought and compromises negotiated.

The environmental knowledge that had developed in the previous years in Berlin had been the discursive mainspring of this policy shift. However, ecology as a planning science differed in striking ways from both amateur research and academic research. Ecological knowledge production did not simply serve the goal of understanding the structural features or the dynamics of urban nature. It was first and foremost a means to bolster territorial claims within the urban policy arena. This had three important implications.

First, ecologists and landscape planners claimed and got accredited authority in matters that touched on normative questions. They dealt with the general issue of what a city should look like, what the place of nature in the city should be, and how such goals should be reached and accommodated to other land-use interests. Sometimes they referred to explicit normative positions, such as the general justifications that they gave for species protection in the city. Sometimes they referred to existing legal regulations that, if taken seriously, in their view required the realization of the measures that they proposed. Furthermore, they developed technical criteria that framed normative questions as technical procedures, such as in the evaluation of biotope types. In all these ways, ecologists were able to achieve normative definition power on urban issues without compromising their status as scientific experts.

Second, Sukopp and his planners became directly involved in the institutional negotiation of a policy. This does not mean that Sukopp and his co-workers had any formal political discretionary power; such power remained the privilege of the Senate, its administrative units, and, of course, the legislative system that eventually had to endorse the new policy. As we have seen, Ermer and his unit in the Senate administration were important mediators as they translated ecological claims into the inner area of the policy system. Though in technical matters they often appeared to be a counter-player to Sukopp, they helped to sustain the basic claims of the Species Protection Program within the political system.

Third, the political context of the Species Protection Program fed back on the kind of knowledge that was produced by ecologists and landscape planners. In their research practice, ecologists had to acknowledge a host of administrative requirements, such as restrictions of budget and time, legal frameworks, and the need to formally integrate their proposals with other policies (notably the FNP 84). Moreover, ecological results and the methods through which they had been achieved had to be communicated and justified to other audiences than the inner communities of ecological field researchers; in other words. they had to become boundary objects. Biotope types, evaluation scores, the maps on which these were represented, and the narratives through which its authors justified the Species Protection Program all had the character of boundary objects. They were based on the observation circuits of ecological fieldwork (and also partly fed back into them), yet at the same time they were constructed according to administrative rationales that ordered urban space as an object of governing. As we have seen, this forced ecologists into cognitive alignments with the logics of administration, and it created tensions with the norms for justification that the ecologists tried to maintain for their academic community of discourse.

Although this chapter has featured the Species Protection Program as the outcome of Sukopp's initiatives, it has also revealed the fragility of the balance of power through which Sukopp and the other ecologists were able to put their stamp on the content of the new policy. Initially, Sukopp was a central figure in the formulation of the policy. He was involved in the legislative process that had paved the way for the program, and he and his working group hammered out the first draft proposal. After 1984, however, the balance of power increasingly shifted from the expert group to the government and its bureaucracy. Although the ecologists' agenda found much support in this context, it became increasingly subject to negotiations and compromises that counteracted Sukopp's initial intentions. Although the growth coalition that had dominated West Berlin's land-use policy could not prevent the Species Protection Program from going through, it demanded a great deal of compromise. This does not mean, however, that the new policy was supported only by Sukopp and his colleagues. As we will see in the next chapter, ecology's role in urban planning gained its momentum from its association with a much broader set of supporting forces, many of which continued to struggle for the realization of ecological land-use claims far beyond the formal endorsement of the Species Protection Program.

4   Building Communities, Forming Alliances

The rise of ecological planning in Berlin was not only due to the dynamics of ecological knowledge production that the city had witnessed since the end of the war. It also reflected the extent to which critical publics within West Berlin had come to acknowledge urban nature as an object of concern and political action. In this chapter I will look at different political communities that coalesced around the regime of urban biotope protection and that supported its central concern against adversaries in the administration, the local government, and the business sector. All these communities represented segments of the new wave of environmentalism that emerged in the 1970s and the 1980s. These regime communities, as I call them here, included the groups of scientific ecologists and professional landscape planners who sought to establish a political mandate in the urban planning arena. They also included a diverse set of citizens' groups (many of them preexisting, others newly emerging) that joined ecologists in their call for urban nature protection and actively promoted this issue in urban land-use conflicts. As a specific aspect of this civic activism, I will also deal with the reverberations of environmentalism in the city's party system, notably through the establishment of the Alternative Liste.

The close combination of scientific-professional expertise and a new civic activism was a salient characteristic of the environmentalism of the 1970s and the 1980s. As we saw in chapters 2 and 3, biological ecology achieved its institutional momentum only in this period, not least because of its claim to contribute to solutions to practical environmental problems.[1] Although many of its representatives, including Sukopp, rejected an overt politicization of their discipline, ecology was often perceived as a Leitwissenschaft (leading science) for a broader reorganization of society and the economy.[2] Ökologiebewegung—the German term for environmentalism—even identified the field semantically with this broader political striving. Ecologists, and even more so the profession of landscape planning, rode on the wave of this environmentalism to extend its institutional basis and to achieve professional "jurisdiction" (Abott 1988) in the planning system. This new environmentalism developed forms of activism that went beyond the elite-based and often politically conservative lobbying of earlier nature-conservation councils and nature-conservation organizations. As analysts have shown, this implied a shift from the often right-wing leanings of traditional nature conservation to the political left (Chaney 2008; Engels 2006; MacCormick 1989). Indeed, many of the activists who in the 1980s joined ecologists in their quest for more wildlife in the city considered themselves left-wing or liberal and adopted the repertoires of contention that had been developed in other protest movements of that period. As Jens-Ivo Engels has shown (2006), participants of environmental protest movements also incorporated elements of political ecology into their lifestyle and identity. Since formerly existing nature-conservation organizations continued to exist in Berlin and also joined the political striving for urban nature conservation, this movement was far from ideologically homogeneous, and tensions between groups of different political shades often became visible.

In this chapter I will describe the institutional structures of these regime communities, characterize their membership, and trace the practices and strategies through which they sought to influence the political game. I want to show how, and to what extent, regime communities claimed, gained, shared, and divided "problem ownership" (Gusfield 1981) with respect to urban biotope protection—that is, to what extent they claimed to legitimately define the problem of urban species protection, and how it should be addressed. It can be said that all of them together shared some ownership of the species-protection issue. Their very existence was justified by their claim to speak for the biotope problem, and they contributed greatly to the way in which it was defined and contested in the local polity. The existence of each group added political weight to the other, so they all benefited from their articulation of the new agenda. They also involved themselves in the same political campaigns and, to some extent, shared individual members. On the other hand, the communities also staked out different roles for themselves. They did not form a coherent nature-conservation lobby. For example, whereas ecologists considered themselves mainly as technical experts whose role was to solve problems in scientific terms, activists saw themselves as procurators of citizens' interests and democratic participation, sometimes also against what they considered an overly scientific or technocratic approach of expertise-based planning.

These groups were as much the product of the discourses and practices of the biotope-protection regime as they were sponsors and operators of this regime. Some of them were newly formed and drew their mission and self-description from the problem claims that guided urban ecology. The new regulatory mechanisms that were institutionalized with the issue of species protection created political opportunity structures that gave rise to specific formations of activism. Some regime communities had a long-standing history that went far beyond their association with the urban biotope-protection regime. Through their embroilment into the new regime, however, they achieved, at least partly, a new identity. Thus, when I use terms such as *community*, *group*, or *organization* in this chapter, I do not refer to any stable predefined structures against which the rise of urban biotope protection can be mapped; rather, I refer to social formations that are co-produced with the new nature regime and to its policy of urban nature conservation.[3]

I map the spectrum of regime communities along a continuum that stretches from institutionalized academia to the more overt forms of political activism. I will start with an analysis of Sukopp's political engagement as an academic ecologist; I will then deal with the emergence of the landscape planning profession and the institutionalization of ecological expertise in the West Berlin planning administration. In the second half of the chapter, I will deal with different strands of activist groups, ranging from classical nature-conservation organizations to neighborhood activists and parliamentary party environmentalism. After mapping the range of regime communities in West Berlin, I will briefly sketch the role of urban environmentalism in East Berlin and discuss how, after the fall of the wall, it became integrated into the entire city.

Ecology and Political Entrepreneurship: The Example of Sukopp

Scholarly ecologists have often been among the active promoters of environmentalist agendas in policy making or in the broader public debate (Cramer 1987; Hannigan 1995; Frickel 2004; Haas 1990). Although Berlin ecologists, unlike environmental activists, have rarely been directly engaged in contentious politics, they have not been merely passive purveyors of instrumental knowledge. Here I will focus particularly on Herbert Sukopp and his role as a political entrepreneur in contributing to the public articulation of the urban nature problem.[4] This does not mean that Sukopp was the only ecologist who promoted the issue in Berlin. Other members of the local research community joined him in calling for political action, and many of his own interventions were actually collective activities in which other ecologists were involved.

As we saw in the preceding chapter, Sukopp and his colleagues promoted their vision of species protection and urban ecology at the national and international levels as well as in Berlin. At the federal level, Sukopp had been involved in many practically oriented commissions, including the Deutscher Rat für Landschaftspflege (German Council for Landscape Care). Notably, he had promoted his vision of urban ecology in the Arbeitsgruppe Biotopkartierung im besiedelten Bereich (Working Group Biotope Mapping in Settled Areas) (see chapter 3). The meetings and the regular proceedings that were published by the group helped to formulate urban biotope protection as a distinct political problem, to make it visible, and to establish contacts with urban policy institutions. The group also kept track of the cities in which biotope mappings had already been carried out and embraced the use of landmark biotope-protection surveys.

Since the 1960s, Sukopp had developed direct working relations with the West Berlin administration. An early example was the evaluation of the environmental impact of a highway cutting through the Tiergarten park that he had carried out with Berger-Landefeld in 1962 (Berger-Landefeld and Sukopp 1962). In the 1970s, the green planning unit at the Senate Department for Construction and Housing received larger budgets for research grants and commissioning expert reports on planning-related issues (Schindler 1971). The inventory of the nature and landscape reserves that Sukopp hitherto had carried out on behalf of the Landesstelle was now officially commissioned by the Senate (ibid.: 78). In 1975 the Institute of Ecology was involved in the evaluation of the effects of the construction of a power station in the Spandauer Forst, next to the bog called Teufelsmoor (Sukopp 1976). This was the starting point of a whole range of evaluation studies that were carried out in the late 1970s and the 1980s. The largest research commission that Sukopp received from the Senate was the work for the Species Protection Program. As we saw in the preceding chapter, it was in this context that his group was able to translate its ideas on nature protection in the city into a comprehensive policy strategy.

The most important pathway through which Sukopp achieved his role in urban policy was the Landesstelle für Naturschutz und Landschaftspflege (Land Board for Nature Conservation and Landscape Care), or, as of 1979, its successor, the Sachverständigenbeirat für Naturschutz und Landschaftspflege (Expert Council for Nature Conservation and Landscape Care), and also especially his own position as a Landesbeauftragter for nature conservation.[5] In the late 1950s, Sukopp had begun to use the

Landesstelle to create public resonance for his research and to feed its results into nature-conservation policy. On the basis of his doctoral fieldwork, he had drafted a memorandum for the commission (Sukopp 1958) in which he had described the deteriorating state of the bogs and made proposals for their more efficient preservation. He had also suggested the creation of new nature reserves in other ecologically significant areas.[6] In 1962, when he was a young assistant professor at the Institute for Applied Botany, the Landesstelle already considered Sukopp "one of the best experts on Berlin's bogs."[7] It was also for the Landesstelle that Sukopp, in the 1960s, began his monitoring of the Berlin nature reserves. In 1971 Sukopp was himself appointed a member of the Landesstelle, and in 1975 he succeeded Otto Ketelhut as Landesbeauftragter.

Ketelhut had occupied the position on a full-time basis since 1955. He had maintained close connections to the Volksbund Naturschutz (People's Nature Conservation Alliance), of which he also had been a member and which he presided over from 1961 to 1970 (Weiß 1997: 722). With the appointment of Sukopp, the Senate paved the way for a more academic model that was meant to provide input for the ongoing reform of nature-conservation law and to accompany its subsequent implementation. In order to allow Sukopp to combine his advisory tasks with his university work, the Senate changed his position into a voluntary one. Furthermore, Sukopp came to be assisted by two academically educated employees who were based at the Senate Department for Urban Development and Environmental Protection. A further step toward professionalization was taken with the Nature Conservation Act of 1979, which replaced the Landesstelle with the Sachverständigenbeirat. Whereas the Landesstelle had included representatives of nature-conservation organizations and of the relevant administration units, the new Sachverständigenbeirat was a scientific advisory committee, appointed for one legislative term. Most of the members of the Sachverständigenbeirat had an academic background in ecological disciplines or landscape planning.[8] The 1979 act also strengthened the position of the Landesbeauftragter and left open the question whether it was to be a voluntary position or a full-time one (Muhs 1979). Sukopp's mandate expired in 1980 and was renewed a few months later; he then retained the position until 2001,[9] when he was succeeded by Ingo Kowarik, who had just taken over Sukopp's university chair. Thus, even after Sukopp's retirement a direct institutional link continued to exist between the Berlin administration and the Institute of Ecology.

Both in commissioned research and in advisory activities, the production of ecological knowledge was inextricably connected to the active promotion of political decisions on the behalf of urban nature. By 1969, Sukopp's lobbying had already led to a special protection act that imposed restrictions on bathing and motorboat traffic to protect Berlin's rivers and lakes.[10] Whereas the Landesstelle had traditionally focused on protecting the countryside and outstanding trees, Sukopp used his influence to promote nature-conservation measures in the inner city. Besides proposing reserves or monuments, the commission voiced opinions about development projects, issued press releases, and organized excursions for administrators and journalists. His position as an advisor was also the formal reason for Sukopp's involvement in the parliamentary hearings that preceded the enactment of the nature-conservation law. And, as we saw in chapter 3, after the proposal for the Species Protection Program had been issued, he used his position to support the realization of its goals. Besides supporting the overall plan, Sukopp and his colleagues at the Sachverständigenbeirat lobbied directly for those areas that the original proposal had earmarked for reserves but where conflicting plans for development remained on the agenda.

Occasionally Sukopp also published articles for broader audiences in which he explained the basic tenets of urban ecology and argued for urban nature-protection measures.[11] But, notwithstanding occasional media coverage, Sukopp did not become a widely known public figure, as other champions of nature conservation in West Germany did.[12] He pursued his goals mainly through institutional negotiations with administration agencies and policy makers.

Sukopp drew the legitimacy of his political role from his status as a scientist, which he carefully defended by keeping distance from the more politicized strands of environmentalism. Both during and after the debates on the Nature Conservation Act, parliamentarians of all parties acknowledged his reputation as an expert advisor or drew on his insights. In one of our interviews, Sukopp retrospectively characterized his style of lobbying as the attempt to convince with "matter-of-fact objections" (sachlichen Einwänden) instead of overt political statements.[14] This has even become a trait of Sukopp's public image, as shown by the characterization given by the former Senator for Urban Development and Environment Volker Hassemer in a Festschrift celebrating Sukopp's 65th birthday. According to Hassemer (1995: 22), Sukopp's political independence had allowed him to be accepted by all political parties and throughout all political constellations. Sukopp actively engaged in what Gieryn (1999) has called "boundary work," explicitly demarcating ecology as a science from its political applications as well as the public activism that surrounded the nature-conservation agenda in the 1970s. As he explained to me, this was also the reason he had declined the presidency of the VBN.[15]

Another source of authority on which Sukopp could base his interventions was the legal and administrative framework in which he operated as a Landesbeauftragter. However, one should not overestimate the power that he and the Sachverständigenrat had in political negotiations within the Senate. It was only in close alliance with other actors in the administration—notably the landscape planners in the Senate Administration for Urban Development and Environment—that Sukopp was able to realize some of his political ideas.

Professional Environmentalism: The New Planning Experts

Besides representatives of academic ecology, academically trained landscape planners (or land-care professionals, as they were also called) represented another important regime community that promoted nature conservation in the city. In the 1970s, with the issues of ecological planning and the protection of biotopes, landscape planners were able to achieve and maintain what the sociologist Andrew Abbott (1988) has called a field of "jurisdiction"—that is, a stable niche within the institutionalized distribution of labor in the planning sector. As consultants or administration officials, they performed the day-to-day tasks of planning and implementing biotope-protection measures. In doing so, they promoted conservation claims and defended them against countervailing powers. Graduates of biology and ecologically educated geographers also found jobs in this area.

The Making of the Ecological Landscape Planner

Since the 1930s, landscape care or land care had been the rallying point of an emerging occupational group that claimed special expertise in the shaping and maintaining of landscapes.[16] The early advocates of land care were professional landscape gardeners who sought to extend their scope of activities beyond the confines of conventional horticulture. During National Socialism they had been able to establish themselves as Landschaftsanwälte (landscape advocates) who coordinated landscaping measures along highway and other infrastructural projects. They were also engaged in the Reichskommissariat zur Festigung des Deutschen Volkstums (Imperial Commissariat for the Consolidation of the German Folkways), a national agency that planned the large-scale engineering of "German" landscapes in the occupied territories of Eastern Europe (Schulz 1991). In the FRG, the occupational self-understanding of land care comprised a broad spectrum of tasks: park and garden design, technological practices of landscape construction (such as the greening of abandoned mining areas, dikes, canal and railway embankments, or streets), nature and landscape protection, and the development of "landscape developing plans" (Landschaftsentwicklungspläne) for localities or regional authorities.

A programmatic 1964 article by the Hannover professor Konrad Buchwald and two colleagues characterized the mission of land care as the "scheduled securing, maintenance, and construction of an environment that is adequate for man and close to nature" (Buchwald, Lendholt, and Preising 1964). By ordering space on the basis of ecological insights, it was to help to maintain a mutual balance (Ausgleich) between nature and the needs of society. Its three related segments were characterized by different primary tasks and spatial foci: landscape care (Landschaftspflege) focused on the "open landscape," nature protection (Naturschutz) dealt with the protection and preservation in open and settled landscapes, and green ordering (Grünordnung) focused on settled areas. With these definitions Buchwald et al. advanced a broad claim that not only extended to settled areas in cities and villages but also encompassed tasks over which other groups had already established jurisdiction. Thus, whereas nature conservation had traditionally been dominated by a semi-professional advisory system, Buchwald and other advocates of land care promoted a leading role for land-care specialists.[17] Designing urban greeneries had traditionally been a job for horticulturalists. By extending its claims to the more comprehensive task of green ordering, landscape care also penetrated into the turfs of other professions, including city planning and architecture.

The constitution of land care (or landscape planning, as it was now also called) as a distinct profession implied the establishment of specialized training programs, research institutions, and occupational associations. Study programs in landscape care or landscape shaping (Landschaftsgestaltung) had been offered by the major horticultural academies (Höhere Gartenbauschulen).[18] The first academic program in horticulture was launched in 1929 at the Institut für Gartengestaltung (as of 1939, Garten- and Landschaftsgestaltung) in Berlin. Landscape shaping was represented by a special chair.[19] After the dividing of Berlin, the program was continued in the agriculture program of the Technische Universität. After the agriculture program was dissolved in 1970, it was replaced by a new program of landscape planning (1972). Comparable programs also existed at the Hochschule für Gartenkultur (Academy of Horticulture) in Saarstedt (since

1947), at the Technische Hochschule Hannover (as of 1952), and at the Technische Universität Munich (Weihenstephan) (since 1957). Another program was launched in 1970–71 at the newly established Gesamthochschule Kassel (Comprehensive University of Kassel). In the early 1970s, the programs which originally were namend horticulture- and landscape shaping (Garten- and Landschaftsgestaltung) (as in the original Berlin and Weihenstephan programs) were relabeled as either landscape care (Weihenstephan, Hannover) or landscape planning (Berlin, Kassel). The offering of these programs and the founding of a number of policy-related research institutes in the 1950s and the 1960s led to the institutionalization of land care or landscape planning as a distinct field of research and professional expertise.[20] This became visible in a steady production of specialist publications, including the journal *Natur und Landschaft* (since 1953) and a pathbreaking textbook by Buchwald and Engelhard (1968). Moreover, several occupational organizations existed through which horticulturalists and landscape-care specialists lobbied in educational and landscape planning policy.[21]

The 1970s brought further changes. First, the new nature-conservation legislation entailed an increasing demand for expertise in the administrative organizations (often newly created) that were responsible for its implementation. Previously, the instrument of landscape planning had been wielded in a relatively selective way.[22] The new legislation foresaw an array of planning categories (Landschaftsrahmenpläne, Landschaftsprogramme, Landschaftspläne) that implemented the concept of landscape planning on different levels and throughout virtually the entire territory. In addition, the impact regulation (Eingriffsregelung) of the German Nature Conservation Act and the introduction of Environmental Impact Statements created strict demands for the assessment of human interventions into the landscape and for mitigation measures. As we saw in the preceding chapter, leading representatives of land care had been very active in promoting these new regulations. Not only did they pave the way for new ecological framings to enter the world of policy making; they also actively extended their own professional jurisdiction.

The second aspect was a rapid quantitative expansion of the education of landscape planners. In 1963 an official of the Federal Ministry for Housing calculated the increasing demand for degreed horticulturalists (at that time, the formal title of landscape-care graduates) at all levels of the administration (Schöning 1963). The overall number of degreed horticulturalists was believed to be 450 at that time. The housing official estimated that the real demand for degreed horticulturalists was already much higher, and that between 630 and 810 would be needed within the next 17 years. Besides the general benefit of rational planning, he referred particularly to the enactment of the Federal Building Code (Bundesbaugesetz), which in 1960 had allowed local authorities to stipulate compulsory landscape plans. Based on these calculations, representatives of the profession urged for the training of more graduates in this discipline.[23] Until the 1970s, the call for more planning experts remained on the political agenda. This was mainly nourished by the expectation that the new nature-conservation legislation, once implemented, would require a competent community of operators. At the same time, such claims dovetailed with a broadly held belief in the problem-solving capacity of rational planning (Metzler 2005) and with a general concern about the low standard of education in West Germany (discussed under the keyword of an "educational disaster"). In the late 1960s and the 1970s, as West Germany witnessed a wave of educational expansion and university reforms, this also entailed a significant extension and restructuring of the teaching capacities of land care and landscape planning. As a result, the amount of academic professorships at West German universities went from 11 chairs in 1966 to 32 in 1976.[24] Besides the restructuring of existing teaching programs (Berlin, Hannover, Weihenstephan) and the upgrading of vocational schools (Gärtnerfachschulen) to polytechnics (Fachhochschulen), it also included the creation of entirely new programs (Fachhochschule Nürtingen, Gesamthochschule Kassel).

The subject of landscape planning fit perfectly with the growing concerns for environmental issues within the young generation. The number of students in West German universities' land-care programs (including West Berlin) had increased from 340 in 1966 to 1,420 in 1976 (Nimmann 1978: 894), and another 1,020 students were in similar programs at the polytechnics (ibid.). The Technische Universität Berlin became the largest center of education in landscape planning. In the winter of 1976–77 the application rate had risen to about 560, and there were only 20 professors (ibid.: 895). This demand exceeded the capacity of the universities, and only about 110 students were admitted (Ermer 1977). In the late 1970s, the number of land-care and landscape-planning graduates exceeded the number of newly created jobs, so many of the graduates had difficulty finding jobs or worked in uncertain positions.[25]

The third aspect was a redefinition of the profession and its cognitive references. Landscape care had traditionally consisted of an uneasy combination of two distinct strands (Eckebrecht 1991). The first was an artistic approach of landscape shaping that had evolved from the tradition of landscape gardening. It defined the work of the land-care specialist mainly in aesthetic terms and cherished the individual designer's intuition and genius. The second strand was managerialist in its approach. It drew on natural-science methods (e.g., biological and landscape engineering) and also on administrative and juridical competences. Notably, concepts from vegetation science (plant sociology) and landscape ecology (the geographically oriented branch of ecology) were regarded as important operational resources in this field.[26] The late 1960s and the early 1970s witnessed an increasing withdrawal of land care from the artistic tradition. It now became redefined as a science-based and technical planning discipline. As such, it sought to strengthen its links with ecology and also to develop links with the social sciences. The first trend led to the creation of a chair for "landscape ecology" at the Technische Universität München and the integration of the Institute of Ecology at the Technische Universität Berlin into the department of Landscape Development (Landschaftsentwicklung). This allowed the new dynamics of ecological research that the 1970s witnessed to feed directly into the teaching programs and the professional self-understanding of landscape planners. At the social science pole, landscape planning saw the emergence of Freiraumplanung (open-space planning), a new approach that was based on quantitative and qualitative studies of users' needs.[27]

The self-definition and the social boundaries of the profession remained in flux. By using terms such as landscape development (Landschaftsentwicklung) and landscape planning, some of its promoters sought to emphasize their claim on directing large-scale planning processes. They rejected the term *land care* because they considered it as too closely associated with a traditional conservation approach to nature protection (Hübler, Kiemstedt, and Sittel 1981: 3). Others still cherished land care as a comprehensive category that, in their view, more realistically characterized the profession's jurisdiction (ibid.: 3). The term *Landschaftbau* (meaning landscape construction), which had become common in the 1960s, emphasized the technical aspect of landscape shaping (in an explicit analogy to similar terms such as *Wasserbau*, *Strassenbau*, etc.) in object-focused projects. Between 1971 and 1972, Landschaftbau was the name of the Technische Universität Berlin's specialist division that was later changed to Landschaftsentwicklung (Landscape Development).

The main conflict within the profession, however, was kindled by the general turn toward science-based planning. Whereas members of the Deutscher Rat für Landschaftspflege largely supported the reorientation, the Bund Deutscher Landschaftsarchitekten (Alliance of German Landscape

Architects)—the main group representating practicing landscape architects
in Germany—argued that this would lead to a loss of the defining qualifi-
cations of the professional landscape shaper (Bund Deutscher Landschaftsarchitekten 1979).[28] At the Technische Universität Berlin, these tensions
led to fierce debates and prevented cooperation in research and teaching.
Later in the 1990s, this conflict led to a differentiation of horticulture
and landscape planning into two different study programs.[29] Within the
renewal movement of landscape planning itself there was a rift between
advocates of an ecological approach and proponents of social-science-
based open-space planning (Freiraumplanung). The advocates of an ecological approach followed Buchwald's understanding of land care as landscape
ecological planning.[30] They advocated ecological methods of research,
such as ecological zoning (Pflug 1973), ecosystem analysis (Tomasek 1978;
Haber 1992), or biotope mapping (Kaule, Schaller, and Schober 1979), as
new methods for landscape planning. In Berlin, Sukopp and his group
represented a botanically oriented version of this approach. Advocates
of a critical social science approach to open-space planning drew on analyses of users' needs and perceptions. They criticized aesthetic shaping and
the imperative of species protection. Both, in their view, undermined
the character of open spaces, which in line with the literal meaning of the
German term *Freiraum* they considered as "free" spaces.[31]

The creation of the Specialist Division Landscape Development at the
Technische Universität Berlin and the launch of a new study program
in landscape construction (Landschaftsbau) were based on the new self-
understanding of the profession. The designers of the program presented
this explicitly as a turn away from the existing program of horticulture,
which they characterized as "hitherto largely aesthetically and artistically
dominated" (Technische Universität Berlin 1972: 29). Instead, the new
curriculum embraced the idea of Landschaftsbezogene Umweltplanung
(meaning landscape-oriented environmental planning). At the same time,
they wanted the program to reach beyond a "purely technocratic function"
(ibid.: 30) and to enable future landscape planners to also understand and
reflect upon the underlying societal reasons "for the increasing environmental damage." The curriculum included technical skills of planning,
substantive contributions from environmental sciences such as ecology,
and background knowledge in social and economic analysis. In addition
to traditional academic teaching methods such as seminars and lectures,
it made use of new teaching formats such as problem-oriented courses and
(sometimes self-organized) study projects.

The program of landscape planning was interdisciplinary and involved
the entire spectrum of disciplines of the Specialist Division, including the

Institute of Ecology. Sukopp's urban ecology figured prominently in the institute's central lecture series on ecology. Practical skills in urban ecology were acquired in the Großpraktikum Ökologie (Large Practical Course in Ecology) that was offered annually by the various departments of the Institute of Ecology. Each year the Großpraktikum focused on a new topic, which was researched from the angle of different ecological disciplines in exemplary fieldwork sites in Berlin (Weigmann et al. 1981). Other practical courses related more directly to ongoing debates in urban planning and sought to provide direct input from an ecological perspective. For example, in June of 1980 one study project evaluated the different wasteland areas in Berlin and made proposals for their future protection (Behrens et al. 1982). By participating in these projects, students not only appropriated knowledge and practical skills of urban ecology, but also tended to share with ecologists a close commitment to the mission of urban nature conservation.[32]

As a result of the aforementioned developments, Berlin had a large number of ecologically orientated landscape planners who conceived biotope protection and other landscape planning measures. In addition to academic ecologists and citizen activists, landscape planners helped to defend conservation claims against other actors within the administration. For example, in the early 1980s, the protest against development projects in Berlin's urban wastelands was to a significant extent sustained by students of landscape planning.[33] Some landscape planners remained active members of citizens' groups or maintained close relations with them.[34] The institutionalization of the nature-conservation movement led to the emergence of several social-movement organizations that provided jobs for landscape planners, often financed via social security funds (so-called employment-creation-measure jobs) or through combining part-time salaries with voluntary work. In these arrangements the boundaries between activism and professional work became blurred. At the same time, the self-organization and institutional success of these planners as an occupational group rode on the wave of increasing concern for urban nature conservation and urban ecology.

Reshaping the Planning Administration

The evolution of new administrative structures allowed landscape planners to increasingly play a professional role in West Berlin's planning administration. In the mid 1970s, only one full-time and two part-time employees were in charge of nature conservation at the Senate Department for Construction and Housing (Senatsverwaltung für Bau und Wohnen).[35] In only five of the twelve districts was a full-time official exclusively devoted to nature conservation.[36] Over the next few years, new staff members were hired and a new working group in charge of landscape planning was created. In the horticultural units of the districts, according to a communication of the Senate in 1979, the new Nature Conservation Act had resulted in the creation of 39 new positions.[37]

At the same time, the tasks of the nature-conservation units were redefined according to the precepts of the Berlin Nature Conservation Act, and a distinct sub-unit for landscape planning was established within the green planning administration of the Senate. New administrative personnel were hired to develop the new legislation, to pave the way for the Species Protection Program, and to implement the other policy instruments of the Nature Conservation Act (notably the "impact regulation" that will be discussed in chapter 5). In 1981, the administrative units in green planning, which hitherto had been located in the Senate Department for Construction and Housing, were moved to the newly created Department for Urban Development and Environmental Protection. Thus nature conservation, landscape planning, and open-space planning were strengthened as components of the field of environmental politics adding to pollution-oriented and public-health-oriented competences.[38] Volker Hassemer, who was appointed as a Senator for the new department, had earlier been the director of a unit on "environmental planning" at the Umweltbundesamt (Federal Environment Agency), and in that function had been engaged in a cooperative project with Sukopp.

Many of the newly created administrative positions were filled with graduates of programs in landscape planning who had studied with Sukopp and other ecologists. These landscape planners played a crucial role in the translation of ecological knowledge into actual policy proposals and administrative decisions and in defending ecological policy goals against opposition from competing administrative units. The main example was the Species Protection Program, which, as we saw in the preceding chapter, developed in close interaction between the Institute of Ecology and the planners at the Senate.

Although landscape planners in the administration shared a general commitment to nature conservation, their relationship was also characterized by tensions and struggles concerning the delineation of administrative competences. Within the Senate Department for Urban Development and Environment, the unit for landscape planning coexisted with a host of urban-planning-oriented departments whose approach to urban problems was guided by economically motivated development goals. Under the roof of their own administrative division, ecologically oriented landscape planners also competed with a horticulturalist unit, which cherished an aesthetic and cultural-heritage approach to park maintenance. As the Kreuzberg landscape planner Martin Schaumann reported from his own experiences, similar conflicts existed in the districts where the competences for both nature conservation (since 1958) and landscape planning (since 1979) had been implemented into a preexisting green administration. In his view, the latter remained in a relatively subordinate position to the horticulturalists, who traditionally had dominated the units. This changed partly after unification, as the newly created Eastern districts decided to shift the competences of landscape planning from the horticultural departments to newly created environmental departments (Umweltämter). This model, however, was not realized in the former West Berlin districts where the horticultural departments (now redefined as Grün- und Naturschutzämter) feared that they would lose their direct say in land-use planning procedures. (See Schaumann 1992: 107–108.)

Besides the Senate Department and the districts, a number of quasi-autonomous non-governmental organizations (QUANGOs) were established in Berlin to implement administrative measures in green planning and landscape planning. In this respect, the Berlin administration followed a general trend in public administration to relocate administrative competences to foundations or private corporations. Landscape planners and a number of biologists found new positions here. In 1981 the Senate established the Stiftung Naturschutz (Nature Conservation Foundation),[39] which in subsequent years launched various programs (often in cooperation with schools, homeowners, etc.) to promote awareness of environmental problems and to mobilize them for action projects. The Stiftung Naturschutz also financed the publication of *Grünstift* (literally, green pen), a monthly journal of nature conservation and environmental policy that focused almost exclusively on Berlin. A second institution was Ökowerk, a state-subsidized environmental education center that promoted environmentally friendly ways of life, gardening practices, and ecological techniques. It was located at a former waterworks in the urban forest Grunewald, a site which itself was shaped as a pioneering project of ecological housing and gardening. It had a permanent staff, which consisted mainly of biologists and cooperated closely with citizens, education experts, and local schools. Two other QUANGOs were concerned with park administration but also became important players in the shaping of nature-conservation policies in the 1990s: Britzer Garden GmbH, which administered the venue of a garden show in the 1980s—and BUGA 1990

GmbH, which had been established to organize a new Bundesgartenschau (federal garden show) in the early 1990s. When the Senate canceled its plans for a 1995 Bundesgartenschau, the two organizations were united as the Grün Berlin GmbH. Since then, that organization has been in charge of various park projects, among them the ruderal park at the Südgelände mentioned in the introduction.

The Consultancy Market

A second way in which landscape planners became involved in Berlin's biotope-protection regime was through their work as private consultants, working as freelancers or in small firms and receiving commissions from the Senate, from the districts, or from public and private developers (e.g., to perform environmental impact statements for their projects). This form of professional work paralleled the work of architects, landscape architects, and horticultural designers. Although their ranks included graduates from other disciplines (e.g., biologists, geographers), they followed the new professional profile of landscape planning, including its strong emphasis on ecological research. Landscape planning consultants conducted impact assessments in the context of the impact regulation and made proposals for measures to compensate for encroachment. (Such a plan was called a Landschaftsplanerischer Begleitplan.) The development of landscape plans at the district level became a task that was typically done by these consultants.

Like their colleagues in the administration, the consultants maintained close working relations with the Institute of Ecology and with Sukopp. Names such as Ökologie und Planung (Ecology and Planning) or Arbeitsgemeinschaft Ökologie und Landschaftsplanung (Working Group Ecology and Landscape Planning) signaled the commitment of these offices to an ecology-based planning approach. The first consulting firm was set up by Christian Schneider and Barbara Markstein, two graduates of the former horticulture program of the Technische Universität. Both had studied with Sukopp, and their firm was closely involved in the early development of the Species Protection Program.[40] The Arbeitsgemeinschaft Ökologie und Planung was among a number of new firms that were established in the subsequent years and which adopted a similar profile. Its founders, Harald Fugmann and Martin Janotta, were landscape planners who studied in Berlin. In 1986, with Schneider, they submitted a concept for planning-oriented nature-conservation research in Berlin that was meant to further substantiate the findings of the biotope-mapping survey

(Fugmann, Janotta, and Schneider 1986). Their professional commitment to ecology, however, did not imply that such firms were always ardent promoters of nature-conservation claims. Notably, encroachment assessments were produced on behalf of the project holder who had to provide the necessary data when applying for building permits.

In the late 1980s, private consultants had taken over many tasks that hitherto had been done by university researchers. Thus, when after the unification of 1990 a new biotope-mapping survey was carried out in the eastern part of Berlin, this was not done by researchers of the Technische Universität but by the landscape planning firm Plantage. (See Plantage 1993.) The transfer of advisory tasks to private consultancy firms was less an indication of decline in authority of the university research than an effect of the successful implementation of ecological expertise that turned many advisory tasks into professional routines which were no longer attractive for university researchers.

Although the success of the study program paralleled the creation of new jobs in the field of practical planning, from the early 1980s on it proved increasingly difficult for graduates in landscape planning to find positions. Working in precarious part-time positions as part of publicly subsidized job-creation measures or in internships became common. In 1983 a Gruppe Arbeitsloser Landschaftsplaner (Group of Unemployed Landscape Planners) was formed in Berlin; it then organized two conferences at which calls for action were launched.[41] The unemployed landscape planners claimed that professional organizations of landscape planners tended to underestimate the extent of unemployment in their field. Most of these planners, they argued, did not appear in the statistics because they tried to survive in under-qualified positions. At the same time, the group claimed that the need for landscape planners was still high. Therefore they maintained that intensive lobbying was required to motivate the public administration to create new jobs or to commission advisory tasks. The unemployed landscape planners also referred to urban ecology as one of the newly emerging potential fields of activity. In order to advance the situation of landscape planners, they advocated public-relations work by professional associations, intensified research at the universities (to maintain and extend the profession's fields of activity), a more open and less discriminatory hiring policy of consultancy firms and administrative bodies, and less rigid admission criteria for state-supported employment grants. This did not keep the market for consultancy office from becoming increasingly competitive.

Civic Environmentalism: The New Urban Activists

Whereas ecologists' and landscape planners' involvement in the biotope-protection regime was part of their academic and professional mission, other regime communities had a more civic character. The wave of environmentalism that shook Germany's public life in the 1970s and the 1980s led to the emergence of a wide range of environmentalist social-movement organizations. Some long-standing nature-conservation associations achieved new popularity in this context.[42] In addition, a host of informal and temporary activist groups—so-called Bürgerinitiativen (citizens' initiatives)—arose to deal with specific land-use conflicts. These groups opposed ecologically contested development projects, campaigned for the designation of nature and landscape reserves, lobbied in legislative processes, coordinated practical amateur work in nature conservation, and promoted conservationist values in the public domain (Andritzki and Wahl-Terlinden 1978; Meyer-Tasch 1985; Schneider-Wilkes 2001; Hager 1995).

At least until the early 1990s, these groups remained an important factor in public life in Berlin. Along with organic food stores, production collectives, and the increasing use of bicycles instead of cars, they formed part of an alternative lifestyle that flourished in Berlin and other West German cities in this period. Some sectors were deeply rooted in the protest milieu that concentrated in the districts of Schöneberg and Kreuzberg. Although after 1989, the convergence with the remains of the East German environmental movement gave some new impulses to urban environmental activism, the momentum of these groups slowed down in the 1990s. This did not mean that environmental activism ceased to exist, but much of the coherent protest milieu eroded, and militant activism gave way to more institutionalized forms of policy making. A striking example was the establishment of the AL, and later the Greens, as a stable factor in the Berlin Abgeordnetenhaus.[43]

Organized Nature Conservation

Among the nature-conservation organizations, the Volksbund Naturschutz continued to play a pivotal role in West Berlin. Largely dominated by older and politically conservative nature lovers, the VBN differed markedly from the checkered assortment of environmental activists that emerged in the 1970s.[44] This association widened its focus only hesitantly beyond its traditional confines of nature monument care and developed a more open attitude toward ecological planning and the activism of the citizens' initiatives.[45] In 1981, under the name of Heinrich Weiß, an advocate of programmatic renewal within the VBN, the association published "20 Primary Demands for Nature Conservation in Berlin" in which it called for the implementation of landscape planning, including comprehensive biotope mapping and strict application of the so-called impact regulation (Weiß 1981). It also called for a more efficient nature-conservation administration and for the cancellation of ecologically problematic development projects. In line with urban ecologists, it demanded a positive attitude toward inner urban wastelands and an acknowledgment of the "ecological and aesthetic value of spontaneous weeds" (ibid.: 656). In the 1970s, many young ecologists and landscape planners who had graduated from the Technische Universität Berlin became members of the VBN. Until the early 1980s, the organization's newsletter, *Berliner Naturschutzblätter*, remained one of the central publications through which nature-conservation activists spread their views. In the long run, however, the traditional conservationism with which this organization was associated was difficult to connect with the attitudes and habits of the new political environmentalism of the 1970s. This led to an increasing isolation of this conservative organization within Berlin's growing environmentalist scene.

Other social-movement organizations tended to focus on specific aspects of nature preservation such as forests (Schutzgemeinschaft Deutscher Wald), trees (Baumschutzgemeischaft), birds (Bund für Vogelschutz), or hedgehogs (Igelschutzgemeinschaft). Some of them had emerged through secession from pre-existing organizations, or by the expansion of nationwide organizations to Berlin.[46] In 1979 an umbrella organization was founded: the Berliner Landesarbeitsgemeinschaft Naturschutz (Berlin Land Consortium for Nature Protection), which represented between ten and twelve local member organizations.[47] It soon developed into the main lobbying institution in West Berlin's nature-conservation policy. In 1985 the World Wildlife Fund set up a local section in Berlin (*Grünstift* 1985), but it did not play a significant role in local land-use policy.

An important newcomer was the Bund für Umwelt und Naturschutz Deutschland (Alliance for Environmental and Nature Protection in Germany), a nationwide social-movement organization that had established a section in West Berlin in the mid 1980s. Since its foundation in 1975, the BUND witnessed a rapid growth in membership and public attention.[48] Its founders advanced the BUND as an alternative to the conventional nature-conservation associations, which they blamed for the alleged ignorance of the new quality of environmental problems that the world was faced with.[49] Its work consisted mainly of thematic campaigns, which were devised centrally and implemented by local actors. The BUND was also the first nationwide nature-conservation organization to explicitly embrace the issue of species protection in the city. In 1985 it even launched a distinct campaign on this topic. The BUND, however, was much less rooted in the West Berlin naturalist scene than the other preservation associations. Its members were often young and had a commitment to a much broader scope of environmental politics. Established nature-conservation associations therefore regarded the emergence of the BUND groups and their professional public-relations work with some suspicion.[50] This did not hinder them from forming strategic alliances, and it did not keep the BUND from becoming more and more popular. Eventually in the 1990s it developed into an important pillar of the Berlin nature-conservation movement, and it became a member of the Berliner Landesarbeitsgemeinschaft Naturschutz.[51]

The Bürgerinitiativen

In the 1970s, West Berlin had also become a fertile ground for the emergence of Bürgerinitiativen (citizens' initiatives).[52] Bürgerinitiativen were at the center of all the emerging environmental conflicts of the 1970s in West Germany, including protests against nuclear energy, polluting industries, and road and airport construction.[53] Other Bürgerinitiativen belonged to the various "urban social movements" (Mayer 2000) that, in the 1980s, opposed the dominant paradigm of urban development. Such groups addressed a wide range of issues that were linked to the living conditions in a neighborhood, including noise abatement, urban renewal schemes, housing policy, and lack of green space. People who squatted in abandoned houses in urban renewal areas formed a significant strand of such urban activism in West Berlin. Many Bürgerinitiativen were pragmatic middle-class-based interest groups that sought solutions for conflicts and problems within a neighborhood. At the same time, some of them were also rallying points for left-wing (including Maoist and anarchist) activists who saw their efforts as part of a broader struggle against the capitalist system. The journal *Umweltschutzforum Berlin*, which was founded by a local church group in 1971, was meant to provide a common platform for the various environmental Bürgerinitiativen that had emerged in the city. In 1972, environmental Bürgerinitiativen of West Berlin and West Germany formed the Bundesverband Bürgerinitiativen Umweltschutz (Federal Association of Environmental Citizens' Initiatives), an umbrella organization that gave them a broader public voice.[54] It was also typical of Bürgerinitiativen that many activists and observers conceived their involvement not only in terms of the material problem that they addressed in the first place, but also, more generally, as a model of public involvement that should complement established procedures of representative democracy with direct citizen participation.[55]

The quests of the many Bürgerinitiativen dovetailed with the agenda of urban ecology or were even explicitly geared to the promotion of similar environmental issues.[56] The typical Bürgerinitiative was organized around a focusing theme or political cause that was expressed by its name, usually a couplet including the abbreviation BI (or, alternatively, AG or AK, standing for Arbeitsgruppe, Aktionsgruppe, or Arbeitskreis). Whereas the BI Westtangente defined its mission by the development project that it opposed, others featured the name of the place that they wanted to protect from interference, often in combination with a concise slogan or call to arms.[57] Along with these project- and place-focused Bürgerinitiativen, a number of working groups existed that had been formed around concerns about more general environmental or conservational problems.[58] Many of these initiatives were rather short-lived and died out once their cause had lost attention or had become obsolete. A few Bürgerinitiativen, however, persisted over longer periods of time. This was true of the BI Westtangente, which was founded in 1973 and which since then has been a permanent opponent of the Senate's highway projects. Another example was the BI Südgelände (later the Förderverein Südgelände), which has promoted a nature park at a former railway area; its work will be discussed in chapter 5. These were also among the few Bürgerinitiativen that had achieved the status of formal associations according to German association law.

Nature-conservation organizations and Bürgerinitiativen often addressed the same political issues. In some cases they formed strategic alliances in which several smaller groups coordinated their campaigns against unwanted projects. The first example of such broadening cooperation was a coalition that formed in the mid 1970s to oppose the Senate's plan to build a new conventional power station in the Spandauer Forst, a popular suburban forest.[59] It was not the safety of the station that concerned these activists. Instead, the protest against the power station (which was originally planned at a site next to a residential area), raised concerns about air pollution, noise, and damage to the scenery. When the Senate decided for the site in the forest, protest focused mainly on the negative effect on this important recreation ground. The activists shared this concern with Sukopp and a group of ecologists who had evaluated the project for the Senate (Blume et al. 1975; Sukopp 1976). The BI Oberhavel, the first group to oppose the project, was soon joined by nature-conservation organizations (including the VBN), other West Berlin Bürgerinitiativen, and individual activists; in

1976 they formed the Aktionsgemeinschaft Oberjägerweg (Oberjägerweg Action Association), named after the road in which the plant was planned. The coordinated efforts of these activists, in conjunction with critical ecological evaluation reports, led to a court decision against the clearing of forests. The project became more and more unpopular, and eventually the Senate canceled it.

Many personal links existed among nature-conservation organizations and Bürgerinitiativen, on the one hand, and professional biologists and landscape planners on the other. Sukopp had been a member of the VBN since the 1960s.[60] In the 1980s, the lists of the newly joined members that were regularly published in the association's journal contained the names of students or graduates of the Technische Universität Berlin, who also were professionally involved in nature-protection policy. The BI Südgelände was founded by landscape planning students and graduates of the Technische Universität Berlin—among them Rita Mohrmann, one of the authors of the landscape program. In particular, the larger nature-conservation organizations, such as the BUND and the BLN, were highly professionalized organizations whose day-to-day business was run by people with academic credentials in nature-conservation-related disciplines. Like consultancy offices, Bürgerinitiativen and nature-conservation organizations benefited from state sponsored job-creation measures. Some biologists and landscape planners worked on these terms for the Berlin nature-conservation groups. This was in striking contrast with the former nature-conservation associations, which were run from the personal homes of their leading members.

Although they all pursued what in a broad sense qualified as urban biotope protection, nature-conservation organizations, Bürgerinitiativen, and professional ecologists differed sometimes considerably in their specific goals and in how they framed the problem of urban nature. In contrast to Sukopp and many landscape planners, traditional nature conservationists remained focused on nature at the fringe of the city and only slowly developed a commitment to the protection of specific urban forms of wildlife. The Bürgerinitiativen, on the other hand, were concerned primarily with the lack of recreation space or with traffic or noise, rather than with endangered species or the loss of biotopes. However, they often opposed the same projects that ecologists attacked because of their effects on nature. Although their own agendas were often heterogeneous, they embraced ecological arguments to bolster their claims. In April of 1973, the Umweltschutzforum Berlin devoted a whole issue of its public newsletter to the topic of "research for the environment" (*Umweltschutzforum*

1973). The issue contained articles in which Sukopp and other scholars from the technical universities presented recent findings on the environmental situation in Berlin. In the cases of the Bürgerinitiativen, which promoted the protection of nature parks at the Südgelände or the Gleisdreieck, concerns for urban nature protection were even the cause around which activists assembled in the first place.

A feature that distinguished much of the environmental activism of the 1980s from the classical nature-conservation movement was its distinctive political framing of the problem of urban nature. This was consistent with the shift in nature protection from a politically conservative and elitist project to an issue of left-wing and liberal politics that other scholars have also observed in this period (Engels 2006). Whereas traditional nature-conservation organizations tended to see nature as a means to promote regional identity and national pride, activists of the 1980s linked their activities to a broader quest for political participation and democracy. In their discourse, conflicts among nature and development projects were often portrayed as clashes between the interests of "the citizen" and an ignorant or capital-dominated bureaucracy. With the notion of "the citizen" as it was advanced in the term *citizens' initiative*, they positioned themselves as a legitimate voice of citizens confronting the institutionalized administrative-political complex.

Allotment Activists

Although they were not nature-conservation organizations themselves, the associations of allotment gardeners have to be mentioned here as important communities of environmental protest. The cause of nature-conservation associations' and BI's opposition, was sometimes supported by allotment gardeners' organizations that had a stake in the same land-use conflicts. They were well organized and could draw on their existing bonds to launch protests against projects.

Between Contention and Cooperation

In comparison with the classic nature-conservation movement, activist organizations of the 1980s made use of a much wider "repertoire of contention" (Adam, Tarrow, and Tilly 2001). Since the mid 1970s, the mobilization of protest campaigns had been a salient strategy through which the Bürgerinitiativen attracted public attention. A landmark event in this respect was the campaign against the power station at the Spandauer Forst. In fall 1976, the Action Association Oberjägerweg organized "information trips" to the envisioned location of the power plant. These soon grew into demonstrations of several thousand participants.[61] In February of 1977, local musicians supported the protest with a public concert. Activists also appropriated the site in a physical sense by placing banners on trees and constructing permanent "watch huts." Often clearance measures that were meant to prepare the beginning of contested development projects kindled public protest, sometimes leading to violent confrontations with the police. In 1982, activists tried to prevent the clearing of trees for a new highway connection between West Berlin and Hamburg. As one of the participants later complained, the actions had failed because of the massive presence of the police, which had choked off any attempt to "secure trees" with chains or to erect barricades or huts in the forest. According to his report, many of the activists were injured and arrested (anonymous 1982). Similar squatting practices also appeared during the conflict around the so-called Lennédreieck, which will be covered in the next chapter.

In the 1980s, the BI Westtangente organized bicycle rallies against contested highway projects, a new type of demonstration that became popular in many cities in that period. Public protest often took on a festive atmosphere, combining the articulation of political claims with neighborhood street parties, including information stands, food stands, and music and theater performances. Sometimes the political authorities behind the development plans were the direct target of contentious action. One morning in 1979, members of the BIW placed apples that they had harvested along the envisioned track of the Westtangente at the Senate Department for Construction and Housing to demonstrate what would be lost if the project went through. In January of 1984, members of the BI Südgelände who were dressed up as plants and animals visited the office of the Senate Department to voice opposition against the plans for the freight station that was planned at that wasteland. Activists also disturbed public ceremonies. For example, in 1980. when the first section of the Westtangente was opened, members of the BIW slaughtered a papiermâché "highway swine" (symbolizing the highway plans of the city) and tore apart the ribbon that a senator was supposed to cut during the opening ceremony. In 1978, about 200 people squatted at the site where the Senate was preparing to cut the first trees for the construction of the Westtangente.[62] There was widespread consensus among Bürgerinitiativen that such actions of civic disobedience were legitimate ways to counter technocratic decision making by the state authorities.[63]

In another strategy, environmental groups filed lawsuits against planning decisions of the Senate. Before 1979, activist groups were not able to sue collectively. Therefore, they had to rely on formally concerned individuals to bring their cause to the bar. During the conflict in the Spandauer Forst, only four (later seven) adjacent residents sued against the Senate's decision to start clearing the forest in order to prepare for the plant's construction. Two court decisions against the beginning of clearing measures (one in 1976, one in 1977) were landmark successes for the protesters. In particular, the second decision, in which the judges explicitly emphasized the need for nature areas in Berlin and suggested an alternative site for the power plant, paved the way for the eventual cancellation of the project.[64] In many other cases, however, litigation failed either because the suit was turned down for formal reasons[65] or because the court did not follow the legal argumentation of the claimants. In 1983 a revison of the Berlin Nature Conservation Act introduced the possibility for formally accredited nature-conservation organizations to institute collective proceedings (called Verbandsklagen) against public decisions that would impact "nature and landscape" (§ 39a). The introduction of this mechanism had long been a main concern of activist groups in Germany, and Berlin was the only Land where the procedure was actually installed. In practice, however, the use of that right faced significant obstacles.[66] In 1986, two lawsuits were issued by local nature-conservation organizations, one against a magnetic train around the ruderal areas next to the Tiergarten and one against the highway in the Tegel Forest; both were stopped by Berlin's Oberverwaltungsgericht (Higher Administrative Court) (Bilzer, Ormond, and Riedle 1990: 79). According to the court, the lawsuits were not legitimate; the regulations on collective litigation had left too much room for interpretation so that it was not applicable in a way that was consistent with the constitution. Not until 1988, when the Bundesverwaltungsgericht (Federal Court of Administration) decided in favor of the nature-conservation organizations,[67] was the way paved for more regular use of association litigation by the Berlin nature-conservation associations.[68]

Litigation was a collective project that went beyond the confines of formally entitled claimants. The groups or individuals that sued were supported by a range of activists who collected money for "support funds." Moreover, the activists drew on the support of a few lawyers who shared their concerns and who regularly acted as advocates of their cause. Raising money for such campaigns and reporting on their results thereby became in themselves strategies for raising public awareness and for mobilizing the public. Even if a legal action failed in judicial terms, it helped to mobilize support for a political decision on the matter.

A third strand of activism involved formal and informal negotiations with the administrative agencies that were responsible for the contested projects. Theorists of social movements have seen such involvement of activist groups as typical of the institutionalization of the protest movement (Brand 1999). It did not only reflect internal reorientation of the activist groups, but also the turn to a governance-oriented approach within the policy system that sought to delegate political power to nonstate actors. Ever since 1971, the German planning law had provided for formal participation procedures, which allowed concerned citizens to launch opinions against ongoing projects.[69] In 1979 the Nature Conservation Act included similar provisions for the stipulation of landscape plans (§ 11), It also introduced the legal status of formally accredited nature-conservation associations which were to be consulted by the administration whenever decisions were taken that impinged on nature and landscape, and a system of financial subsidies to support the work of nature-conservation organizations.[70] As we saw in the preceding chapter, many individuals and associations used such possibilities to raise their concerns about the Species Protection Program and the FNP 84.

The establishment of the BLN was a direct reaction of nature-conservation organizations to these newly emerging opportunity structures. In 1977, nine nature-conservation organizations that had already been involved in the hearings on the new Nature Conservation Act created a working group to coordinate their lobbying activities (Kalesse 1979). Following a proposal by the VBN member Heinrich Weiß, ten associations created the BLN as a permanent umbrella organization to strengthen their position in formal and informal negotiations with the Senate. Although the BLN did not achieve accreditation itself, it was meant to coordinate the work of its accredited members and to make sure that the nature-conservation lobby would talk with one voice. In its annual report it explicitly mentioned that, despite in some respect contrasting opinions, good relations had developed between the BLN and the Senate (in contrast to the districts where the report diagnosed a more distant relationship).[71] From 1982 on, the BLN received direct financial support from the Senate. A considerable increase in subsidies in 1984 allowed the BLN to develop into a professionalized movement organization, with office space, a paid managing director, and a number of additional part-time or internship-based co-workers. Because of its professional base, the BLN took on a position within the West Berlin nature-conservation sector that was very similar to the Landesbeauftragter and the Sachverständigenbeirat. Notably, the first years of the BLN's existence were characterized by a latent competition with the Sachverständigenbeirat.[72] The BLN's request for a permanent place on the Sachverständigenbeirat was rejected by the Senate (which was responsible for their appointment). In the mid 1980s, both organizations developed a more cooperative style; Axel Auhagen, the managing director of the office of the Landesbeauftragter, was even appointed vice-president of the BLN.

Another site of common institutional involvement of nature conservationists and the administration was the Stiftung Naturschutz Berlin (Berlin Nature Conservation Foundation). It administered a fund of 2 million Deutsche Mark, with which it was supposed to promote nature-conservation measures in Berlin. The foundation became an important source for activist groups to acquire funding for nature-conservation projects. At the same time, the accredited nature-conservation organizations were members of the council of the foundation, so that they had a direct influence on the funding agenda.

The action repertoire of nature-conservation groups and Bürgerinitiativen was not restricted to contentious politics and lobbying. It also included the public promotion of their concerns through publications, lectures, and guided tours. Organizations such as the BLN or the Bürgerinitiativen issued press releases and maintained personal contact with local journalists in order to put their cause on the public agenda.[73] Until the 1980s, the journal of the VBN, *Berliner Naturschutzblätter*, had been the main forum through which Berlin conservation activists exchanged their views. In the early 1970s, *Umweltschutzforum Berlin* featured the debates of the broader spectrum of environmental Bürgerinitiativen in West Berlin. The monthly magazine *Zitty*, a cultural and political landmark of the emerging alternative milieu in Berlin, also offered space for Bürgerinitiativen to present their agenda.[74] In the 1980s, *Grünstift* became the main mouthpiece for a politically oriented nature conservation in Berlin. Published by the Foundation Nature Protection and edited by a professional journalist, it covered the whole spectrum of environmental debates in Berlin. For more than a decade it remained a site of critical debate among professional landscape and nature-conservation experts as well as citizen activists. It was more politicized in style than *Naturschutzblätter*, featuring sharp critical comments, political manifestos, and cartoons that mocked ongoing development policies. *Grünstift* also included a calendar in which events were announced and a list of activist groups and other civic centers of nature conservation in the city. After unification, another journal, *Der Rabe Ralf*, appeared in Berlin. It grew out of the tradition of the East Berlin environmental movement.

Many groups were also involved in practical projects that pioneered alternative techniques or land-use practices (e.g., ecological gardening, roof greening). Experiments with roof greening and courtyard greening were launched at the UFA-Fabrik, an alternative culture center at the site of an abandoned film studio. Likewise, the federal campaigns by the BUND for "More Nature in Village and Towns" (see *Grünstift* 3/85) and "The Garden without Poison" sought to promote ecologically sound practices of using private land. Such public campaigns were often joint endeavors between activists and government organizations or QUANGOs. Both the Ökowerk and the organization that prepared the International Construction Exhibition (IBA 1985) stimulated such greening projects.

Appropriating Ecological Knowledge
Social movement activists also typically engage in what Eyermann and Jamison have dubbed "cognitive praxis" (Jamison 2001; Eyermann and Jamison 1991). That term refers to the fact that activists who are making political claims actively draw on, and thereby also contribute to, the production of knowledge. As we saw above, knowledge production had always been involved in the activities of nature-conservation groups. Together with botanical associations and university academics, the latter contributed significantly to the observation circuits through which ecological knowledge about the region was assembled. Knowledge production remained an important function of nature-conservation organizations. In the 1970s and the 1980s, Bürgerinitiativen normally did not share this general commitment to the study of nature. Nevertheless, they were also forums through which action-oriented ecological knowledge was selectively appropriated and put to political use by their members. For example, the BI-Westtangente activist Norbert Rheinländer later noted how in the early 1980s his group had benefited from the arguments Sukopp had put forth in his expert report on the Tiergarten tunnel (Rheinländer 2000). The reference to science provided these activists with a socially robust justification to bolster their alternative land-use claims. Rather than just emphasizing putative aesthetic needs, or being associated with romanticist sentiments, they could deploy arguments that were on an equal footing with the economic facts policy makers had used to justify their projects. In this vein, a landscape planner and member of the BI who wanted to preserve the agricultural areas in the South-Western outskirts maintained that his cause "did not only emerge from nostalgic longings, but instead, is becoming evident as an ecological necessity" (Kalesse 1979).

Ecological knowledge was processed at the regular meetings of activists of the Bürgerinitiativen, or at special working groups that were created on certain topics. Also the publications of the activist groups helped to sustain a specialized public in which critical debates on planning processes were connected to the interpretation and distribution of ecological knowledge. In contrast to the traditional activities of nature-conservation groups, activists' cognitive praxis was not confined to the understanding of the ecological features of certain contested sites. A large part of the knowledge making consisted of the critical analysis of the dominant justification of the projects and the development of alternative solutions. Activist groups thereby benefited from the participation of various specialists, such as traffic planners or landscape planners. As has already been noted, sometimes a close interaction evolved between student project groups at the university or landscape planning offices, and activist groups, and helped them in developing alternative planning concepts.

The production and circulation of activist knowledge developed along other vectors than those in former naturalist groups. As we have seen before, the latter developed a spatially extensive but thematically narrow knowledge of particular species or other aspects of the natural environment. Their organizations and their networks of knowledge exchange followed an almost disciplinary logic. Although such knowledge remained an important part of activists' knowledge practices, their orbit of attention was broader. Their interest was strategic and not intellectual: knowledge, for them, was mainly a political resource. Thus, in principle, everything that was of potential use to bolster their cause could also become a reason for activists' quest for knowledge. This included knowledge claims about the economic necessity, the juridical status, or the technical feasibility of projects as much as claims that helped activists to champion alternative visions to administrative planning practices.[75] Moreover, activists developed skills that helped them to interact with the administration and to critically monitor the latter's intentions and actions. Besides official policy declarations and media coverage, this included informal contacts and the procurement of insider information. For example in the campaign against the freight station at the Südgelände, activists received much political leverage from an internal strategy paper that had been passed to them whose content did a lot of damage to the legitimacy of the Senate administration.[76]

In some instances, activists also pursued knowledge claims that clashed with those made by academic ecologists. An example was a dispute between Sukopp and citizen activists about the clearing of a forest next to an airport runway in the suburban district of Gatow. Whereas the local citizens' groups opposed the clearing, Sukopp posited that there was no ecological reason to object to the clearing. He maintained that the forest could not be compared with the old oak tree forest at the power station as the activists seemed to imply. In Gatow the clearing would result in a type of heath, that according to Sukopp, would have presented a much higher species diversity than the original forest, and that therefore, also provided a better aesthetic quality for visitors (Sukopp 1979). Similar conflicts emerged around the Fritz-Schloss-Park and an old citadel in Spandau.

The very concept of urban ecology was also interpreted in different ways by academic ecologists and activists. This became visible when, during a "Summer Academy on Urban Ecology" in 1985, the ecologist Wolfgang Erz criticized what, in his view, had become a widespread misunderstanding of urban ecology. (Erz is quoted in Schwarzer 1985.) His concern was mainly about biotope-promotion measures in gardens and courtyards that according to him were far too individualistic to live up to the real complexities of the species-protection problem. On the other hand, the philosopher Heinrich Weiß—one of the leading figures in the campaign against a power plant in the Spandauer Forst, and the founder of the BLN—opposed the guiding principles of Sukopp's urban ecology (e.g., Weiß 1989). Weiß maintained that urban ecology had mistakenly been reduced to species protection. The type of urban ecology, that Weiß envisioned, should be concerned much more with the viability (Lebensfähigkeit) of the city than with the protection of certain species. Weiß' position was in no way representative of the entire nature-conservation scene in the city, and similar arguments could have been raised by academic proponents of a broader version of urban ecology (e.g., by environmental scientists promoting a comprehensive systems approach to the city). It exemplifies, however, the extent to which members of the activist groups developed their own understanding and even claimed epistemic authority independent from credentialed experts.

From Activism to Party Politics
Public concerns about the lack of nature in Berlin also found their way into local politics via the political parties. There were members in all parties of the Abgeordnetenhaus who had sympathies with the conservation lobby and who regularly advocated issues of nature conservation. Throughout the 1970s, the parliamentary debates that accompanied the new nature-conservation legislation showed that a widespread consensus existed among all parties about the need for a more efficient nature-conservation law. Thus, in 1978 the Nature Conservation Act was passed with a great majority, which included the opposition parties. Political conflict centered mainly on the details of the regulation and the scope to which the problem should be regulated at the competing levels of federal government and

states. Both the liberal FDP (the coalition partner of the governing SPD) and the opposition CDU criticized the power plant project in the Spandauer Forst, and welcomed the 1977 court decision to stop the envisioned clearing of the forest (*Der Tagesspiegel* 1977). CDU and FDP members of parliament also voiced their doubts about the plans for the renovation of the newly reopened Teltow Kanal, which entailed the replacement of the embankments and their vegetation by a metal wall.[77]

It was notably through the efforts of the AL that the concern for urban nature conservation achieved a solid place on the parliamentary agenda in Berlin. Founded in 1978, the AL represented the concerns of a broad set of protest movements.[78] Although it became aligned with the newly established federal party Die Grünen in 1980, it retained its distinct name until 1993. Not wanting to be identified exclusively with "green" issues in the narrow sense, it put equal emphasis on social equality, grass-roots democracy, gender, and other subjects. Environmental issues were framed in the context of a broader critique of capitalism. In its 1981 program, the party blamed "a form of economy that considers everything—nature and man—exclusively under the perspective of profit-making" (reprinted in Buhnemann 1984: 149) as the fundamental cause of the environmental crisis.

The AL used its parliamentary position to keep nature-conservation issues on the agenda and to exert pressure on the Senate. In 1984, when the Landscape Program still lingered in the process of administrative finalization, the AL launched two series of parliamentary motions in which it urged immediate action.[79] In order to prevent competing land use, the AL proposed to immediately designate the critical areas that were earmarked in the species-protection report as protected reserves. In justifying the motions, the AL repeated the arguments given by Sukopp and his colleagues in the species-protection survey almost verbatim.

The AL played a major political role between 1989 and 1990, when it formed a coalition government with the SPD. Owing to the influence of the AL, the coalition committed itself to a policy of "ecological urban renewal." It attempted many of the biotope-protection projects that had been discussed in previous years. Notably, it planned to launch a federal garden show that would have included pioneering nature parks on urban wastelands. The coalition, however, was too short-lived to realize many of these goals. After unification, conflicts between the AL and the SPD about a police action against squatters brought the coalition to an end. After the subsequent elections, Berlin came to be ruled by a coalition of the CDU and the SPD that again gave more priority to classic urban development goals. The AL, which now worked together with an alliance of ecologically oriented parties in the former East Berlin (formed by Grüne Partei, Bündnis 90, and Unabhängiger Frauenverband), remained a constant element of the parliamentary opposition. In 1993, it changed its name to Bündnis 90/ Die Grünen, the name that the German Greens adopted after their fusion with an East German civil rights coalition.

Beyond the Wall: Environmentalism in East Berlin before and after 1989

A remarkable strain of environmental activism that had developed in East Berlin in the 1980s continued to play a role in the unified city after the fall of the wall. Although this book focuses on developments in West Berlin and unified Berlin, it is important, at least briefly, to have a look at this tradition and its relation to the trajectory of West Berlin's urban ecology.

Nature conservation in the GDR relied largely on the active participation of citizens—even more so than in FRG, where the local nature-conservation administration played the leading role. Individually, or in small groups, amateur Naturschutzhelfer (nature-preservation assistants) took care of areas for which they had been given responsibility. An overarching organizational structure was provided by the Gesellschaft für Natur und Umwelt (Society for Nature and Environment). The GNU had been founded in the 1970s as a successor to the Natur- und Heimatfreunde (Friends of Nature and Homeland), and was itself a section of the state-sponsored Kulturbund der DDR (Culture League of the GDR).[80] The GNU organized inventories of nature reserves, educated amateur assistants, and launched petitions (Eingaben) in which it alerted state authorities of critical developments in nature.[81] In contrast to the generally pervasive state control of private and economic life in the GDR, direct state involvement with the day-to-day technicalities of nature conservation was relatively weak, and Naturschutzhelfer could often operate quite independently on the individual patches of land for which they had been assigned responsibility. This relative autonomy also contrasted strikingly with the formally administrated nature-conservation system of West Berlin.

In the late 1980s, several locally based Interessengruppen Städtökologie (Urban Ecology Interest Groups) were founded by members of the GNU (Nölting 2002). These groups consisted mainly of younger people, and, in contrast to the rest of the GNU, considered themselves as political organizations similar to the Bürgerinitiativen in the West. As with the environmental movement in West Germany, they were broadly concerned with the impact of industrial development (notably air pollution and chemical pollution as a result of intensifying production). This brought them into direct opposition with the GDR regime, which, having followed a relatively active environmental policy in the early 1970s, sought to prevent any public debate on environmental conditions.[82] Although political repression and continuous supervision by the state security agency restricted the work of the urban ecology groups, the Interessengruppen Stadtökologie were able to engage in the collection of data, to launch petitions, and to organize smaller public protest events.[83]

Another institutional niche in which ecological protest in East Berlin thrived was the church. As of 1986, an Umweltbibliothek (Environmental Library) operated in the basement of a church in the district of Prenzlauer Berg. More than just a library of (largely censored) publications, it provided a gathering ground for opposition activists and published a semi-legal periodical. Two years later, the Ökologisches Netzwerk Arche (Ecological Network Arche) was founded to coordinate the work of church-based environmental groups throughout the GDR, one being based in East Berlin. Even under church protection, however, environmental activism in East Berlin remained personal risky, and activists suffered from various repressive reactions.[84]

Although the East German environmental groups reacted to quite different political circumstances, they were not isolated from West German and West Berlin activism. Not only did they draw on similar discursive repertoires, including some landmark publications about the environmental crisis; they also followed to some extent the conflicts in the Western part of the city, and supported positions of West Berlin activists. In 1988, the undogmatic leftist West Berlin newspaper *Die Tageszeitung* published a declaration in which a church group expressed solidarity with squatters who were protesting against the construction of a highway by occupying an abandoned lot along the wall (that at that time still belonged to the GDR). The East German activists also criticized their own government for selling the area in which the highway was being built to West Berlin, thereby supporting the "reactionary politics of concrete" of the West Berlin Senate (*Die Tageszeitung* 1988).

Environmental activist groups were not only a major segment of the opposition movement in the GDR (Nölting 2002). After the fall of the wall, the lifting of former repressive measures led to an enormous growth and vitality in the East Berlin activist scene. In many districts, local citizens' committees (Bürgerkommitees) emerged that pursued ecological goals.

Among them was also the citizens' committee that in the early 1990s called for the creation of a nature park at the former airport in Johannisthal (see chapter 5), or a group, that, according to a report in *Der Rabe Ralf,* launched grass-roots activities to renaturalize a local pond in the district of Marzahn that was not formally protected and whose property status was unclear to the activists (anonymous 1992). At the same time, West German nature-conservation organizations such as the BUND and the DBV expanded to the GDR, providing a new institutional framework for existing local groups. A Green Party was founded that formed a common list of candidates with Bündnis 90 (an alliance of former opposition groups) and successfully participated in the first federal and Berlin elections that were held after the unification. With the Grüne Liga (Green League), which was founded in 1989, a new nature-conservation organization emerged that was explicitly committed to the heritage of the East German environmental movement. Its founding members stemmed from the urban ecology groups. After struggles with the other members of the GNU, who they criticized as being too much aligned with the former system and unwilling to reform, they formed the Grüne Liga as a decentralized network of local groups.[85]

Very soon, however, the environmental groups that had existed in the former East Berlin lost much of their public support, and hence also much of influence on policy making. This was also the case with the individuals or local groups who, as volunteers, had taken care of individual nature reserves relatively autonomously. Their tasks were now taken over by administration officials, often landscape planners hired from the West, who operated in a more legalistic and bureaucratic framework.[86] This led to much frustration on the side of the nature volunteers, who felt disenfranchised. A 1992 article in *Der Rabe Ralf* (Bringmann 1992) expresses some of the frustration of East Berlin nature amateurs who felt reduced to mere passive "onlookers" with respect to their former territories. The author even describes how local activists tried to prevent an excursion team of the Senate from entering the area they still felt responsible for.

It was mainly through the Grüne Liga that the specific East German tradition of environmentalism survived in the unified city. Since the early 1990s the Grüne Liga has also been a member of the BLN. It became involved in specific East Berlin conflicts, such as those concerning the future of the wastelands along the former wall, or the fate of Berlin nature-conservation reserves that had lost their protected status as a result of a mistake in the reunification treaty.[87]

Conclusion

It was only through the support of various regime communities that biotope protection achieved a stable place on the political agenda of West Berlin, and later, the reunified Berlin. Although external developments such as federal legislation or national and global debates on species conservation had paved the way for such a policy, it was given shape and momentum by the concerted efforts of such communities. The ecologists who had studied nature in the city and who had decisively been involved in the making of the Species Protection Program were such a community. As a political entrepreneur, Herbert Sukopp took an active role in the promotion of this program in the administrative context. For ecologists such public involvement was as much a matter of moral commitment to its research objects as it was a strategy to establish their discipline as a socially relevant domain of expertise. This applies also to the landscape planners who used the issue of biotope mapping to redefine the scope of their competences and to claim a space within the administrative system and the emerging consultation market. Allied both with ecologists and with activists, they promoted concern for nature in the city and, in turn, benefited from the options that this policy offered for their professional success. All these projects would have been futile had they not been backed by the variegated culture of civic activism that existed in the city, with its well-established nature-conservation organizations and its innumerous Bürgerinitiativen. As we have seen, these developments also had a complement in East Berlin, where a different strand of activism matured under different political circumstances.

I have tried to show that the communities were themselves dynamic constructions, whose self-understanding, membership and political strategies were co-produced in close interaction with the development of the biotope-protection arena. Thus, the profession of landscape planners of the 1970 and the 1980s, differed markedly from those of the preceding landscape-care specialists and horticulturalists. Their constitution as a community was the result of study reforms in which a new, ecologically defined profile for the landscape planners was articulated. Likewise, the activist communities that dominated the civic protest in this period differed from the earlier lobbyism of traditional nature-conservation organizations. As we have seen, even the long-standing VBN was challenged by the joining of new members and the competition with other groups in the field of protest. The successful institutionalization of the biotope-protection problem was another factor that provoked adaptations of the communities.

Thus, activist groups became increasingly involved in organized procedures of the political-administrative system, which required cooperative styles of negotiation instead of the contentious forms of protest that they had otherwise articulated.

My account of the emergence and political positioning of environmental groups has focused mainly on the 1970s and the 1980s, an especially dynamic period both on the professional end of the spectrum and on the civic-activist side. It was in conjunction with the cognitive innovations in urban ecology that the emergence of these communities shaped and sustained the biotope-protection regime in Berlin. The existence of a coherent set of problem claims and a critical constituency that supported them, however, did not guarantee the success of the actual biotope-protection policy that these communities promoted. In the following two chapters I will examine some of the political struggles through which the problem claims of the regime were contested, rejected, or partially implemented.

5 Places in the Making: From Wastelands to Urban Nature Parks

The Species Protection Program provided a blueprint for the reorganization of Berlin's urban space on the basis of ecological knowledge. It was through more specific site-focused projects, however, that the goals of the program actually became implemented. In this chapter I will use the example of urban wastelands or ruderal areas, as ecologists also called them, to zoom in on the dynamics through which the biotope-protection regime materialized in the cultural and material order of urban space. More then any other of the biotope types that ecologists had distinguished in their surveys, wastelands became the focus of long-lasting and socially extended campaigns for preservation. As I will show, the campaigns for wasteland protection crystallized around a complex set of meanings of these places—or "place images," as I call them here. (See also Shields 1991.) I will trace how promoters of wasteland protection constructed these place images, how they mobilized them to undermine dominant policy schemes, and how they translated them into alternative projects of building "nature parks." Activists and other political stakeholders in urban nature conservation, however, appropriated ecological schemes of meaning selectively and aligned them with others that were equally important for their political practice. What in the end became institutionalized as a nature park was a social and material result of these practices and the place images they articulated.

After giving a short overview of the development plans against which the nature-park idea was articulated, I will show that these plans relied on a very negative framing of the wastelands. I will then show how critics of the development assembled an alternative place image, one that aligned ecological representations (species numbers, biotope type category) with images of wilderness, historical memory practices, and new modes of visual and physical apprehension. These images, however, did not remain stable throughout the trajectory of the political process that led to the designation of nature parks. They became accommodated with further concerns, which resulted in different interpretations of the concepts of wastelands and nature parks. At the same time, tensions that existed within the original concept of the nature park led to different interpretations of the park, giving special emphasis either to recreation issues, to garden aesthetics, or to the site's significance for species protection. In the case of the park at the Gleisdreieck, this led to complete abandonment of the idea of a nature park that originally had motivated the project.

The Quest for Redevelopment: Urban Planning in the 1980s

It was only because of their marginal locations that vast areas of the former city center had remained undeveloped for such a long time. Whereas other parts of West Berlin had been widely built up in the 1950s and the 1960s, and a new city had evolved in Charlottenburg, the broad zone running along the wall remained largely untouched. In the east and the south of the Tiergarten park, huge vacant areas had been left by demolition done under the National Socialist government (as a preparation for the creation of a gigantic new center) and by the bombing raids during the war. The railway tracks that traversed the former center, including the Anhalt passenger station and Anhalt freight station, had been severely damaged during the war but were used to some extent by the East Germany company that operated the railway network of the entire city. Further east along the wall was the Friedrichstadt (as the northern part of Kreuzberg was called), a working-class neighborhood south of the former city that had been severely damaged during the war. Until the late 1970s, reconstruction of these areas proceeded in a rather limited and piecemeal manner.

Since the 1960s, chunks of the Tiergarten, the adjacent wastelands, and the partly abandoned railway tracks that connected the center with the south had been earmarked for the so-called Westtangente, a newly envisaged element of Berlin's inner-city highway system.[1] Although the Senate renounced them in 1981,[2] road construction plans for the area reappeared on the agenda of the newly elected Christian Democratic Senate (*Der Tagesspiegel* 1982). After the signing of the Grundlagenvertrag (foundation agreement) in 1972, a more cooperative relationship emerged between the GDR and the FDR; this helped the Senate to launch development projects on the various no-man's-lands that existed between the two parts of the city. Since 1972 the Senate had been in (secret) negotiations with the GDR to purchase the Potsdamer and Anhalter Güterbahnhöfe (freight stations) in the so-called Gleisdreieck, an area of about 63 acres that separated the densely populated districts of Schöneberg and Kreuzberg. In 1979 an agreement was reached and plans for its development were devised.[3] The Senate and the GDR also agreed upon plans for a new freight station at the Südgelände, the abandoned former shunting station about 7 kilometers south of the Gleisdreieck. Although geographically located in West Berlin, the area belonged to the railway system that was run by the GDR railway company and therefore had remained beyond the jurisdiction of Western planners.[4] In 1988 West Berlin also purchased the so-called Lennédreieck, a smaller territory at the wall that before the war had been a part of the shopping and cultural center around the Potsdamer Platz. It formed a triangular extension of the East Berlin border beyond the confines of the wall. Ringed by a fence, and not systematically supervised, it formed a kind of no-man's land between the two parts of the city. The West Berlin Senate intended to built the projected Westtangente across this site.

A major impetus for the redevelopment of the destroyed inner areas came from the Internationale Bauausstellung 1984 (International Construction Exposition 1984) and the planning procedure for the Zentraler Bereich (Central Area). The plan for the Internationale Bauausstellung (IBA) had been launched by the SPD/FDP Senate in 1978. Besides the rehabilitation of old housing estates, the rebuilding of the destroyed Friedrichstadt became one of the two main foci of the IBA. Committed to what the architect Joseph Paul Kleihues dubbed "critical reconstruction," the IBA also aimed at the inclusion of green spaces (courtyards, roofs, facades, etc.) into residential areas (Bodenschatz 1987). Political and technical intricacies delayed the completion of the project until way beyond the official exposition year, which had already been postponed from 1984 to 1987.

The planning procedure for the Zentraler Bereich covered a vast space that extended from the Großer Tiergarten to the districts in the north and the south. The inauguration of a separate procedure for the Zentraler Bereich was partly a result of a critique of the IBA (Schneider, no year). In contrast to the comprehensive and therefore, presumably unduly complex approach of the IBA, the new plan focused on "major points of high priority" (ibid.: 9). Organizationally, the design of the Zentraler Bereich was clearly distinguished from the IBA.[5] Its city-planning ideal took a somewhat different tack. Rather than turning the city into a livable place (the official goal of the IBA), its advocates sought to provide the center with a new representative function (Jahn 1983).

These development projects were supported by an actor coalition that included planning officials, architects, public housing companies, and private developers and investors. Since the period of postwar reconstruction, the Senate Department for Construction and Housing and the Senate Department for Traffic, which were both formally in charge of large-scale development policies, had operated in close coordination with the public housing cooperatives and construction firms in the city (Bodenschatz 1987).[6] It was in these networks that plans for road and railway construction in the city emerged and were continuously kept on the agenda. With regard to the planning of the Zentraler Bereich, the initiative came from the Architekten und Ingenieursvereinigung (Association of Architects and Engineers), which was rather influential in Berlin and which also had many members in the Senate departments (Stimmann and Machule 1980). Pivotal roles were played by IBA GmbH (IBA limited-liability company) and by the steering team of the planning process for the Zentraler Bereich, which were commissioned by the Senate to draft the plan (Hoffmann-Axthelm 1980). Many of the members of those two bodies were architects and urban planners.

In the discourses that guided the redevelopment plans, wastelands as such did not have any positive place image. Typically, development advocates considered these areas according to their official designations in existing planning schemes. Even if a formerly residential area had been bombed, it was still represented as a residential area. Likewise, the Gleisdreieck and the Südgelände were formally still designated as railway sites. In addition to the formal planning categories, redevelopment advocates employed aesthetic characterizations that framed the wastelands as signatures of urban disfigurement and decay. What made these sites problematic in the eyes of planners was not only the lack of ordinary land use, but also the negative reverberation that this had on the adjacent districts. For example, Kleihues bemoaned that the "emptied-out area of the former Postdam Personen Bahnhof acted as barrier between, on the one hand, the largely reconstituted West, and on the other, the southern Friedrichstadt, the Luisenstadt, and SO 36, which were partly devastated by misdirected planning and run down due to lack of investment" (1981: 288). He maintained that a "decisive and artistically ambitious development" was needed to demonstrate the ambition to change the situation.

At the same time, planners attempted to create a positive image of the sites, which they derived from reference to their history, layout, and former function and from their location in the wider setting of the city. Thus, the architects' concept for the Zentraler Bereich drew upon metaphoric attributes to articulate a new symbolic function of the inner city (Jahn 1983). The Tiergarten was dubbed the "central garden of the city," the Kulturforum the "real core," the Reichstag area a "place of free speech," and the open spaces between Potsdamer Platz and Kulturforum the "gate area that keeps memories alive." The symbolic meanings attributed to the sites were interpreted not only with respect to the West but also with respect to to their function as symbolic connections between West Berlin and East Berlin (ibid.: 4). Although this was primarily a redevelopment program, these planners also considered a role for open spaces in their projects. In contrast to the IBA advocate Kleihues, the architects of the Zentraler Bereich opposed the redevelopment of the Gleisdreieck and appealed for a park that would integrate elements of the existing ruderal vegetation (Jahn 1983).

Toward an Alternative Place Image

The first proposal to protect ruderal areas in the inner city had been made in 1974 (Kunick 1974: 46). Worried about the loss of many urban wastelands, he had argued that at least some permanent observation fields should be preserved as study objects for ecologists. It was only in reaction to the various development plans of the 1980s, however, that these calls for wasteland protection received broader public attention in Berlin. Local activist-planner coalitions emerged—notably around the Südgelände and the Gleisdreieck, but also around other smaller wastelands—that called for the designation of these areas as nature reserves or publicly accessible open spaces. These campaigns revolved around an alternative place image that featured the actual biophysical features of the spaces and endowed them with a number of positively evaluated properties. From this perspective, redevelopment was anything but an innocent measure of urban renewal. It was essentially a destructive act, a sacrifice of Berlin's most extraordinary pieces of urban nature.

Places of Nature

At the most general level, these place images materialized in categories and names that advanced alternative use values and identities for the sites. Generic categories such as "wastelands" (Brachflächen), "ruderal areas" (Ruderalflächen), and "nature parks" (Naturparks) had a positive ring for those who articulated them and helped to place them on an equal footing with other forms of land use or development in the city. The ecologist Ingo Kowarik (1991, 1992) coined the term Natur der Vierten Art, meaning Nature of the Fourth Kind, to align urban wastelands with three other forms of nature in the city: the remnants of the prehistoric landscape, the remains of landscape used for agriculture, and the artificially shaped urban greeneries. Kowarik argued that "nature of the fourth kind" had evolved through the spontaneous adaptation of vegetation to the urban environment and thus was the most characteristic form of nature in the city. He suggested it should be valued as highly as the other three types and should be actively promoted in urban planning.

Promotors of wasteland preservation drew on vernacular or newly invented place names that gave the sites an identity and functioned as symbolic markers of the political campaigns that surrounded them (e.g., in the name of activist groups that acted on their behalf). Terms such as "Gleisdreieck" (Railway Track Triangle, for the Anhalter and Postdamer Güterbahnhof), "Südgelände" (Southern Area, for the Tempelhofer Rangierbahnhof), although not new, became buzzwords of their alternative political projects. During heated conflicts around the Lennédreieck in 1988, left-wing activists renamed the site after a fellow activist who had committed suicide in prison. (See Frank 2000: 261.) After unification a contested piece of the Gleisdreieck received much attention under the name Wäldchen (diminutive for grove), a name that aligned the area with the forest, an almost archetypical icon of nature in German culture (Brück 1999). The diminutive expressed a particular affectionate relationship with the area and underlined its vulnerability. Ecologically minded planners also advanced the notion of the Grüne Mitte (Green Center), which referred to a coherent complex of open spaces in the center of the city—various wastelands as well as the Tiergarten (Sukopp 1982; Maas 1982). The term Grüne Mitte allowed them to articulate an alternative meaning to what hitherto had simply been referred to as the Zentraler Bereich.[7] Whereas orthodox planners had considered the large open center as a barrier that divided the city and therefore should be developed, they maintained that these large open spaces with their various belts would actually connect otherwise distant neighborhoods and offer the chance for a more environmentally friendly urban development (Maas 1982: 25).

A pivotal claim of these alternative discourses was that wastelands represented a form of nature that was extraordinary or even (at least in the case of the Gleisdreieck and Südgelände) unique. In the species-protection survey and in various other reports, ecologists associated wastelands with high numbers of species and with the presence of some rare species. Thus, according to a survey conducted in the Südgelände in 1980, the area contained about 334 species of ferns and flowering plants, about one-third of Berlin's total (Asmus 1981). Among these species were various plants of foreign origin. Anti-development activists also featured the presence of

Figure 5.1
Nature and railway technology at the Südgelände. Photo by Uta Steinmetz or Konrad Wita; courtesy of Grün Berlin GmbH, Berlin, from Grün Berlin brochure "Natur-Park Südgelände," 1995.

foxes, falcons, and a rare species of spider that hitherto had only been known to exist in caverns in Southern France and was supposed to have been brought to Berlin by freight trains during the war (ibid.). In 1981, three hitherto unknown beetle species were discovered in the area. These findings, reported in the press (*Berliner Morgenpost* 1981), contributed to the image of the area as a spectacle of exuberant wilderness. The Alternative Liste quoted extensively from ecological reports (Alternative Liste Berlin 1984: 29–32), describing the Südgelande as an "oasis of nature with a variegated and diversified flora and fauna" (ibid.: 59). Other activists referred to the Lennédreieck as a "wild, undeveloped metropolitan paradise . . . with about 161 plant species" (Kubat Dokumentation, quoted in Frank 2000: 261).

The claim that the sites were extraordinary was closely coupled to the claim that they were essential for the progress of ecological research (Kunick 1974: 46). The argument was particularly relevant when Hans-Peter Blume and other ecologists at the Institute of Ecology sought to prevent the redevelopment of the Dörnbergdreieck.[8] The claim was also supported by the

Alternative Liste members of the Abgeordnetenhaus.[9] In line with the general problem framing of urban ecology and the Species Protection Program, promoters of wasteland preservation also maintained that wastelands had a positive influence on the urban climate.

Nature for the People: The Rise of the Park Concept

The wastelands and their new place images would not have received as much public attention as they did if they had not also been connected to the quest for recreation. For decades, planners and citizens had criticized the lack of green space in the central districts that were affected by the development projects. (See chapter 1.) Actually, such alternative uses of the areas were already developing informally, since residents used to go to the Südgelände or the Gleisdreieck to sunbathe, to barbeque, or to walk dogs. In order to address the perceived shortcomings in recreation areas, the BI Westtangente advanced the concept of a Grüntangente (green tangent) as an alternative concept to the Senate's plan. The concept, which was developed further in cooperation with a project group of landscape planners (in 1980), comprised large parks at the Gleisdreieck and the Südgelände and a green strip along the closed-down railway track, which connected the Gleisdreieck and the Südgelände with one another and to existing parks in the north and the south (Grüntangente, no year). In its first concept for a Grüntangente, the BI even suggested covering the railway track in order to create new green space on top of it. It even accepted that with such an encroachment "an absolute protection of the wildly grown ecology is not possible" (ibid.: no page). And in 1979 the Kreuzberg chapter of the SPD called for a "large urban park" with a swimming pool, adventure playgrounds, sporting grounds, as well as the newly planned Museum für Verkehr (museum for traffic) at the Gleisdreieck (Gembitzki 1980).

Conventional understandings of recreation parks, however, soon gave way to the concept of a "nature park" that would preserve the ruderal vegetation and make it a significant element of a new recreation environment. In 1980, a Symposium on Urban Ecology held in Berlin called for the protection of the Südgelände as a nature park.[10] In an article in *Natur-schutzblätter*, Ingo Kowarik (1980) invoked the idea of a "nature garden" as a model for the future of the Gleisdreieck. The BI Südgelände, established in the fall of 1980, campaigned specifically for a nature park at the Südgelände. In 1988, the Arbeitsgemeischaft Gleisdreieck began to pursue the same goal for the Postdamer and Anhalter Güterbahnhof.

The concept of the nature park or nature garden was inspired by the Dutch horticulturalists Ger Londo (1977) and Louis Le Roy (1973), both of whom had cherished spontaneously developing vegetation as an alternative to planted greeneries. In their own projects in Dutch cities they had prepared plots on which they allowed a wild vegetation to develop. LeRoy and Londo had found followers in Berlin who tried to replicate their ideas.[11] The promoters of nature conservation maintained that the Gleisdreieck and the Südgelände were already such nature parks, albeit informal and hidden from public consciousness. Their goal was therefore not so much to create anything new at the sites as to preserve their current state, to acknowledge their function as green spaces, and to make them accessible to the public.

Not only did the framing of wastelands as nature parks align nature-conservation issues with the general quest for open space in the city; it also advanced a new definition of the recreation needs of the urban dweller. According to one observer (Maas 1982: 26), extreme overuse of the Tiergarten had led to intractable conflicts among recreation demands and those of gardeners. "For this reason," Maas concluded, "the wastelands at its fringe are so important, where without control by its shaping and predefined regulations, new forms of use have established themselves spontaneously." Promoters of the new place image were enthusiastic about the wild scenery of the areas, which, in their view, offered a more authentic and unregulated experience of nature than the ornamentally designed parks. They maintained that the wastelands displayed a peculiar beauty, and that they allowed urban residents to experience nature more authentically than they could at the overregulated spaces provided by conventional urban greeneries.[12] In 1980, the AL and a student project from the Technische Universität Berlin published a joint press release that featured the Südgelände as a place for alternative experiences: "One can collect mushrooms, harvest fruit trees, pick wild flowers, enjoy a wide horizon and the warmth of the dry meadows, simply really rest in proximity to the apartments. At the same time children and youngsters are able to romp and to play, without the well-known, uniform and expensive playground equipment."[13] The new place image contrasted the supposed wilderness of the wastelands with a conventional park in which "recreation would only be possible on park benches" (Alternative Liste Berlin 1984: 22). Likewise, the BI Schöneberger Südgelände characterized the area as "an extraordinary landscape that no garden architect would be able to shape" (BI Schöneberger Südgelände, no year: 2). As one BI Westtangente activist I interviewed put it, there was also a politically motivated wish to escape the neat and well-ordered garden culture of the provincial towns from which many of the protesters had come.[14]

Learning to See: The Beauty of Wilderness

By re-imagining wastelands as potential nature parks, advocates advanced an aesthetic sensibility that overturned conventional evaluations of wastelands as ugly or empty. For example, an amateur botanist who |had strolled through the former Potsdam Güterbahnhof area wrote the following:

Around the place, in particular at the southern part, busy traffic dominates, quite the busiest that we have in West Berlin. . . . Only few steps further, behind the remains of the external walls of the station, behind bricked doors one finds at some points almost the quietness of the wood. High birch trees, poplars, maples, willows, at some sites already enlaced by hop and clematis, have taken possession of the railway tracks and roadbeds which were lying next to each other in different altitudinal belts. So densely that at some spots due to the lack of light not a single herb grows, and that the dense coverage of leaves at the bottom almost makes you forget that everything is growing on the most scrawny sand and on gravel. (Stricker 1974: 36)

The discourse on wastelands did not simply erase the differences between nature and the city. It rather drew on their contrasting qualities and put them into a new aesthetic relationship. With its close proximity of noise and quietness, cars and clematis, and leaves and gravel, the Postdam station was an entity opposite to the city yet at the same time, evidently located within the city. Moreover, it was associated with images of excessive sprawl, which presented nature as a power that slowly takes over human artifice.

In the 1980s, similar statements appeared in countless political pamphlets, petitions, expert reports, and brochures in which activists and ecologists promoted preservation of the wastelands. The subtitle of a volume of photographs of the Südgelände that the Bürgerinitiative Schöneberger Südgelände published in 1982 described the area as a "landscape in the city." In another photo book, titled *The Hidden Green of Schöneberg*, the reader was invited to take a virtual tour of the site:

From the diffuse light of the copse we enter into a bright and wide landscape. During hot summer days the glittering sunlight will blind us und the temperatures that dominate here will make us sweat. Memories of southern landscapes are awakened, an effect that is enforced by the loud chirring of innumerable cicada and the buzz of the bees and bumble-bees. And besides this, no sound is noticeable, a peculiar quietness rests on the landscape. (Bürgerinitiative Südgelände 1985: 6)

For one of the photographers who contributed to the aforementioned volume, the Südgelände also evoked memories of his childhood, when he had strolled "in the search of adventure" through meadows, woods, and abandoned lots (Krauke in Krauke and Schnell 1982). Based on his experiences at the site, the photographer claimed to have access to a truth of the place that was inaccessible to the "politicians, planners, and technocrats," who did not bother to visit the site: "For them the Südgelände is still a railway area, very simply because on planning maps it is indicated as such" (ibid.).

For many opponents of the development projects, the Südgelande and the Gleisdreieck were not only places of nature but also places of urban memory. They recalled the most significant periods of urban history, such as the rise of industrialization, the transport system, and the destruction of the city during the war. Booklets and other publications contained historical overviews or chronologies that presented the history of the sites in terms of significant events and distinct periods. These historical narratives found their visual complement in remaining artifacts such as water towers, railways tracks, or railway sheds. According to these historical narratives, nature had "reconquered" human artifice. The abandoning of the station had created special conditions that allowed nature to gain a foothold in a place that had hitherto been exclusively human. The juxtaposition of the dense vegetation with the remains of urban-industrial culture made the sites metaphors for of the vanity of human existence.[15] For many observers, it also revealed the hubris of modern technology. It showed, as one of the photographers put it, "how volatile technology is" (Krauke and Schnell 1982). According to the botanist Walter Stricker (1974: 37), this even implied a positive message, namely that nature was also able to reestablish itself in areas that had been completely destroyed by man.

Such descriptions were typically written in the form of itineraries that took the reader on a virtual excursion to the areas. Images and texts were thus complementary to the actual excursions that promoters of nature parks regularly organized. By guiding visitors along a selected route, with stops at significant landmarks, and by means of verbal comments and clues, they helped draw the participants into a common aesthetic sensibility, which was at the same time visual, auditory, and corporal. Photographs and a film complemented the textual descriptions of the site. In the introduction to one photo book, the BI Schöneberger Südgelände stated that, rather than present "'hard' facts, numbers and plans" about the freight station project, it wanted to convey a "sensual impression of the vastness and beauty" of the area (Krauke and Schnell 1982: no page). In the introduction, the editors suggest that the reader take a look at the landscape, at least virtually via the photos, and then "think about how much we urban dwellers are able to see and experience there, if we do not destroy this nature with excavators" (ibid.). Photos taken by two Berlin photographers and publicly exhibited by the Bürgerinitiative Schöneberger Südgelände featured groups of trees, blossoming meadows, snowy winter landscapes, decrepit buildings, and a large water tower, mostly in close interplay with the material remains of the railway station. Here and there the camera spotted small details such as fallen trees, the railways tracks, a switch, or a discarded bathtub. All of this resulted in an image of jungle-like wilderness, abundant with secrets to be discovered by the adventurer who made it to this site. The water towers at the Gleisdreieck and the Südgelände became popular icons of the sites, and were reproduced in countless photographs that appeared in the leaflets of activist groups and nature conservationists.

An Emerging Conflict

In the early 1980s there was a fierce confrontation between the claims for redevelopment by the Senate and its various supporters, on the one hand, and the advocates of the nature parks, on the other. The plans for the freight station had already been officially endorsed in 1980, and one year later the Senate was already about to begin with clearing measures to prepare the area. This decision spurred parliamentary motions, an intervention by the Landesbeauftragter, demonstrations, and public-relations campaigns. Formal mistakes in the Planfeststellungsbeschluss (planning procedure) became obvious and eventually forced the Senate to resubmit the plan in 1982. The new plan at least formally acknowledged the area's potential nature-conservation value and included a landscaping scheme (Elvers et al. 1982). When the plan was publicly exhibited, about 3,000 objections were raised.[16] Various expert reports, some produced by activist groups, led to a critical evaluation of the project's impact on the climate, its noise effects, and the underlying traffic-management concept. In the Abgeordnetenhaus, meanwhile, both the AL and the SPD called for cancellation of the project. In 1985, the Senate, reacting not only to the increasing unpopularity of the project but also to a reestimation of the future traffic needs in West Berlin, decided to cancel the project. In 1988, in line with that decision, the Landscape Program earmarked the Südgelände as a landscape and nature reserve and the FNP designated the area as an open-space.

When the conflict at the Südgelände reached its apogee, the development plans for the Gleisdreieck were still in the making. When the Senate renounced the Westtangente plan in 1981 it kindled much optimism among the anti-development activists. Paradoxically, the project of a nature park might have even benefited from the realization of the Süd-gelände project. As ecologists and landscape planners maintained, the protection of the Gleisdreick was the only possibility for the Senate to fulfill the mitigation requirements that, in their view, were legally required for the Südgelände project (Senator für Bau- und Wohnungswesen 1983; Elvers et al. 1982: 105–106). When the plans for the Westangente soon reappeared on the agenda, the idea of a nature park had already found much public support. Even the architects of the Zentraler Bereich opposed the redevelopment of the area and pleaded for a park that would integrate elements of the existing ruderal vegetation (Jahn 1983). When the first devel-opment construction project—the new Museum für Technik und Verkehr (Museum for Technology and Traffic)—was built at the fringe of the area, protection measures included incorporating adjacent ruderal vegetation into the new architectural ensemble. These debates also motivated the production of further evidence about the area and of inspired plans for wasteland nature parks (Schaumann 1986; Behrens, Hühn, and Karbowski 1982).

Sometimes also bombed plots in the city received wider public atten-tion. One example was the so-called Dörnbergdreieck. In 1980, to prevent development of the site, Hans-Peter Blume and Herbert Sukopp demanded

Figure 5.2
Protest against the development of the Gleisdreieck: Fixing a banner on a street near the area. (photo by Bernd Latzel)

Figure 5.3
Protest against the development of the Gleisdreieck: a construction trailer on the area serves as a citizen information center. (photo by Helmut Mehnert)

the designation of the area as a nature reserve.[17] Although this demand was taken up by the nature-conservation department at the Senate, it was eventually turned down, and the western fringe was developed with residential buildings.[18] The main threat for the area, however, emerged in 1985, when it became public that the Senate was making plans to construct a hotel on the site. Ecologists' protests against the hotel were also supported by a group of residents of recently constructed residential buildings. The call for preservation was also supported by the local district administration and by the parliamentary faction of the Alternative Liste.[19] All this, however, could not prevent that the hotel project was eventually built in 1986.

Another conflict concerned the Lennédreieck. From May 26 to July 1, 1988, the area at the wall (which then still belonged to the GDR) was squatted by activists (see Frank 2000: 260). The squatters included nature conservationists, activists from the BI Westtangente and the Alternative Liste, and anarchists. At the site, they erected a tent village, which soon grew to a settlement of wooden huts. The West Berlin police closed the area (access and supply remained possible through a narrow path), and occasionally attacked the squatters with water cannons, tear gas, and loud music. When the Lennédreieck was formally transferred to West Berlin, the police cleared the squatters.

An important issue was the extent to which the provisions of nature conservation actually had to be applied to wastelands. Whereas ecologists claimed that valuable "nature" was also an integral part of the city, from a juridical standpoint this was much less clear. Since the new nature-conservation law was no longer restricted to the protection of extraordinary natural areas, Sukopp and nature-conservation activists claimed that its provisions would apply for the entirety of West Germany and, respectively, West Berlin territory, including the inner center. (See chapter 5.) Accordingly, they considered development projects on ruderal areas as a clear case for "impact regulation," which would have provided for strict conditions and probably compensatory claims. (See chapter 6.) It was the opinion of the Senate for Housing and Construction, however, that these sites, owing to their official designation either as railway areas or residential areas, were exempted from the provisions of the Nature Conservation Act. Activists and nature conservationists mocked the formalism and ignorance of the Senate, which, in their view, simply denied the existence of nature at these sites: "After all, it has developed without having asked the administration for permission and hence, in legal terms does not exist at all." (BI Schöneberger Südgelände, no year: 2)

During these conflicts, the contested sites or their direct surroundings became venues for public action. In a study of British road protests, Andrew Barry has argued that activists, by physically appropriating the ground of contested areas, were not just symbolically voicing their public opinion, but in a more direct sense pointing out—or demonstrating in the literal sense of the word—a specific truth about the site. Rather than just being a site for potential traffic use, they were shown to be a site where the "existence of humans, animals and the land were, in whatever way, mutually implicated" (Barry 2001: 183). In a similar way, the activists who mobilized against the projects at the wastelands of Berlin, by their very action, created an alternative meaning of those places. During a neighborhood party, members of the BI Westtangente built a gate of plants to mark the entrance to the railway track that they wanted to become part of the greenbelt (Bürgerinitiative Westtangente 2010). Likewise, guided tours on which BI members explained the value of the area to citizens, and a spectacular waste-clearing action that the BI Westtangente carried out at the ruderal areas along the envisioned greenbelt (in 1981), directed public attention to the site. Around 1988, the BI Gleisdreieck organized tours to the Gleisdreieck, and placed a former construction trailer in the area that functioned as an information center.[20] Voicing protest with these activities merged with the ongoing informal appropriation of the wasteland by residents, who often found holes in the fences around the area.

The squatting of the Lennédreieck was a general manifestation of an alternative use of urban space. The squatters' village was supported with food and money by citizens and tourists. The BI Westtangente held regular meetings there. Potatoes and vegetables were grown there. The place was not only the object of concern; for many participants it was also a symbol for a broader political alternative. According to one leaflet (quoted in *Die Tageszeitung* 1988: 17), the squatting was "not only directed against the destruction of a green space for highway construction, but against the system that required such swinishness." In this sense, it was not different from other sites in the city (residential buildings, the Kreuzberg district, etc.) in which alternative culture flourished. It was no accident, however, that an urban wasteland became the gathering ground for this activism. Besides its legally ambivalent status at the border, the presence of plants and the wilderness character of this inner urban area achieved a particular political significance. The activists referred to the place as a "green space" (see quotation above) and decided to keep the spontaneous greenery clear of huts and tents (*Die Tageszeitung* 1988).[21] The natural qualities of the space, for these activists, however, did not only reside in the ruderal vegetation. They had no qualms in also using parts of the area to grow vegetables and potatoes, although that might have actually violated the vegetation that ecologists wanted to protect. Although the squatting was widely supported by the left-wing public of the city, this form of activism also created tensions within the nature-conservation scene. "As soon as it was accessible," one nature conservationist later remarked critically about the squatting of the area, "squatters established themselves there, demanded conservation of the area, and battered it with a tent village, vegetable patches, and actionism." (anonymous [Heinrich Weiß?] 1988: 3)

Nature Parks as Exhibits: The Rise and Fall of BUGA 1995

Bundesgartenschauen (federal garden shows) had been held regularly in German cities since 1951. Traditionally this meant that a park-like open-air exhibition area was shaped with colorful flowerbeds and other ornamental horticultural arrangements. In 1985, Berlin had arranged a Bundesgartenschau (abbreviated BUGA) in the Southern district of Rudow, which took a much broader approach to landscape architecture. In 1985, Senate officials began planning the project of a second BUGA to take place in 1991. This idea originated from the ongoing debate about the Zentraler Bereich, and was meant to complement the urban shaping of this area.[22] In May of 1988, the Senate presented a first proposal for the exhibition, which now was scheduled for 1995.[23] In October of 1989, after debates in the public and in the planning community, the newly elected red-green Senate presented a revised proposal.[24] The Bundesgartenschau Berlin GmbH, a QUANGO that had been founded by the Senate in October of 1988, was formally in charge of the further development and implementation of the project.

    With the BUGA plan, the vision of a Green Center moved from the realm of critical discourse to the heart of the Senate's policy. This did not mean, however, that the Senate simply followed the claims of urban ecology. Earlier BUGAs had been heavily attacked by ecologists for their ignorance of ecological conditions.[25] Moreover, like the larger planning concept of the Zentraler Bereich, to which it was tied, the early BUGA plans were largely geared to the creation of historicized, representative green spaces.

    The further development of the BUGA plans was characterized by an increasing acceptance of ecological motives. In 1988, Hendrik Gottfriedsen, the director of the BUGA GmbH, which was responsible for the plans, announced that the protection of some "ruderal vegetations" should be included in the plan.[26] Ecologists and conservationists, however, remained skeptical toward the BUGA. The alternative Berlin newspaper *Die Tageszeitung* quoted Ingo Kowarik as having argued at a protest meeting that the realization of the BUGA plans would even reduce the available green space. Alternatively, Kowarik had suggested the BUGA should be used as a means to preserve the spontaneous vegetation that existed in the center (*Die Tageszeitung* 1989). During the debate about the FNP of 1988, Frank Kapek of the Alternative Liste claimed that the BUGA plans were meant only to attract tourists to the city and did not take "the chance to create an ecologically valuable area."[27]

The 1989 elections that resulted in a majority of SPD and Alternative Liste in the Abgeordnetenhaus paved the way for a revised version of the BUGA plan which put more emphasis on the protection of the ruderal vegetation. The Gleisdreieck and the Südgelände (which was now included into the BUGA project) were now both intended to become nature parks. For the railway track that connected the two areas and some other sites (the Anhalter Bahnhof, the Moabiter Werder, and the Postdamer Personenbahnhof) the plan envisioned more conventional parks. Even at these sites, extraordinary pieces of nature such as the black locust grove at the Moabiter Werder were to be preserved. Although the new Senate canceled the plan for the Westtangente, the Alternative Liste was not able to get the protection of the Lennédreieck accepted by its coalition partner. Instead, the BUGA concept earmarked the area for housing projects. In line with the Species Protection Program, the individual green projects of the BUGA were seen as elements of a larger greenbelt stretching from the area of the first BUGA in the south to the Tiergarten and the Moabit.

Ecologists and activists now became also closely involved in the planning process. Members of the BI Südgelände consulted regularly with representatives of Bundesgartenschau Berlin GmbH (Federal Garden Show Berlin LLC) to nail down the specific outlines for the Nature Park. The appointment of one activist by the QUANGO strengthened the links between the two groups. In October of 1989, the BUGA GmbH and the Kreuzberg district organized a symposium at which planning officials, members of the AG Gleisdreieck and the BI Westtangente, ecologists, and landscape planners discussed the future of the Gleisdreieck. When new urban planning projects conflicted with plans for green space, the BUGA also became an important coalition partner for activists. It opposed the plans of Senator for Housing and Construction Wolfgang Nagel to construct large new buildings that would have destroyed the grove of black locust at the Moabiter Werder (*Die Tageszeitung* 1990).

The increasing acceptance of the nature-park concept implied that the meaning of the place shifted from a target of nature conservation to an aesthetically designed park. In 1991 the BUGA GmbH invited various landscape architects who presented concepts for the shaping of the park area (Bundesgartenschau Berlin 1995 GmbH and Bezirksamt Kreuzberg 1991). In contrast to the earlier landscape plans of the district of Kreuzberg (Schaumann, no year), which restricted the modifications to the area mainly to the construction of a pathway system, some of these plans envisaged far-reaching modifications to the site and its vegetation. Ecologists criticized this and called for a more modest design. As we will see later, this remained an important issue in the realization of the park projects in the 1990s.

Notwithstanding these contested issues, the BUGA project was one of the most successful steps toward an implementation of the Species Protection Program. The policy, however, was very much an outcome of the specific policy constellations in the walled city of Berlin. When the wall came down, only a month after the Senate had presented its revised BUGA plan, this changed the terms of the political debate. In principle, it meant that the wasteland areas in the undeveloped areas to the east of the former wall could be integrated into a much wider Green Center. Indeed, the Senate began to revise its plans for the BUGA by including further projects in the East. However, the program of keeping large areas in the center open as green spaces was no longer realistic under the new political conditions. The first reason was a change of government. Only a few months after the fall of the wall, a conflict among the coalition parties about police action against squatters led to the dissolution of the AL-SPD coalition. With the formation of an SPD-CDU coalition, ecological planning lost one of its main supporting parties. Second, the fall of the wall and the unification of the city had resulted in enormous economic pressure on the urban areas. Lots that formerly were wastelands in fringe locations, suddenly turned into "Filetstücke," (literally fillets) as excellent sites in central locations with high real-estate value were called. International developers and real-estate firms—which had played only a minor role in the pre-unification period—took over and became important players. Third, policy making itself gave priority to the stimulation of urban development (Rada 1997; Strom 2001). This was not only due to the Senate's commitment to prepare Berlin for its future function as the political capital. In order to position Berlin in an increasingly competitive European and global economy, the Senate sought to attract investment capital, often from international firms. In the case of the Postdamer Platz, the Senate, which owned the site, sold it for only a symbolic sum to the auto company Daimler-Benz. Urban planning was now largely conceived of as a means to create attractive conditions for investors, and closely coordinated with the latter. Although the hype of urban development kindled a debate on architecture and urban form, issues of ecology and nature conservation became widely marginalized.[28] After attempts to relocate the BUGA in a suburban district of East Berlin failed, the project was given up entirely by the Senate.

New Projects in the Unified City

Claims for ambitious wasteland-protection schemes could no longer be realized under these circumstances, notwithstanding their promotion by ecologists.[29] As with the Diplomatenviertel, many wastelands of the former West Berlin, as well as those that ecologists had hoped to be able to protect in the East, were eventually developed. Such backlashes went hand in hand with the resurgence of park projects at other sites. This was due in part to the need to provide mitigation for encroachments into urban nature that were caused by the ongoing development of the city. (Measures of mitigation and compensation were provided by the Nature Conservation Act and will be discussed as a theme in its own right in chapter 6.) In addition, institutions and constituencies that had formed around this cause were still in place and still promoting their ideas. When it became clear that compensation needs had to be fulfilled for other projects, the nature-park idea achieved new momentum. When planners were searching for possible compensation areas for the development projects in the city, it was most obvious to do this by reviving the nature-park projects.

The first project was the nature park at the Südgelände. After the BUGA had been canceled, activists as well as Grün Berlin, the former BUGA GmbH, which was already responsible for the administration of various green spaces in Berlin green space, continued to pursue the project. For a while, however, the Eisenbahn-Bundesamt (Federal Railway Authority) and the Senator of Housing and Construction even considered rebuilding the station (*Die Tageszeitung* 1991). The way for the park was eventually paved by the need of the Eisenbahn-Bundesamt to provide compensation for the loss of "nature and landscape" that was caused by the newly built railway projects in the center, notably a tunnel under the Tiergarten. Since the Südgelände was officially designated as a railway area, its formal change to a green space was legally considered a fulfillment of the compensation requirement. The park project found financial sponsorship from a private foundation and was meant to be Berlin's contribution to the world exhibition EXPO

2000 (which was mainly located in Hannover). Under the jurisdiction of Grün Berlin, the area was widely reshaped in the following years until it was opened to the public in 2000. Furthermore, in 1999 the Senate approved a decree that designated the most valuable areas of the site as a nature reserve.[30]

A second nature park was realized at an abandoned airport in the former East Berlin district of Adlershof. Opened in 1908, the airport at Johannisthal had once been the first airport for motor aviation in Germany. It was closed down after the war, and since that time used as a military training ground. Fenced off from the neighborhood, it was, much like the Südgelände, inaccessible to the public. After unification several proposals were made for the future use of the site. The idea of a park was initially propagated by two neighborhood activist groups who were mainly interested in the recreation potential of the site.[31] From 1991 on, Berlin ecologists— notably the office of the Landesbeauftragter—argued that it was a remarkable dry-land habitat, which, because of its size, was extraordinary for the region. In order to strengthen the conservation claim and to influence the planning process, the Landesbeauftragter and the BLN issued surveys on the flora and fauna of the site which added further evidence to this claim (Gerstberger 1992; Saure 1992; Kielhorn and Kielhorn 1993; Köstler and Stöhr 1993). Although the number of species was much lower than at the Südgelände, the site proved to be a worthy habitat for various birds, beetles, and wild bees. Ecologists attributed this to the extent and plenitude of the area, as well as to the character of the vegetation, which showed many features of a steppe habitat.

In contrast to the scenery of the Südgelände, the open-meadows of the airport were much less impressive to most visitors. Promoters of a "close-to-nature recreation park," however, emphasized the "spatial vastness and emptiness" that was extraordinary in the midst of an otherwise densely built-up urban environment (Becker Gieseke Mohren Richard 1995: 22). Various remnants of aviation history existed at the site, such as research facilities, hangars, and next to the place a former "tailspin tower" for experiments in aviation medicine. As with railway history in the case of the Südgelände, these visual remnants, as well as the narratives that linked the wasteland with its past as Germany's pioneering airport, helped to ascribe historical meaning to the site (ibid.: 15–16; Interessengemeinschaft Flugplatz Johannithal 1991).

The Berlin Senate viewed the area around the airport as a disturbing example of social and industrial decay in the former East, but also as a site that provided space for potential development projects. In 1991, it launched plans to turn the area into a "science city" with university and industrial research institutes and a few residential areas. As with other science cities or science parks that already existed elsewhere in the world, the project aimed at strengthening the ties between science and industry (Tischer 1996). Although the project was a potential threat to the claim for conservation of the park, the latter was eventually integrated into the plans of the science city. On the one hand, this was the result of the successful campaign by ecologists, who found a powerful resource for their claim in the Berlin Nature Conservation Act. Dry meadows, as they were supposed to cover large parts of the site, were among the few categories, which were explicitly listed in the law as valuable habitats. Given these legal restrictions, consensus was easily achieved among planning participants that agreed at least, that these ecologically most valuable parts of the airport had to be preserved.[32] Moreover, the construction of the science city would have entailed various encroachments on the natural environment, which had to be compensated for. The airport turned out to be a suitable territory on which to provide space for these compensation measures. On the other hand, developers themselves began to see such a park as a means of creating an attractive urban environment for the science city and to boost its image.

The shape and the design of the park emerged from a long process of negotiation and mutual adaptation among the proponents of the science city and those of nature conservation. This debate was formalized in a series of hearings and mediation meetings, organized by a local landscape planning office, that brought together a wide range of stakeholders.[33] Among the participants were departments of the Senate's and the local district's administration, the nature conservation counseling agency, the semi-private agency that was in charge of the development of the site, investors and scientific institutes that were engaged in the science city, representatives of the scientific institutes, and various external experts, among them landscape architects, ecologists, and urban planners.

After the fall of the wall, conservation experts and activists of the Bürgerinititative for a nature park at Gleisdreieck continued to lobby for a park at that site. However, they were unable to keep the area from being used as a logistics center for the construction work at the adjacent Potsdamer Platz. This meant that large parts of the ruderal vegetation were removed, which diminished the sites conservational value. In March of 1993, the Senate, the developers of the Potsdamer Platz, and Eisenbahn-Bundesamt (the formal owner of the site) agreed that a major section of the Gleisdreieck should be turned into a park.

The park was to be financed by the investors and thereby to provide the compensation for the rebuilding of the Potsdamer Platz. The realization of this project was stymied by the reluctance of the Eisenbahn-Bundesamt—which meanwhile saw the chance to make a higher profit by selling the areas to potential developers—to fulfill its promises. Moreover, large parts of the Gleisdreieck area were not included into the compensation agreement and were thus earmarked for development projects.[34] Only after intricate legal debates, a (failed) suit by the AG Gleisdreieck, and continuous lobbying was a contract signed between the Land of Berlin and the representatives of the Eisenbahn-Bundesamt—the firm Vivico, in September of 2005, which paved the way for the actual realization of the park. In contrast to the other two parks, however, the concept of a nature park had meanwhile lost its function as an all-encompassing guideline for the site's development. The site had already witnessed a piecemeal appropriation by different activist groups who used the area for different purposes, among them a dog run, a nature-experience path, and an intercultural community garden (run by a group of exiled Bosnian women). Activists and local districts promoted a "citizens park" that would be run partly by neighborhood groups and would continue the activities that had already developed spontaneously on the site. The Senate, by contrast, promoted the idea of a representative city park that would be artistically designed and that would attract tourists and provide recreation space for the people working in the adjacent office areas. Although an agreement was reached that the design and the use of the park should be consistent with the protection of the grove, this did not lead to a designation of these areas as nature reserves.

Nature Parks: Between Wilderness and Landscape Architecture

Basically, the notion of a nature park can be seen as an alignment between two pre-existing concepts: a nature reserve (which aims at protecting a valuable piece of flora and fauna) and a shaped public recreation space. Initially, the plasticity and comprehensiveness of this concept served to create alliances between groups of actors who tended to prioritize either of these two aspects. However, during the process of planning and designing the park, the tensions between these aspects became apparent.

A first bone of contention concerned the balance between the concern to protect the flora and fauna of the wastelands and the attempt to open these spaces for public use. While many activists and landscape planners advocated the unrestricted accessibility of the area, ecologists and nature-conservation officials feared that the presence of too many visitors might have a detrimental effect on the flora and fauna.

The second issue concerned the extent to which the parks should be shaped by means of landscape architectural design. In the first nature-park proposals, design activities consisted almost exclusively of the construction of a pathway system.[35] Later, however, more ambitious attempts to design the nature parks emerged. Design processes and the relative weight given to ecology and aesthetic considerations differed considerably among the three sites.

At the Südgelände, the design process was organized under the guidance of Grün Berlin. In 1991 the organization commissioned two landscape planning offices, one of them led by Ingo Kowarik, to develop a plan for the site (ÖkoCon 1992; Grün Berlin 1995). Thus, from the onset, ecological considerations had a strong voice in the design process. Other conservation experts, such as the office of the Landesbeauftragter for nature conservation, gave further support. This entailed a relatively restrictive policy with respect to public access, as well as a rejection of any attempt toward aesthetic design. As the landscape planners argued, nature as it already existed at the two wastelands should be made perceptible for the visitors. The creation of some kind of "eco-Disneyland" (ÖkoCon 1992: 129) should be avoided.

When the landscape planners presented their first conception, in 1992, it met with harsh resistance from Green Berlin, the Südgelände activist group, as well as from the foundation that provided financial support.[36] These issues were debated by a consulting panel in which all these groups were represented. Although the participants agreed that the existing features of the flora and fauna of the site should be protected and made the major theme of the park, design plans show a constant move from a relatively rigid conservation scheme toward a concept of a park that was more accessible to the public and, at least partly, artistically designed. According to the original plan, the meadows in the center of the area were to be turned into a nature reserve and fenced off to the public. Only a small viewing platform at the fringe of the site would have allowed visitors a view of this inaccessible area. A revised plan which was presented by the two offices in 1995 assigned for a wooden foot bridge for visitors in the nature reserve (Grün Berlin 1995). For conservation reasons, the foot bridge was to be of a height of about one meter, and fitted on both sides with a railing. With the track which was eventually realized, however, the design of the park took a considerable step forward toward landscape art. The alternative design was proposed by a group of metal sculpture artists who had their workshop in an old railway shed at the Südgelände, and immediately received the support of Grün Berlin and the Allianzstiftung. The "walkable sculpture" (begehbare Skulptur), as the artists named it, was made of rusty steel and consisted of a narrow trail which rested on low metal rollers and, at some points, was marked by larger steel sculptures. The rusty material and the wheel-like roundness of the rollers were an explicit allusion to the history of the site as a railway station. The track allowed people to walk into the core of the nature reserve without having any physical contact with soil or vegetation, while being as low as possible to the ground and not restricted by a railing. More than being just a pathway system, this "sculpture" turned the Südgelände into an artsy place. Nature and art became closely connected to one another, one providing the background for the other. This process of artsification continued after the opening of the park, when further items of steel art were added. Among early promoters of the nature-park concept this led to an increasing disappointment. When three of them evaluated the site in 2004, they complained that it was overburdened by art and had moved too far away from the original idea (Kowarik, Körner, and Poggendorf 2004).

At Johannisthal, aesthetic motives have been influential since the very beginning of the planning process. Whereas the core region of the airport

Figure 5.4
Urban wilderness as a park: a trail in the Südgelände. (photo by the author)

was to be kept untouched as a nature reserve, planners of the science city called for a more intensive design regarding the remaining area. In contrast to the Südgelände, the uniform meadows of the former airport were not considered as aesthetically appealing unless they were framed in the context of some form of landscape design. As was stated in one of the planners' meetings in 1993, the "ruderal area" in the center "had to be put on a stage."[37] In 1993 landscape architects were invited to present proposals for the design of the area.[38] Eventually consensus was reached that the site should be designed according to the proposal submitted by the Berlin landscape architect Gabriele Kiefer. Rather than completely reshaping the area with newly planted woods, as other architects had proposed, Kiefer suggested a landscape park that retained the vastness and openness of the former airfield and integrated parts of the existing flora and fauna.[39] The park also provided an open field of vision that was sharply bounded by the edge of the science city. The nature reserve in its center was to be surrounded by a narrow wall of "gambions" (baskets of wire filled with stones), which was intended to make the place attractive for birds.

At both sites, ecological considerations and the attempt to aesthetically design the park existed in a relatively uneasy relationship with one another. However, aesthetic design was not intrinsically at odds with ecology. At the Südgelände, the processes of negotiation that developed between ecologists and conservationists, on the one hand, and garden architects and other proponents of aesthetic design, on the other, resulted in mutual alignments between these perspectives that most of the participants were willing to accept. To some extent, ecological concepts and definitions were also incorporated into the design schemes and thus became part of the visual scenario the designers aimed to emphasize at the two sites. This was even the case in Johannisthal, although aesthetic considerations were much more dominant there.

A very different tack was taken at the Gleisdreieck. When in 1991 the BUGA invited design proposals for a future nature park at that site, the six participating landscape architects provided much more ambitious ideas for shaping the site (Bundesgartenschau Berlin 1995 GmbH and Bezirksamt Kreuzberg 1991). This was at odds with the ideas of the AG Gleisdreieck, which gave much more priority to the conservational aspect (AG Gleisdreieck 1991). After the site had been transformed by its use as a logistics center, design issues further came to dominate conservational considerations. In 2006, the Senate organized a landscape architectural competition and, on the basis of the winning concepts, began to reshape the site as a park. Activists were also included in the jury for the landscape planning competition, and a leading activist even contributed his own proposal. This did not, however, prevent many activists from seeing the prize-winning concept of the landscape firm Atelier Loidl, and its implementation in the hands of Green Berlin, as an assault on the grass-roots initiatives that had developed in the park.

Although the Atelier Loidl concept was much less invasive than some of the competing concepts, it distanced itself explicitly from an emphasis on the "myth of nature" as well as railway history.[40] The architects rather emphasized the potential role of the park as a stimulus for new developments which would reach beyond these former points of orientation (Atelier Loidl 2007: 14). In this way they hoped that the park would become similar to what, in their view, was characteristic for Berlin at large: "What Berlin really is; multicultural, chic, without swank, modern, flexible, fun-oriented, and most importantly, sensual." (ibid.) The architects conceived the park as a wide plain defined and structured by planted woods, pathways, and preexisting (sometimes renovated) railway facilities. Much of the plan provided for the horticultural creation of entirely new landscape features. It also devised for sporting grounds and a café. Of the grass-roots activities that had developed here, only the intercultural gardens and a "nature experience area" were included in the park. At one spot the architects gave priority to the "subjects of ecology." This was the grove and the strip along a railroad track. As the designers put it, "they will be put on stage as ecologically sensitive areas and will sensitize the visitor to the nature of the city" (Atelier Loidl 2007: 26). In comparison with the original idea of the nature park, however, ecological goals have become largely marginalized in their relation to aesthetic goals.

Nature Parks: Between Wilderness and Management

While open-space planners sought to give the parks an aesthetic form, conservationists advocated a range of management practices that aimed at the optimization of the ecological qualities of the site. Such measures were proposed in many statements, and later, they were carried out under the guidance of the responsible nature-conservation authorities (which at the Südgelände and Adlershof were formally committed to developing maintenance schemes for the areas that were designated as nature reserves).

In a study of a recent Chicago conservation campaign, Reid Helford (1999) has shown that this campaign also aimed at actively shaping the vegetation. (See also Gross and Hoffmann-Riem 2005.) The goal of these activities, in particular burning the existing coverage of vegetation, was to reconstitute the type of savannah that had existed in the region before the coming of European colonizers. Management was thus meant to remedy the human influence of settlement that had presumably resulted in a deterioration of the landscape. Berlin wasteland management took a different tack. Rather than reconstituting some allegedly more natural ancient state of the site, it aimed at promoting the forms of flora and fauna that had actually developed on the wasteland. These were not considered as being free of the influence of human activities. In contrast, in cities, according to the concept of "nature of the fourth kind," these influences were to be accepted as a peculiar feature of these sites. Thus, estimations of the previous state of the sites could not serve as criteria for evaluating its present state or for defining maintenance goals. At the same time, however, not everything that happened to the flora and fauna of the place was deemed optimal. It was throughout the practices of fixing and negotiating maintenance schemes that new boundaries between what was to be deemed as "natural" and "not natural" in these places were locally produced.

At the beginning of the controversy around the Südgelände, active management of the site was not yet considered necessary. In the early 1990s, however, the spontaneous process of greening, or what ecologists called "succession," was increasingly considered to be a problem. The authors of a vegetation survey (ÖkoCon 1991) concluded that the woody vegetation had increased dramatically since 1980. As the authors of the survey estimated, it would not have taken much time until the wasteland were entirely covered with a dense wood. Although in principle this could have been seen as the quite normal result of a natural process, it was considered to be a serious form of deterioration. Many of the rare species that populated the site depended upon the existence of meadows and were now threatened. Moreover, the meadows consisted of various plant societies that were deemed worthy of conservation. The juxtaposition of woody and open sites was one of the most appealing visual features of the Südgelände and therefore was considered worthy of conservation. Based on these considerations, the design plan in 1992 included a first maintenance scheme (ÖkoCon 1992). In 2000, an official maintenance protocol was issued by Berlin's nature-conservation administration; it was further elaborated in close cooperation with administration representatives (Planland 2000). In the following years, trees that had spread into the meadows were cut. Single plots of the meadows were also mowed regularly, and individual plant species, which tended to abound in the meadows were removed. These interventions aimed at maintaining or even increasing the existing diversity of plant communities, including those at a very early stage of succession.[41]

At Johannisthal, active biotope maintenance was called for from the very beginning of the concern for conservation. Thanks to the military use of this place, as well as the continuous application of herbicides, the succession of the vegetation had been blocked, and according to ecologists it was just this that made the place as valuable as it was. Maintenance activities were thus deemed indispensable if the site were to be kept in its current shape. In 1994 a provisional maintenance program was submitted and, after some negotiation on this issue with government officials and planning agencies, a revised version appeared in 1999 (Saure and Steinlein 1994; Becker Gieseke Mohren Richard 1999). Since 1997, sheep have been temporally grazed on alternating sites of the dry meadows. Additionally, since 1999 some plots of the area have been ploughed annually in order to reinitiate succession processes. Park maintenance has also included digging pits into the sandy soil, in order to promote the development of a special form of plant community, the silver meadow (Silberrasen). Further measures of design and maintenance aimed directly at promoting the fauna, for example, putting stakes in the areas which are meant to make the place more attractive for birds.

The need for maintenance activities was far from self-evident. Some participants deemed the very idea of fighting succession as an illegitimate intervention into the spontaneous development of nature. What had long impressed so many people at the Südgelände had indeed been the supposed wilderness of the vegetation and the related imagery of nature having "reconquered" this place in the city. Members of the Bürgerinitiative, the representative of the sponsoring foundation, and the artists who constructed the track were rather hostile to intensive maintenance measures.[42] As one of the artists put it in an interview, making decisions on which kinds of plants should be extinguished was "a kind of fascism."[43] Although ecologists succeeded in convincing most skeptics that a complete dominance of wood in the area was to be avoided, more fine-tuned interventions in the meadows, such as mowing, still face harsh criticism. The issue of maintenance has long been a bone of contention between Grün Berlin, which was in charge of the park, and the nature-conservation authority that was responsible for the reserve at its center. The conflict escalated in 2004 when the nature-conservation authority begun to use the meadows in the Südgelände as pasture for sheep. As the project manager of Grün Berlin complained, the sheep had grazed only at the bottom of the bushes, so that their physiognomy had been disfigured to an extent that was unacceptable. Since the nature-conservation authority was formally in charge of the reserve, Green Berlin had no way to prevent it from pursuing this policy.

Ecological scientists took an ambivalent stance with respect to this issue. On the one hand, they were expert advisors for park maintenance, and in that position they had actually been the ones who had initially promoted the idea of maintenance. On the other hand, for them the spontaneous process of "succession" on these sites was also an important research object. In that respect, they had a strong interest in keeping the sites free from intervention. At the Südgelände, the conflict was largely pacified by a division of the park into two different areas in which both of these policies now coexist. In the wood of black locust, which contributed to the image of wilderness and on which the research interest of ecologists has focused, succession has been allowed to proceed spontaneously. In the open meadows, however, a quite rigid conservation scheme has been applied.

In many nature-conservation reserves, the removal of alien plants had been a primary target of maintenance measures. This was also the policy of the Berlin nature-conservation administration in the fens and woods at the margins of the town. Given the urban character of these two nature parks, alien plants, however, were seen as a typical feature of "nature of the fourth kind" and thus not, per se, prey to a policy of ecological triage. Only those plants that tended to abound in the area, and therefore tended to disturb the overall balance of the ecosystem, were removed (Planland 2000).

Managers conceived their schemes as provisional and in need of being continuously adapted according to new experiences gained on the site. Thus, rather than applying a full-fledged body of knowledge to the problem of wasteland maintenance, this knowledge only took shape throughout local processes of negotiation and material interference with the two sites.[44] The design and maintenance of the wasteland parks was a kind of experiment through which new strategies of urban nature promotion were to be explored. Given this uncertainty, the beginning of maintenance activities went hand in hand with monitoring surveys which aimed at evaluating their effects (ÖkoCon and Planland 1998; Köstler et al. 1999). At the Südgelände and Johannisthal, the surveys were repeated regularly every year, sometimes leading to smaller adaptations of the maintenance schemes. For example, at the Johannisthal meadows the first two monitoring surveys led to an adaptation of the technique of sheep-grazing. As the surveys (Köstler et al. 1999; 2000) made clear, sheep were not always willing allies in the maintenance program. They had trampled the newly constructed sandy pits and hills to such an extent that these had lost their original shape with the consequence that none of the expected pioneering plants had settled there. Moreover, they had grazed so intensively on the site that the meadows had become considerably damaged from overgrazing. It was proposed to fence the pits and hills off from the sheep. In order to avoid overgrazing, the size of the herd was reduced from 250 animals to about 100.[45] Thus, rather than changing from sheep grazing to mowing, the planners of the management scheme aimed at adjusting—or "calibrating"— the technology of sheep-grazing according to what they considered the adequate development of the vegetation.

Conclusion

Only a few of the wastelands that ecologists wanted to protect in Berlin were eventually excluded from urban development: the Südgelände and the Johannisthal airport, and for a while also the plans for the BUGA at the Gleisdreieck. These became the focal points around which a new model for the provision and use of open space crystallized: the urban nature park. In contrast to existing parks that had been designed in terms of horticultural aesthetics, and with respect to the putative recreation needs of residents, the urban nature park was based on a place image that blended the generic ecological category of "ruderal biotopes" to site-specific sceneries and experiences.

As the story of the sites has shown, however, the concept of the nature park and its guiding place images were not a stable base from which the projects evolved. Rather, it took shape throughout the process and remained a matter of conflict and constant negotiation between the participants of the planning processes. As we have seen, in all of these three respects, quite different approaches were taken to the wastelands, each reflecting the peculiarities of the locales and the trajectories of the conflicts that evolved around them.

The attractiveness of the category of the nature park resided in its ability to redefine the protest against development as an alternative project of urban land use. Whereas the category of the ruderal area only emphasized the ecological features of the sites, their framing as nature parks legitimized their existence through their actual or potential significance within the human fabric of the city. It was mainly through this broader conceptualization of the projects that actors other than ecologists, such as activists, planning authorities, or the sponsoring foundation, embarked on the project. Such an alignment, however, was not without consequences for the ecological framing of the place. When the park concept was first articulated, in the 1980s, it merely emphasized the actual status of the sites and their legitimate use by visitors. The project of a nature park was mainly a matter of formally acknowledging the function of these areas and making them accessible to the public. When the projects matured, and the park idea increasingly became aligned with notions of an active shaping of these areas, tensions between the ecological framing and the park framing became apparent, and led to the conflicts about the extent and kind of design. The 1990s witnessed an increasing drift from the nature-park concept toward a renewed emphasis on horticulture, and grass-roots activities were increasingly motivated by the idea of the New York community gardens, rather than by urban nature protection in the sense as it had been articulated by urban ecology. When issues of nature protection, in the end, became widely marginalized in the plans for the Gleisdreieck, this was not only due to the physical impacts that the Potsdamer Platz construction had left on this site; it also reflected the increasing significance of these discourses.

One might see the appreciation of unregulated wildlife at the nature parks as a continuation of the quest for natural authenticity that had already inspired the landscape park model. Likewise, the emphasis of nature-park promoters on the value of unconstrained physical appropriation of urban spaces reflects some of the earlier ideas of the people's park and open-space planning. Although these continuities are apparent, the ecological framing of the sites gave such earlier motives a very different twist. In contrast to landscape parks, where horticultural scenery was meant to put some kind of rural landscape on the stage, the wastelands that existed on these sites were considered to *be* nature. Turning them into a park therefore mainly meant preserving them, basically in their present state, or enhancing existing visual features to make them more significant to the visitor. On the same basis, the content of experience of nature in the city was defined differently than in earlier Volkspark models. Not only did nature parks provide visitors with fresh air and space for physical exercise; if competently used, according to park promoter's scripts, they were supposed to allow them to experience a deeper truth of nature.

Although the two parks were not realized before the unification of the two Berlins, and one of the parks was located in the former East, it was from the background of West Berlin's urban ecology that claims for nature protection were raised and successively promoted. The local network of ecological expertise, which extended from the Technische Universität to landscape planning offices and parts of the local administration, not only provided a considerable amount of knowledge and practical experience with inner urban nature; it was also able to use its position to feed its concepts and values into the political debate of the city. Moreover, the Nature Conservation Law of 1979, which after unification was extended to the East, proved to be a valuable resource on which proponents of urban nature conservation could draw, even when faced with powerful interest groups acting on behalf of alternative land-use projects. It was due to these local knowledges and policies that wasteland parks became the hallmark of urban nature conservation in Berlin. Although other cities have followed the example of Berlin, it is by no means self-evident that attempts at ecological planning or species protection in cities always entail the same kinds of policies. In contrast, ecologists favoring "dense cities" have argued for building up urban brownfields in order to avoid the destruction of nature due to urban sprawl. Moreover, projects for greening other cities have often taken a quite different tack. Vienna's Biogärten, London's William Curtis Park, Toronto's renewed waterfront, and the ecologically shaped environments of the French science city Sophia Antiopolis are all further examples of recent attempts to ecologize the city.[46] However, they differ in what is considered as nature, and they reflect quite different local political constellations and knowledge cultures. For example, William Curtis Ecological Park in London (1976–1985), as with the two Berlin cases, was also an attempt to re-naturalize (though only for a limited time) an inner-city area, and it was also based on a conception of promoting urban wildlife. However, this ecological park was created on an area almost totally denuded of vegetation and, in striking contrast to the Berlin concept of promoting "nature of the fourth kind," was planted with a diverse mixture of native plants (Emery 1986: 83).

6 From Conservation to Mitigation: The Management of Urban Encroachments into Nature

Even if they were supported by broad alliances of activists, many of the conservation claims that ecologists had pushed forward in the Species Protection Program failed because in planning policy, priority was given to other development goals. Exempting spaces from development, however, has been only one strategy through which urban nature conservation has sought to realize its goals. Another strategy took the potential impact of a development project as a starting point. In this case, planners provided for technical means of landscape shaping which were meant to mitigate the projects' deteriorating effects or even to create alternative ecosystems and habitats that would substitute the nature that was destroyed by the project. Notably, in the aftermath of unification, when conservation claims were confronted with strong land-use interests by developers, investors, and state agencies, such mitigation packages were often the only way to accommodate conflicting goals. As we saw in the preceding chapter, all the large park projects that were realized on Berlin's urban wastelands, owed their eventual success to the need to provide compensation for ecologically damaging projects that went through elsewhere. These compensation policies were a far cry from the ambitious concerns that had characterized the urban biotope-protection regime as it had evolved in the 1980s in West Berlin. On the other hand, the shift to mitigation allowed the original regime to adapt to the politically unfortunate circumstances of the 1990s and to continue, albeit in a weaker form, to maintain a role in the urban planning system.

In this chapter I will trace some of the discourses and institutional practices that surrounded and shaped this transformation of the biotope-protection regime. As I will show, mitigation was more than simply replacing a given piece of nature by an artificially created equivalent. What counted as an encroachment, what counted as an improvement and, finally, how an improvement was made to qualify as an equivalent to an encroachment—all of these have been highly negotiable matters that have been characterized by various interpretive uncertainties and dilemmas. It was also the very principle of compensation that was at stake in these negotiations. Since legal regulations were the basis of this policy, a great deal of the environmental mitigation discourse was concerned with the interpretation of legal requirements and their relation to the ecological knowledges and practices through which mitigation measures were realized. Whereas it was the original intention of biotope protection in the city, to prevent the loss of biotopes in their given status, the result of such policies can be seen in the proliferation of what might be called phantom biotopes. These are real spaces or options for future improvements of such space that, on the basis of the informed judgments of environmental consultants, are valued as functionally and symbolically equivalent to the biotopes that are sacrificed for the sake of urban development.

Legal Regulation and the Politics of Mitigation

Designing and shaping landscapes in order to mitigate the negative impact of urban or technological interference had been promoted by gardening architects and landscape planners for more than 100 years. The environmental historian Marcus Hall (2005) has argued that the attempt to "repair" human damage to the earth had already figured in the programs of nineteenth-century conservationists such as George Perkins March in the United States, and late-nineteenth-century Italian landscape gardeners. In the nineteenth century, the Prussian horticulturalist Lenné viewed gardening as kind of aesthetic packaging of the effects of utilitarian rationality of modern industry and machinery (Wise 1999). For example, in 1835 he designed a garden around the new Borsig factory in Berlin, which placed the plant in the middle of a flowery landscape scenery. Since the 1920s, champions of land care had sought similar compromises between industrial development and the landscape scenery. This led to the greening of slag heaps or open pits in mining areas or, as with the Teufelsberg in Berlin, of rubble heaps which were erected during the process of reconstructing destroyed cities. (See chapter 1.)

As of the early 1960s, the call for a systematic use of mitigation measures figured prominently on the agenda of the German nature legislation reform debate. In 1961, the Green Charta of Mainau considered both the prevention of "avoidable encroachments," as well as the "reparation of unavoidable encroachments" as pillars of future landscape protection (Grüne Charta von der Insel Mainau 1961). After these principles had been introduced into the planning laws of some Länder, they also became included in the Federal Nature Conservation Act of 1976, and the subsequent Berlin Nature Conservation Act.[1] In contrast to earlier forms of gardening or land care, the so-called impact regulation (Eingriffsregelung) was not merely about maintaining an aesthetically pleasing environment. What the regulation defined as encroachment (Eingriff) encompassed all kinds of modifications of the shape or the land use of an area that "to a significant and lasting extent" interfered either with the landscape scenery or with the "performance of the household of nature."

The regulation stipulated a cascade of decision criteria that should help public authorities to minimize or prevent the impact of such encroachments. In the first place, it obliged the originator of an encroachment—for example, the developer of a housing project or the organizer of an outdoor event—to do everything to prevent such impairments. If this was technically impossible, he or she had to provide for compensation measures (Ausgleichsmassnahmen) that, within a given time frame, would fully rectify the damage. Encroachments that were of such a severe kind that it was impossible to compensate their impacts, were inadmissible. Under these circumstances it was still possible for public authorities to approve the project, but only on the condition that it was justified by "interests of the generality" (§14, 5), which overrode the goals of the Nature Conservation Act. Even in that case, however, the regulation stipulated that impairments which could not be compensated, had to be remedied by alternative measures of landscape improvement, the so-called substitution measures (Ersatzmassnahmen). A final option was reserved for the case that even suitable substitution measures could not be realized, for example because no space was available for the realization of such measures. Public authorities could then levy a charge from the bearer of the project, which had to be used to finance measures of nature protection or landscape protection. For a long time, however, this remained a purely formal option since the Berlin Senate had not issued the decree that would have been required to implement such a charging procedure.

Another strand of legislation that spurred the policy of impact mitigation was the introduction of compulsory environmental impact assessments. Following the example of the US National Environmental Policy Act of 1969, the West German Government decided in 1975 that projects of federal institutions should undergo a systematic evaluation of their impact on the environment.[2] In 1978 the Berlin Senate endorsed similar principles for its own jurisdiction (Senat von Berlin 1978). The goal of such a procedure was on the one hand, to determine if a project had negative effects on the environment, and if so, how this could be prevented, compensated, or minimized (Art. III). In 1985, the European Community issued a directive on environmental impact assessment that extended this procedure to a number of large-scale private construction projects.[3] It took until 1990, however, for the Environmental Impact Assessment Act (Gesetz über die Umweltverträglichkeitsprüfung) to implement this directive into German law. Although there is much overlap between the goals of impact regulation and the goals of environmental-impact assessment (EIA), notable differences existed between these two legal instruments. One was the broader scope of the EIA which not only focused on "nature and landscape" in the sense of the Nature Conservation Act, but generally on all kinds of "harmful effects" on the environment. At the same time, it was also weaker since it only applied to a restricted class of interventions, usually large-scale projects, which were explicitly listed in the law. It was also weaker since it had only a consultative function, and hence, had no binding effect on the decisions of public authorities.[4]

As we saw in chapter 3, the Species Protection Program already included measures of maintenance and development of nature in the city (e.g., watering of wetlands, roof greening, and so on). In many cases, these were meant to mitigate the effects of prior human interference with the preexisting shape of nature. The impact regulation provided for similar measures in order to compensate for environmental impacts of ongoing projects. It was not uncommon that provisions of the Species Protection Program and compensation measures overlapped, and, as with the nature park at Südgelände, that measures that had long been promoted for an area eventually only went through when they also served as mitigation for some other development project.

As a result of the new legislation, building promoters and permission authorities were increasingly confronted with demands for the assessment and management of environmental impacts. Such concerns were raised by the nature-conservation agencies and the formally accredited nature-conservation associations, which both had to be consulted before the responsible authorities could give their approval to such projects.[5] These groups tended to be rather critical of the projects and therefore often asked for extensive evaluation studies and—if they were not able to prevent the projects entirely—at least called for far-reaching impact compensation. They were supported by other opponents of the projects such as local activist groups, political parties, or the wider public. Be it the loss of forest areas due to the construction of the new highway connection to Hamburg, the infliction on the Tiergarten of the projected Westtangente, or the various construction projects on the closed-down railway areas—in all these cases, compensation measures were called for, suggested, or even implemented. Besides buildings and infrastructural projects, also temporary activities such as air shows, open-air concerts or technical modifications of the landscape by new agricultural or forestry activities were evaluated and regulated on this basis.

Regarding most regular city planning, however, the impact regulation played a more minor role than many nature conservationists had wished. It did not apply to urban planning schemes since these had only a preparatory character and therefore were not considered an actual encroachment. Another limitation was the notorious unwillingness of significant parts of the Berlin building authorities to apply the regulation in the inner area (Innenbereich) of the city (those parts of the city that were already completely developed).[6] Only when districts interpreted the rule more broadly and reached local compromises with other investors and the building authorites, were building projects accompanied by compensatory measures. Such an example was the decision of the district of Tiergarten to create a small park to compensate for the encroachment caused by the hotel at the Dörnbergdreick (see chapter 4).[7] In addition, the Kreuzberg district was relatively successful in asking the Museum for Technik und Verkehr that was under construction in the Gleisdreieck to implement relatively far-reaching mitigation measures that would integrate the spontaneous vegetation.

The public authorities that were responsible for the project were often less concerned with its environmental impact than the nature-conservation agencies or associations. Permission authorities of the districts or the Senate, which were in charge of approving private building, were often also themselves committed to the development of the respective projects. They tended to shy away from possible debates on compensation and substitution measures, which might have jeopardized the smooth realization of the project and enjoined the investor with additional financial burdens. Their position as formal leaders of the regulatory procedure gave the responsible authorities many opportunities to circumvent or to protract the involvement of nature-conservation representatives.[8] The permission authorities wielded final decision-making power concerning the permission for the project and the stipulation of compensation or supplementation requirements. Although the nature-conservation agencies had to be asked for their formal consent (Benehmen), this did not give them any actual veto power against the project. If evaluation studies had revealed that a project would cause a significant encroachment, the reference to "overriding public interests" allowed the responsible agency to pave the way for the project. Moreover, they were able to split projects into fragmented planning packages—which often were motivated by administrative rationales—that impeded a comprehensive evaluation of the entire project.

This situation was further complicated by various interpretative uncertainties that were left by the formulation of the impact regulation. The meaning of terms such as mitigation, compensation and supplementation, as well as the conditions under which the impact regulation was to be applied, became the topic of intricate conflicts among lawyers and planning experts and litigation. As we have already seen in the analysis of the Südgelände case, it was contested if the impact regulation also could be applied to vegetation stocks that had developed spontaneously on areas that were formally identified as railway areas. The unwillingness of the Senate to apply the impact regulation in the inner zone, a practice that was also notorious in other West German Länder, caused further legal disputes (Gaßner/Siederer 1987: 22–24). This issue remained controversial until it was eventually settled by a change of legislation in the early 1990s.

In general, it can be said that nature conservationists and ecologists took a rather ambivalent position concerning the issue of mitigation. In the Species Protection Program, Sukopp and his co-workers posited that preference should always be given to the preservation of ecosystems rather than to enhancement or rehabilitation measures. Given the "complexity of landscape ecosystems" and the uncertainty that this entailed, they deemed it impossible to prove that any encroachment could actually be completely compensated for (Arbeitsgruppe Artenschutzprogramm Berlin 1984, I: 56). The Arbeitsgruppe Eingriffsregelung (Working Group Impact Regulation), a federal committee that in 1988 published procedural rules for the implementation of the impact regulation (Arbeitsgruppe Eingriffsregelung 1988), claimed that it was impossible to provide a scientific definition of the "functioning [Leistungsfähigkeit] of the nature household" in scientific terms (ibid.: 4), and that therefore, from a scientific point of view it was practically impossible to achieve compensation (ibid.: 9). When Sukopp and Markstein discussed the "possibilities and limits" of compensation for the impact on biotopes, they also referred to "uniqueness"—both with regard to their meaning for local identities and with regard to their long-standing evolution—of an ecosystem as a principal limit of reproducibility (Sukopp and Markstein 1984: 33).

At the same time, however, these ecologists and landscape planners did not hesitate to define standards for the stipulation of compensation measures that would guarantee a minimum quality for the compensation measures that were required by the law (Sukopp and Markstein 1984; Markstein and Palluch 1981; Arbeitsgruppe Eingriffsregelung 1988). The distinction of scientific versus juridical terms—a form of rhetorical boundary work between these two domains of practice (Gieryn 1999)—allowed them to maintain higher standards of academic knowledge (according to which compensation was practically impossible), while at the same time, offering their expertise for the technical operationalization of the legal regulation. In legal terms it was supposed to only require that an impairment was reduced to an "insignificant" amount (Arbeitsgruppe Eingriffsregelung 1988: 9).

It was not only the political unwillingness of the responsible authorities that hindered the search for adequate compensation measures. In particular, in an urbanized environment, it proved difficult to find sites on which compensation measures for valuable biotopes could be implemented. As Markstein and Palluch (1981: 68) estimated, all large biotope types in Berlin—forests, fields, and extensive ruderal areas—were practically irreplaceable. Interventions into those areas were thus likely to cause more damage than they were supposed to remedy (Weiss 1987: 78; Markstein and Palluch 1981). This problem became clear in 1984, when ecologists discussed possible substitution measures for the envisioned freight station at the Südgelände (Sukopp and Markstein 1984). On the basis of a comparison of the ecological conditions and floristic and faunistic patterns, only the Anhalt- and Potsdam Güterbahnhöfe (Gleisdreieck) and the Diplomatenviertel came out as being amply equivalent to the Südgelände. Since the Diplomatenviertel was not available for such purposes, they suggested rehabilitation and development measures at the Anhalter- und Potsdamer Güterbahnhöfe as the only possible substitution strategy in the case that the freight station be built (ibid.: 35).

The ecological evaluation of the encroachment played a pivotal role in the implementation of such measures. Estimating the type and the scale of impairments, which were likely to be caused by a project, was necessary to decide if a project could be approved at all. If it was approved, it hinged upon the outcome of this evaluation what kind of compensatory or substitution measures had to be demanded. Besides smaller encroachments which were evaluated directly by the responsible nature-conservation officials, evaluation studies for environmental impact assessments and impact evaluations of larger encroachments, were carried out by scientific institutes or professional landscape consultants on behalf of the bearer of the project.

Construction Site Berlin: Capital Planning as Compensatory Conflict

It was only in the context of the enormous transformation of the inner city of Berlin in the 1990s that mitigation policies begun to play a crucial role in urban planning. After the unification and the subsequent decision to re-designate Berlin as the German capital, the city witnessed a building boom that widely changed its appearance (Strom 2001; Rada 1997; Blais and Spars 1990). Many open spaces, such as the wastelands along the wall that were considered valuable biotopes, and significant chunks of inner urban parks became earmarked for shopping malls, corporate offices, governmental buildings, or new railway stations. Capital planners also devised for high-speed railway tracks that would impinge on the vegetation that had grown along the smaller tracks that formerly connected the divided city with the rest of the country. The impact regulation, as well as the existence of an extraordinary active local nature-conservation lobby, resulted in a host of compensatory measures that added to the ongoing reshaping of the urban landscape.

Notably around the Tiergarten and in the vast open spaces that stretched along its eastern edge, redevelopment projects conflicted with ecological goals. To a large extent, these construction works were connected to the Senate's efforts to renew the city's traffic infrastructure system. In December of 1991, the Senate decided for the construction of eight tunnels for roads, railway tracks, and a subway line under the Tiergarten.[9] Directly connected to these plans was the creation of new railway stations, some of them in direct proximity to the Tiergarten.[10] Although the tunnel project was heavily attacked for its supposed ecological effects, notably on the vegetation of the Tiergarten, this did not hinder its realization. After extended debates about its environmental risk and the necessary compensation measures, construction work on the tunnels began in 1995. In 2006 the last of the tunnels was eventually opened to traffic.[11]

Other large-scale construction sites existed along the eastern fringe of the Tiergarten. Most significant was the redevelopment of the area around Postdamer Platz. The area is located in the south east of the Tiergarten and had been a vivid cultural and shopping center until it was completely destroyed during the war. Owing to its critical location on the border between the East and the West of the city, it had remained undeveloped since that time. In December of 1991, the Senate approved the plans of a number of international firms, among them Daimler-Benz, Sony, and ABB, to construct high-rise office buildings and a shopping mall on the area around the Postdamer Platz. The construction work, begun in 1994, transformed this former wasteland area into one of the salient architectural landmarks of the new Berlin (Schwambach 2006).

More development projects around the edge of the Tiergarten soon followed. A significant part of the new government district emerged along the Spree River at the area in the north of the Tiergarten, the so-called Spreebogen area. Apart from a few significant buildings, such as the Bellevue castle, a convention hall, and the former German Reichstag (the parliament), this area had been widely dominated by vast lawns and trees. Most of this green space was lost. Beginning in 1995, several new buildings were constructed in this area: an office for the chancellor (Bundeskanzleramt), buildings for the Bundesrat, and various offices and accommodation for parliamentarians. Berlin's new role as a German capital also paved the way for the development of the southern fringe of the Tiergarten, a landscape of ruins that botanists cherished for its biological diversity. In the mid 1990s, the countries that still owned most of the lots in this area began to rebuild them with new embassy buildings.

These projects represented a radical departure from the plans of the West Berlin Senate in the 1980s to develop the area around the Tiergarten into a Green Center. (See chapter 4.) The Anti-Tunnel AG, an alliance of activist groups and political organizations that was formed in 1993, organized protest campaigns, launched caveats, and also sued against the tunnel project (Berliner Landesgemeinschaft Naturschutz and Anti-Tunnel AG 1994). As the critics maintained, the construction of the tunnels could have been avoided if the Senate had opted for a more decentralized railway concept and aimed at reducing car traffic in the inner city. Building the tunnels, in contrast, would mean that parts of the Tiergarten were about to be lost to the gateways and that no healthy vegetation could survive on the surface above the tunnel. They also maintained that the lowering of the groundwater level during the process of constructing the tunnels would cut the water supply for the vegetation in the Tiergarten, so that most of its trees would die. Similar worries about groundwater lowering in the Tiergarten were caused by the development projects in the adjacent areas, notably at Postdamer Platz. Here the planners envisioned the construction of a basement shopping area and an underground station. Opposition against these projects was not able to prevent their realization. It played a significant political role, however, since it paved the way for an extended debate about ecological compensation management.

The ecological effects of the projects were scrutinized in various expert reports and accompanying landscape plans were developed to minimize or compensate for these effects. The tunnel project formed part of a binding planning approval procedure for which an EIA was compulsory. The study was carried out by the Berlin landscape consultancy firm Neumann + Hoffmann and various subcontractors which also developed the accompanying landscape plans.[12] As a reaction to mounting concerns about the ecological effects of the Potsdamer Platz project, the responsible district administration decided in April of 1992 that an EIA should be conducted on the area.[13] In May of 1993, the Federal Investment Facilitating Act provided that encroachments should be evaluated and compensated, for any new development planning scheme (Bebauungsplan).[14] The development of the Tiergarten area was laid down in various separate schemes that therefore all had to be checked for potential compensation needs.

The ecological surveys corroborated many of the worries of the critics. At the same time they proposed a host of minimization and compensation measures which were then also provided by the plans.[15] Both with regard to the tunnel projects, as well as the foundation works at the Potsdamer Platz, interference with the groundwater was seen as a serious threat. Accordingly, the construction process was organized in such a way as to minimize the impact on the groundwater as much as possible. Parts of the tunnels were constructed by a so-called shield method that did not require the lowering of the groundwater. In order to avoid lowering of the groundwater during the foundation works at the Potsdamer Platz, the pitch was sealed with a special concrete while still under water. The resulting tub of concrete prevented any direct interference of further construction work with the groundwater. The plan also suggested that "negative wells" be constructed in the surrounding of the construction sites through which groundwater should be continuously pumped backed into the ground. They were part of a complicated control and monitoring system that was meant to make sure that the groundwater level did not fall short of a certain minimal level, and that no polluting substances from the construction work would end up in the groundwater.[16]

Moreover, the various projects in Berlin's new center implied a severe reduction of the vegetation area that had to be compensated for by new green spaces. Where it was possible, green spaces were integrated into the project area, such as the Spreebogen, or by creating a park at the Postdamer Platz area. Since adjacent areas were themselves earmarked for other construction projects, it was difficult to find suitable areas in the neighborhood where more substitution measures could be realized. It was at the Südgelände and the Gleisdreieck that the planners eventually sought to fulfill their compensation requirements.[17] Accordingly, the railway company would release the Südgelände from its status as a railway area to compensate for the impact of the tunnel and other railway areas in the center. Likewise, the Senate charged the investors of the Potsdamer Platz for creating a park at the Gleisdreieck, an area which had been severely impacted by its use as a logistics center for the construction work at the Postdamer Platz. As was noted in chapter 4, the quick realization of this park failed because the railway company that owned the site withdrew its former approval for the project. Besides providing space for plant and animal species, the newly created green spaces were also meant to compensate for the negative effects that the project had on the climate, the regeneration of groundwater and the landscape scenery.

Planning schemes for the Potsdamer Platz were also modified in such a way as to preserve the lime trees of a former avenue that had survived at the site. Moreover, the plan provided for the greening of roofs in order to remedy the negative impact on the urban climate. Other proposals that were made in the environmental impact statement (AGU Arbeitsgemeinschaft Umweltplaung Berlin and Büro für Kommunal- und Raumplanung, no year) and a related encroachment-evaluation report (AGU Arbeitsgemeinschaft Umweltplaung Berlin and Büro für Kommunal- und Raumplanung 1993) failed because of resistance from developers and the Senate. These were, for example, modifications to the size and the shape of the buildings (to make them more climate friendly), a greenery-promotion scheme for the adjacent neighborhood, and the conservation of a grove of birch trees, which had grown spontaneously on a rubble site in the planning area. Although their ecological value and connection to the original riches of urban nature was heavily contested, these compensatory landscapes can be regarded as phantom biotopes in which the former riches of urban nature survived in the form of minor attempts at urban greening, or even only in the form of financial resources to realize such measures in the near future.

Negotiating the Scope of Encroachment

Although environmental impact statements or encroachment studies had been carried out on all of these projects, their results were not always accepted by their opponents. Such a clash became particularly visible during the controversy around the Tiergarten tunnel. In their evaluation study, Hoffmann + Neumann reconsidered data from earlier studies of the area (e.g., the expert report on the Tiergarten that Sukopp's group had presented in 1979). Additional data—for example, an up-to-date inventory of the flora and vegetation—was contributed by subcontractors (the landscape planning firm Plantage). The environmental impact study (the scientific report that is required for a EIA was based) was the basis of the planning proposals that were submitted by the Eisenbahn-Bundesamt in 1994.

During the consultations of the "bearers of public interests," however, the environmental impact statement was heavily attacked by the critics of the project. A crucial role was played by the BLN and the Anti-Tunnel AG, which for this purpose cooperated with kubus, a public scientific information facility at the Technische Universität. With the help of kubus scientists, the BLN and the Anti-Tunnel AG drafted extensive opinions against the project (Berliner Landesgemeinschaft Naturschutz and Anti-Tunnel AG 1994, 1995). In order to better evaluate its effects, scientists also carried out alternative studies for the BLN that focused on issues such as the health of the Tiergarten trees, the water regime, and chemical pollution (Barsig et al. 1995; Barsig and Klöhn 1997). A conference of citizens and experts that was organized by kubus in January of 1995 provided another important forum for such critical views on the management scheme (kubus 1995). According to the tunnel critics, the EIS underestimated the remaining ecological risks. With respect to biotope protection, they posited that neither the direct loss of biotope areas nor the indirect effects of the project on the vegetation of the Tiergarten had been fully acknowledged in the environmental impact study.

First, the critics pointed to the lack of ample empirical inventory of the present state of the biotopes that would be affected by the project. Since the high value of the flora and fauna of the Tiergarten was well known, they argued that a detailed survey of these aspects was needed.[18] They argued that the environmental impact study was to a large extent only based on results from already existing and large-scale studies, or from studies which themselves had only drawn on data from previous research.[19] They also argued that the corridor in which floristic and faunistic data was gathered was far too small, and that it was collected in a very short observation period. The critics thereby referred to the UVP-Förderverein, which had proposed standard requirements for good practice in an environmental impact study (Berliner Landesgemeinschaft Naturschutz and Anti-Tunnel AG 1994: 53).

The critics also referred to the environmental impact study that Sukopp and a group of scientists from the Technische Universität had carried out in 1979 for an earlier highway tunnel project in the Tiergarten that, back then, was not realized (Sukopp et al. 1979). For the critics, the former study served as a yardstick against which the current impact study was exposed as methodologically flawed (kubus 1995: 6). The reference concerned only the methodology and the content, not the conclusion that the authors of the report had drawn at that time: They actually had pleaded for a tunnel as an alternative to the highway through the park that was discussed as the planning alternative.

The critics also based their arguments on alternative studies that they had carried out in the Tiergarten. In 1995 Michael Barsig, a biologist from the Technische Universität and a member of the Baumschutzgemeinschaft (Tree Protection Community), had made an inquiry into the state of health of the Tiergarten trees (Barsig et al. 1995). The researchers claimed that the trees of the Tiergarten already showed serious impairments, which they attributed to earlier changes of the groundwater level and the stress caused by ozone and air pollution. They concluded that the danger for the Tiergarten was much more severe than expected. In 1997 a second tree vitality study (Barsig and Klöhn 1997) established a further decline in tree health and warned of the air pollution that would be caused by the increasing traffic that would be created by the tunnel.

Second, the critics claimed that a realistic appreciation of the full impact of the project on the flora and fauna of the Tiergarten was missing in the environmental impact study. This concerned, notably, the question of to what extent the health of the vegetation of the trees depended on direct contact to the groundwater. The proponents of the project denied this, and claimed that the trees would tolerate any fluctuations in the groundwater level caused by the tunnel project. The critics by contrast citied evidence from Sukopp's earlier study, according to which, at least some larger trees had direct contact with the groundwater (kubus 1995: 8). They also complained that corresponding evidence from a subcontracted study had not been considered in the final evaluation of the environmental impact study (ibid.: 8). As Barsig explained this problem, it was likely that groundwater fluctuations would have an effect on the trees (ibid.: 16) and that above the tunnel no trees but only bushes would be able to survive (ibid.: 13). The critics also doubted that the monitoring system for the groundwater would allow sufficient control of these effects (kubus 1995). They even predicted that minimization measures themselves were likely to cause negative impacts on the Tiergarten. For example, to avoid the removal of the vegetation, parts of the tunnel should be drilled underground. To prevent the influx of groundwater it was necessary that the ground would be artificially frozen during this procedure. According to the critics, however, this was likely to cause damage to the deeper roots of the Tiergarten trees (Berliner Landesgemeinschaft Naturschutz and Anti-Tunnel AG 1994: 39). For the critics, all this was not only due to the lack of systematic attention by the evaluators, but bespoke the fundamental uncertainty inherent in this project. As they claimed, the project was of such extraordinary complexity and so unprecedented, that it exceeded any previous experience from which its effects could be judged (kubus 1995: 12). One even doubted the very possibility that "construction sites of this incomparable, gigantic size" (ibid.: 10) could become the object of effective reckoning.

Third, the critics rejected the way in which the compensation needs had been established in the study. Besides simple tree counts, Hoffmann + Neumann had made use of the so-called bio-volume method. This meant that the loss of vegetation had been estimated solely in terms of biomass expressed in cubic meter volume. On this basis they had calculated how much new bio-volume had to be created (surface area x height of the vegetation) in order to compensate for the damage. This was a far cry from the detailed ecological survey for a sound evaluation of encroachment effects and compensation needs as had been suggested in the 1980s (Markstein and Palluch 1981; Sukopp and Markstein 1984). Tunnel critics dismissed these purely quantitative methods as being too crude (kubus 1995). By only balancing the bio-volume or the number of trees, it lacked the necessary attention to ecological quality, both of the vegetation lost, as well as at the site where the compensatory measures were deployed (kubus 1995: 7). According to the critics, the situation was even worse, since other projects in the surroundings of the Tiergarten were evaluated separately from the tunnel project and therefore a general picture of the overall impact on the park was lacking.

Fourth, critics of the tunnel project cast doubt on the ecological competence of the evaluators. As one participant put it during the public colloquia, the approach taken by the impact study represented a "planners' perspective," not an "ecological view" (kubus 1995: 7). Another participant went so far as to claim that an EIA should be monitored by nature-conservation organizations, and that in general, it should not be carried out by private institutions but by university institutes that he considered more independent (ibid.: 14). Critics also complained that in contrast to earlier occasions, the Senate had not systematically involved the ecologists from the Technische Universität or specialists of the local environmentalist groups in the evaluation of the project (ibid.: 16).

Finally, the critics bemoaned that a systematic consideration of alternative solutions for the traffic problems in the city was missing. In their view this had had the effect that the authors of the EIA had provided uncritical support for a conventional model of traffic planning (kubus 1995: part 2).

In contrast to earlier planning conflicts, the critics got little support from the ecologists at the Technische Universität. An exception was Barsig, the author of the alternative studies for the kubus, who was a biologist and worked as a researcher at the TU. Sukopp and other established representatives of the urban ecology, however, were not present in the debate. This was not simply due to the unwillingness of the planners to involve them in the process, as the tunnel critics maintained. Indeed, the ecologists from the TU tended to be much less hostile to the tunnel project than the environmental activists who called for their expertise. In their former evaluation study on the highway in the Tiergarten, Sukopp and his colleagues had actually proposed a tunnel as an alternative for the highway that had been planned at that time. During the new controversy, Wilhelm Ripl, at that time professor for limnology at the Institute of Ecology, even defended the tunnel project. In his view the tunnels were the best solution for traffic problems and many of the critics' claims about the negative environmental impact of the project were not justified (kubus 1995: 8). Ripl even suggested that the critics misused ecological arguments as a false pretense for their opposition to a traffic problem and urged them therefore also to discuss it in those terms (ibid.: 18).

Although their engagement resulted in minor adaptations of the management scheme, the critics were not able to challenge the project. This can be seen as a consequence of the general shift of the urban planning agenda to capital development, which left much less room for ambitious ecological claims than in the 1980s. At the same time, it also reflected the relative weakness of the opposition movement itself. Although anti-tunnel activism had started off as a relatively broad coalition of various political groupings, it did not sustain a larger mobilization power over time. This was exacerbated by internal conflicts among the anti-tunnel activists, notably between the BI Westtangente and the so-called Baumpaten, which in early 1996 led to a de facto dissolution of the Anti-Tunnel AG. When in March of 1996 the foundation stone for the tunnel was set in place, protest had suffered from a dramatic loss of momentum. As one architect observed, only a "sad spectacle" of a few activists accompanied the opening ceremony (Hoffmann-Axthelm 1996).

The Quest for a Standard Method

Although expert reports occupied a central place in the impact compensation procedure, no consensus existed in the planning community about the way in which environmental impacts and the demand for

compensation measures should be assessed.[20] This methodological plural-
ism, however, was increasingly seen as an obstacle, and in the 1990s
Germany witnessed efforts by policy makers, as well as landscape planners,
to develop standardized procedures for the evaluation of encroachments
and compensation measures.[21] In Berlin the first proposal for such a standard evaluation scheme was presented by the landscape planners Becker
and Giseke (1993). They had developed their scheme in a very short time
and regarded it as provisional. It was meant to provide an ad hoc tool to
coordinate the implementation of the impact regulation in the ongoing
large-scale building projects in the center, notably around the Postdamer
Platz. By this time, the standardization of evaluation methods was also
actively promoted by the Berlin administration. After a hearing with
stakeholders in 1991, the Senate Department for Urban Development and
Environment commissioned Axel Auhagen to develop an encroachment-
evaluation method for Berlin. Initially, the method was only meant to help
the Senate to implement a levy for encroachments that could not be fully
compensated by compensation measures (Auhagen 1994). Independent of
its original purpose, however, the Senate promoted Auhagen's system as a
general quality standard for the establishment of encroachment needs in
Berlin (Senatsverwaltung für Stadtentwicklung 1999: 36; Senatsverwaltung
für Stadtentwicklung und Umweltschutz and Gruppe F 2001). From the
late 1990s on, extended debates with stakeholders and experts led to significant revisions of the original method and the introduction of an alternative procedure for the evaluation of smaller development plans in the
developed area (Köppel and Dreiwick 2004).

Generally, standardization was seen as a way to replace uncertainties
and arbitrariness in the application of the impact regulation by a transparent and accountable procedure. According to Becker and Giseke (1993: 4),
only a "universal line for Berlin" would guarantee that the bearers of different projects would be treated equally, that the administrative procedure
would have a clear structure, and that "objectivity and transparency"
would be achieved. In the same vein, Auhagen (1994: 10) expected that
his method would increase the transparency and acceptance of encroachment compensation procedures. The Senate Department for Urban Development and Environmental Protection, which commissioned Auhagen's
work, also hoped that a methodological standardization of the encroachment
evaluation would allow it to take a more consistent, and therefore more
successful, position in negotiations within the Senate (ibid.: 12). As Auhagen
explained, the lack of "professional and procedural standards" could have
disastrous consequences not only for the legitimacy of planning decisions

but also for the professional status of landscape planners. As long as any consultancy firm used different methods and arrived at different compensation needs, it was "easy for its opponents to represent them as 'investment obstacles'" (ibid.: 33). Auhagen also warned that the lack of a unified method allowed investors and their consultants to manipulate the outcome on their behalf (ibid.: 37). The concern for such problems was not confined to Berlin. All over Germany, planning experts in the 1990s worried about the fading public acceptance of the impact regulation (Kiemstedt 1995).

Although it was inspired by parallel developments in other parts of Germany, the "Auhagen method" was greatly affected by the intellectual and political circumstances under which it was produced. To a considerable extent it reflected Auhagen's own methodological views and ideas on the implementation of the impact regulation. Auhagen also had to meet a number of demands by the Senate Department for Urban Development and Environmental Protection, which were defined beforehand and further negotiated after he had submitted his first draft.[22] The method was also discussed in various evaluation meetings with relevant stakeholders who all held their own opinions about encroachment compensation (Senatsverwaltung für Stadtentwicklung und Umweltschutz and Gruppe F 2001).

First, it was critically important for Auhagen and the Senate that the evaluation system would include the full range of factors covered by the impact regulation. Auhagen criticized that the widespread biotope-value method (Biotopwertverfahren) was based on the assumption that the quality of all environmental qualities of an area were sufficiently indicated by the evaluation of its biotope features (Auhagen 1994: 21–31, 24; see also Becker and Giseke 1993: 9). Auhagen's own method focused on 18 different functions which he supposed to represent all "subjects of protection" on which the impact regulation applied (Auhagen 1994: 31). Besides the biotope function, they also included non-biotic aspects of the household of nature (water, soil, climate) and the natural scenery (Landschaftbild).

Second, the evaluation method was meant to establish a monetary equivalent to the encroachment and, therefore, had to arrive at a quantitative output. Since the late 1990s, quantification had already been a common trend in the German debate on impact evaluation. In its crudest form, as it was represented by the biotope-volume method, complex ecological phenomena were simply reduced to their outwardly measurable dimensions. Subtler methods such as utility analysis and the biotope-value method worked with scoring systems that allowed a stepwise translation of qualitative data into values on a quantitative scale.[23] Following these examples, Auhagen developed a scoring system that allowed him to express the value of each relevant factor and, on this basis, to calculate an aggregate value for the area under consideration. The difference between the value points given to the original state and those for the state after the completion of the project (including the realization of impact reduction and mitigation measures) yielded a quantitative expression of the remaining compensation needs. Although the precise determination of the price index remained a matter of constant debate, this scoring system allowed Auhagen to represent the compensation needs of different kinds of encroachments on the basis of a common quantitative standard. Such quantifying methods were not uncontested within the planning community. Critics maintained that translating ecological data into metrical values, which was at best available at an ordinal scale, was highly problematic. They also disapproved the computation of these values as not justified by scientific reason and claimed that it produced "deceptive certainty" (Kiemstedt 1995: 62). It was notably with respect to the landscape scenery that quantifying methods were received with much skepticism. Many landscape planners—even if they had no qualms about the quantification of biological and physical aspects of the environment—insisted that the landscape scenery was intrinsically subjective and qualitative, and therefore not amenable to such methods. Back in the early 1980s, Auhagen and his colleagues (who had evaluated Berlin's biotopes for the Species Protection Program) had refrained from any attempt at quantification. Although from an academic perspective Auhagen in 1993 still seemed to share some of the critical doubts on quantification, he did not see any alternative to quantification with respect to the practical needs of an operational levy scheme (Auhagen 1994: 7).

Third, impact-evaluation methods also varied with regard to their requirements for research and data input. Whereas academic ecologists, landscape planning scholars, and nature-conservation activists had often tended to call for an extensive input of science, Auhagen took a more pragmatic stance. He explicitly rejected the proposals of the working group impact regulation (Arbeitsgruppe Eingriffsregelung), an expert group that had evaluated the impact regulation (Kiemstedt and Ott 1993). According to Auhagen, the working group had been far too ambitious in its suggestions for the extent and quality of encroachment-evaluation studies (Auhagen 1994: 29, 31; 1995: 37). Although he agreed with the commission's proposals from a scientific (fachlich) point of view, he considered them as unrealistic and not in line with the actual legal requirements of the impact regulation. Moreover, the Senate had explicitly demanded from Auhagen that his method should be sufficiently applicable to the data that was actually available in Berlin (Senatsverwaltung für Stadtentwicklung und Umweltschutz and Gruppe F 2001: 2). Auhagen therefore designed his method in such a way that it could operate on the basis of the *Umweltatlas*, a collection of environmental maps and statistics that had been published by the Senate since 1988 (Auhagen 1994: 31). Only for the evaluation of the biotopes did it require additional site-specific data that had to be produced in special evaluation surveys. On behalf of the Senate (Senatsverwaltung für Stadtentwicklung und Umweltschutz and Gruppe F 2001: 2), Auhagen made sure that the requirements for additional research were as minimal as possible. Moreover, Auhagen's method included the option for an investor to choose in favor of a general estimation of the site according to its biotope-type characteristic as an alternative to a more costly and time-consuming site survey (Auhagen 1994: 33–34).

Fourth, Auhagen did everything to adapt his system to the local conditions under which the impact regulation was implemented in Berlin. This problem became particularly visible with regard to the biotope-type classification on which parts of the encroachment evaluation was based. The existing list of the Species Protection Program was designed for a general appraisal of the entire Berlin territory, and hence, not specific enough for the smaller scale on which encroachments were evaluated. Becker and Giseke (1993) had therefore already drawn upon a biotope-type list that had been created earlier for encroachment evaluations in Hesse and Hamburg. Auhagen rejected all pre-existing schemes as inadequate for his purpose and presented an entirely new classification of about 155 Berlin biotope types. On behalf of the Senate, Auhagen constructed his method in such a way that it was able to deal with the large-scale building projects that were underway in the center of the city.

Auhagen considered his evaluation scheme as purely conventional. Its legitimacy was not based on empirical facts, but on its ability to produce the expected practical results. The method was soon taken up in many evaluation studies. A main factor that led to the broader use of the method was the active promotion by the Senate (Senatsverwaltung für Stadtentwicklung 1999: 36). Although the method was not formally compulsory, it has become increasingly difficult for individual evaluators to justify the use of any less rigid evaluation procedure.

This does not mean, however, that Auhagen's method remained unchallenged in Berlin. In the day-to-day practices of administering encroachment cases, individual evaluation studies diverted from his prescriptions or adapted them in case-specific ways. One of the most significant revisions that his method underwent in the 1990s, was the inclusion of the so-called recreation function. Auhagen had always argued that recreation was not a subject for protection according to the legal meaning of the impact regulation (Auhagen 1994: 11; Senatsverwaltung für Stadtentwicklung und Umweltschutz and Gruppe F 2001: 2). This implied that measures, which improved the recreation function of an area, did not receive any positive scores in the compensation balance. If they modified the features of a site, they could even qualify as impairments to the household of nature so that extra compensation measures or fees were required for them (ibid.: 11). This became increasingly inconsistent with the Senate plans in the late 1990s to set up a compensation pooling system, and thereby, to realize compensation measures mainly through the provision of recreation space. The Senate, which drew on the master's thesis of a landscape planning student to justify its own new legal interpretation, now sought to integrate the recreation function as a further factor into Auhagen's system (Senatsverwaltung für Stadtentwicklung und Umweltschutz and Gruppe F 2001: 8–9; see also Köppel and Deiwick 2004: 6; Senatsverwaltung für Stadtentwicklung, 2006: 72).

At the same time, the significance of the Auhagen method for the inner urban area was increasingly reduced. When Auhagen had developed his method, the Senate demanded explicitly that it should be used primarily for large-scale construction projects, as they dominated the planning agenda at that time. In the late 1990s, however, many of these projects had been finished so that this criticism lost much if its relevance. Moreover, the workshop of the Senate on the "Auhagen method" in 1991 had revealed many complaints by users about the method, which in their view was too complex (Senatsverwaltung für Stadtentwicklung und Umweltschutz and Gruppe F 2001). In 2003, the Senate therefore introduced a simplified alternative to the "Auhagen method" that was applied to building schemes in the built up area. Instead of a separate evaluation of different environmental factors, it relied on a calculation of the balance of impact and compensation on the basis of a blanket sum of the fictive costs of the mitigation measures (Köppel and Deiwick 2004: 6). The revisions after 2000 represented a significant lowering of the ambitions and the thematic focus of the impact regulation. As such, they formed part of a wider shift in the political function of the impact regulation from an instrument of species protection and nature conservation, to a much broader and flexible form of open-space planning.

From Nature Conservation to Recreation Planning: The "Flexibilization" of the Impact Regulation

The standardization of evaluation systems dovetailed with a more general shift of the goals and procedures of encroachment compensation. Originally, the impact regulation required that compensation and substitution measures were connected spatially and temporally to the project whose damage they should mitigate. Even if it was often not possible to realize them directly at the site of intervention, they were meant to be as effective as possible in remedying the actual damage caused by the project. These priorities were encoded in the strict sequence of decision rules that authorities had to follow when deciding about a project (minimization, compensation, substitution, levy). The late 1990s, however, saw a change to a more "flexible" implementation of the impact regulation. Authorities were now relatively free with respect to the location, timing, as well as the kind of compensation measure that they stipulated for a project. In many other places in Germany, this led to the development of pooling systems in which the compensatory duties of various projects were combined in a larger development project. Instead of reconstituting the status quo before the encroachments took place, such measures were meant to realize more general development goals of open-space planning. In Berlin, this quest for "flexibilization" had the effect that encroachment compensation measures became increasingly tied to goals of recreation planning, and thereby, lost much of their original function of biotope protection. The original biotopes were thereby converted into options that could be used flexibly for different purposes, but which still embodied the supposed value of the biotopes that had been lost. Whereas the original biotopes were lost, they survived as derivative valuables (or phantom biotopes, as I call them here).

This shift in encroachment mitigation was the result of widespread discontent with the established practices of the impact regulation. Among investors, urban planners, as well as parts of the wider public, it was not uncommon to see the regulation merely as a bureaucratic obstacle and an unnecessary financial burden. In line with a widespread call for "deregulation" and "flexibilization" in environmental and other public policies, the impact regulation was increasingly portrayed as a piece of outdated state-centered regulatory law. For different reasons, nature conservationists and landscape planning professionals were also unhappy with the results of the impact regulation. Since the late 1970s, they had bemoaned the notorious lack of space that was available for compensation areas. Too often even the willingness to remedy encroachments by compensation or substitution

measures was hampered by existing physical structures and private property rights. The protective function of the impact regulation that conservationists and ecologists had prioritized, and that was encoded in its
cascade of decision making, had thus become widely eroded in practice.
Moreover, as a Berlin nature-conservation official noted, an opportunistic
"pondering down" (Brandl 2000: 18) had become notorious in the day-to-
day practice of the administration. He also complained that this resulted
in a patchwork of simple substitution measures such as greening roofs or
the planting of trees, which according to critics, were too small and inco-
herent to have any ecological effect (ibid.: 19). Another point of criticism
concerned the lack of implementation of the impact regulation. Owing to
the unwillingness of the bearers of a project to fulfill their compensation
requirements and to weak enforcement by public authorities, many compensation and substitution measures were never realized.[24] The Working
Group Impact Regulation (Arbeitsgruppe Eingriffsregelung), a committee
of landscape planning experts that in the mid 1990s reviewed the "practice
of the impact regulation" in Germany, arrived at the same dismal conclusion.[25] In its three reports (Kiemstedt and Ott 1994; Kiemstedt, Mönnecke,
and Ott 1996a,b) it diagnosed shortcomings and made proposals for new
minimum standards for the evaluation of impacts and compensation
needs. In order to strengthen the ecological soundness of the regulation,
it suggested that compensation measures should be better integrated with
the broader planning goals for the landscape as they were stipulated in
landscape programs.[26]

    These different strands of criticism were also reflected in the legislative
changes that the impact regulation underwent in the 1990s. The Federal
Investment Facilitating of 1993 was primarily geared toward deregulating
planning law. It relocated the implementation of the impact regulation
from the level of the single building permit to the preparatory planning
stage, thereby paving the way for a more comprehensive approach to impact
compensation but, on the other hand, also keeping public authorities from
enforcing the regulation on single projects. The revision of the Federal
Building Code of January, 1, 1998 created the legal preconditions for
a systematic decoupling of encroachment and compensation. The new
version of the impact regulation that was included in this act no longer
distinguished between compensation and substitution and did not provide
any restrictions on the location of these measures.[27] It was also not required
that mitigation measures remedied the actual impact caused by the encroachment to which it was connected. They only had to be consistent with
"sustainable urban development and the aims of spatial ordering, nature

conservation, and landscape maintenance" (ibid.). This Building Code applied only to projects that were based on formal building schemes, as the majority of planning decisions in the developed parts of the city were. The 1998 regulation only applied for entire urban building schemes, whereas for individual projects outside the inner area the original regulation of the Nature Conservation Act remained in place. Only the revision of the Federal Nature Conservation Act in 2002, and the Berlin Nature Conservation Act in 2003, extended this flexibilization to all kinds of encroachment.

In Berlin, this legislative change led to a systematic policy of compensation pooling. Between 1998 and 2001, the Senate began to develop a Comprehensive Mitigation Concept, which was included as a special plan in the existing Landscape Program.[28] According to the concept, the future compensation needs for urban planning in Berlin were to be met in 43 focus areas. The selection of these areas was partly justified by their location along two belts and two crossing traverses which the Senate foresaw as the pillars of the city's future green-space system. Another motive that guided their selection was their suitability to fulfill the needs for recreation in the city. Therefore, all the areas were either parks or areas meant to be transformed into some sort of public greeneries.

It was notably the shift to recreation planning with which the Senate provoked much debate. Nature-conservation stakeholders such as the BNL and the Sachverständigenrat welcomed the new policy of pooling which in their view would allow a more effective approach to compensation than the earlier piecemeal policy.[29] At the same time, however, they complained that the concept gave too much priority to recreation planning, and thereby, marginalized ecological goals. Many of the proposals made for the focus areas were just improvements of the recreational infrastructure which had no value for biotope and species protection. The critics even maintained that from an ecological perspective, such measures had a negative impact and therefore had to be counted themselves as encroachments. The BLN also bemoaned that measures that would really have improved the ecological quality of the city, such as a systematic de-sealing of inner city areas, were lacking in the program. Further caveats were directed against the selection of the greenbelts and transects along which the Senate wanted to distribute the measures. According to the BLN these were purely conceptual constructions which were not related to the existing structure of the city, and therefore unlikely to be realized in this form.[30]

At the base of this conflict was a crucial tension between the goals of recreation planning and biotope protection. At the Senate, compensation measures were increasingly seen as a resource to realize better recreation planning. In particular, on the background of reduced public financial resources, the compensation measures were meant to realize goals of recreation planning that the city could otherwise not afford (Cloos 1999; Bruns, Köppel, and Wende 2000). Nature conservationists, on the other hand, insisted on the role of the impact regulation as a means of protecting the household of nature, notably biotopes and species. They admonished that the policy goals of recreation planning and of compensating environmental encroachments had to be kept clearly distinct from one another.[31] Both groups thereby drew upon different interpretations of the legal requirements of the impact regulation. The Senate based its program on a very wide interpretation of the regulation that also included the goals of recreation planning. The critics, however, argued that the only measures that effectively contributed to improvements of the household of nature fulfilled the requirements of the impact regulation.[32]

Parallel to the mitigation concept, the Senate developed new administrative procedures with which it aimed to direct the compensation resources from the encroachment areas to new targets. The first was the replacement of direct responsibility of the bearer of a project by financial levies. The formal possibility of charging levies had already existed in Berlin since the Nature Conservation Act of 1979. It was only with the new Berlin Nature Conservation Act of 2003, that such a levy scheme was eventually introduced.[33] The second procedural innovation was a so-called eco-pool that allowed the Senate to realize specific measures in the target areas, which later could be redistributed as compensatory requirements to the investors in a planning project (Senatsverwaltung für Stadtentwicklung 2006: 20).

The criticism by nature conservationists did not hinder the general shift in encroachment compensation policy. Although losses of biotope qualities still figure prominently in determining the need for compensation measures, it has become increasingly difficult for proponents of nature conservation to ensure that they will actually be used for biotope improvement measures. At the same time, flexibilization has led to a convergence between encroachment compensation and landscape programs: Compensating impacts has thus become less a planning instrument of its own, than a way to mobilize financial resources for realizing broader goals of the landscape program.

Conclusion

Environmental mitigation has been the most enduring form in which the urban biotope-protection regime has survived after losing its initial momentum. When political support by activists and ecologists had become weaker, and when powerful alliances advocated the redevelopment of the new capital, the regulatory practices that had been established by the impact regulation required at least a minimal consideration of issues of biodiversity. Thus, it was the institutional results of previous struggles for urban nature in Berlin more than the actual power of urban environmentalism that kept this policy on the agenda. The institutionalization of the impact regulation in nature and construction legislation, and the organizational procedures through which this regulation was implemented, provided an infrastructure in which newly upcoming planning issues were constantly monitored, evaluated, and subjected to mitigation policies. More recently, the impact regulation underwent legal recodification in Germany, but it remains an important feature of land-use planning.[34] However, the period of large-scale urban reconstruction measures and, hence, extensive mitigation conflicts, may be over. In any case, examining the locally and historically entrenched dynamics of contestation of the 1990s enables us to understand dilemmas and political tensions that, in one way or another, are likely to reappear in any systematic mitigation policy.

Although the impact regulation provided a clear legal framework, it rarely determined the content of the mitigation policies. The actual implementation of the rule relied on contextual interpretations and negotiations that accommodated the provisions of the impact regulation to the case, as well as the political circumstances in which it was instantiated. Again the production of ecological knowledge played a crucial role in the operation of this policy. Ecological surveys were used to identify the impact of the projects, to determine their impact, and to select appropriate mitigation measures. Struggles among supporters and opponents of projects were increasingly carried out in terms of different interpretations of study results, or even, in a few cases, led to alternative research initiatives. As we already have seen in the context of the Species Protection Program, the conditions under which such knowledge was produced differed considerably from those in academic fieldwork. Thus, issues of evaluation, the need to operationalize legal criteria and definitions, and a tendency to standardization, largely reshaped the practices of observation and the circuits in which ecological knowledge about Berlin emerged. Far from being a neutral technical input for compensation measures, ecological knowledge making was itself the arena and the medium, in which, and through which the struggle for adequate compensation unfolded.

One of the striking features of the trajectory of mitigation policy in Berlin was the constant drift from the restriction of development projects to a flexible compensation that was only loosely connected with the original projects. Whereas the inventors and early promoters of the regulation had hoped to use the impact regulation to derail many critical development projects, or at least to direct them onto more environmentally friendly paths, the regulation soon proved to be relatively weak in this respect. In the 1990s, it was mainly the reason for substitution measures, often in spatially disconnected areas, and even pooled into new programs. This proliferation of phantom biotopes kept green planning initiatives moving, that otherwise would have suffered from a lack of financial resources, or the lack of political concern. Paradoxically, the ignorance of nature in the rebuilding of the inner city resulted in a new wave of green planning in other areas.

The disconnection from project and compensation measures, however, opened up much space for different interpretations of the adequacy and scope of such measures. Was the improvement of the recreation function of a biotope an adequate measure of encroachment compensation, or was it in itself a new encroachment? And how much was to be done in order to fully balance the loss that it was meant to remedy? As we have seen, standardized systems of evaluation and accounting thereby increasingly replaced project-specific proposals by which impact statements have sought to minimize the effects of a certain project. Notably, the conflicts about how to define the loss and the value of equivalent natures, brought basic understandings of nature in the city to the fore: What should count as nature? What is its value? What is its function? How should it be maintained? As with the specific conflict in wasteland design, this did not only concern different prioritization of biotopes and ecosystem types (which ecosystem could stand for the one that had been lost?) but also a more fundamental conflict between nature as a site for biodiversity and as a recreation space. The extent to which a serious biotope-protection policy will continue to exist in Berlin will depend greatly on the ability of its proponents to defend its position in these compensatory conflicts.

Conclusion

Cognitive and social innovations in the field sciences have often been connected to specific places. One might think here of the Galapagos Islands and evolutionary biology, the slopes of the Andes and vegetation geography, Chicago and urban sociology, or the villages of the Nuer and anthropology.[1] Berlin has played a similarly pivotal role for the emergence of an ecology of the city. Although such research has also been conducted elsewhere, it developed here with an intensity and relative constancy that has probably been unparalleled in the world. In this book I have focused on two aspects of this evolution: the establishment and maintenance of a sustained circuit of observation through which knowledge about urban flora, fauna, and ecosystems was generated, and the mutually constitutive interaction of such knowledge-generation processes with a newly emerging politics of nature that focused on the preservation of species in urban spaces. As I have argued, these two aspects were co-produced as integral features of an urban nature regime that exceeded the formal boundaries of institutionalized science. In the 1950s this nature regime was sponsored by only a few naturalists and ecologists; later it took shape in formal planning policies, protest activities, and a variety of regime communities.

It was no accident that this nature regime developed in West Berlin. It was largely determined and shaped by the particular circumstances of this city. One important factor was the entrenched history of urban nature protection and green-space planning that had developed in Berlin since the late nineteenth century. Many of the spaces that ecologists later defended as valuable biotopes were actually the output of earlier policies of greening the city. Moreover, urban nature conservationists benefited from the existence of an institutional infrastructure that had developed in that period, as well as from the attention that greening had received in West Berlin. This also included the tradition of naturalist research, which, in close alliance with earlier strivings of nature conservation, had provided the basis on which the later circuits of ecological observation could thrive.

Second, the biotope-protection regime was to a large extent a product of the marginal situation of West Berlin. Slow economic recovery, which prevented a rapid reconstruction of the center, provided local ecologists with the ideal conditions to develop an ambitious ecological research program. Moreover, the lack of a hinterland motivated many ecologists and naturalists to focus their research on their own city. It also was invoked in public-policy discourse to call for preservation or development of nature in the city. The pursuit of nature conservation also thrived in the strong left-wing and ecological counterculture that had developed in West Berlin, a city whose marginality made it attractive to many politicized young West Germans.

Last but not least, Berlin's relative marginality made it much easier for ecologists and activists to defend conservational claims in spaces that in other cities would have been much more threatened by competing land-use interests. This is clearly suggested by the fact that after the fall of the wall newly developing dynamics of urban development resulted in a radical restriction of the nature-conservation policy that had developed in West Berlin.

Although the tradition of urban ecological research is still alive in Berlin, and although significant parts of the earlier policy initiatives have survived in the city owing to regulations that date back to the 1980s, Berlin has lost its relatively exceptional status. How to make cities more environmentally friendly, how and in what form to integrate wildlife in urban spaces, and how thereby to enhance the livability and social justice of the city for its inhabitants are issues that, meanwhile, have achieved worldwide scientific and political significance. Around the globe, and increasingly in developing countries, many cities have developed programs for sustainable development that have sometimes also included programs for urban wildlife promotion. In addition, international policy institutions have identified cities as hot spots of the environmental crisis and have called for a shift in urban planning policies.[2] The problems that are discussed in this new urban environmental discourse have often led to a dialog between urban planning and ecology. Of course, the local conditions under which such dialogs are developing in other cities, and the extent to which these dialogs manifest themelves in actual policy change, might vary considerably from city to city. The long-standing record of Berlin in pioneering such a discourse, however, yields insights into some basic problems and dilemmas that might also be relevant for initiatives elsewhere.

Reinventing Nature and Reinventing the City

Throughout modernity, cities have always been critical sites in which and around which symbolic borders between the natural and the artificial or cultural have been drawn, renegotiated, and undermined. At least since early modernity, the urban has been identified with the artificial, and nature has been located outside of the city (Kaika 2005; Böhme 1989). Whereas the enlightenment discourse of the eighteenth century had featured the presumed artificiality of the city generally as a positive attribute, the nineteenth century witnessed an increasing negative evaluation of the city as an "unnatural" space (Böhme 1989: 60–61). As we saw in chapter 1, from nature protection to urban planning, the artificiality of the city was seen as one of the main reasons for the ills of the city. Nature conservation, in all its different shades, was an attempt to exempt spaces that had not yet been affected by urbanization and industrialization from these tendencies, and thereby re-constituted them as counterparts of the allegedly un-natural and hence socially unhealthy spaces of the city. At the same time, the city witnessed various attempts to approximate it to nature, either by bringing nature to the city or by bringing the city to nature (ibid.; Kaika 2005). The first tendency led to the creation of parks and greeneries, the second to the creation of garden cities or suburbs that located the city in existing natures. Both tendencies culminated in the idea of the city landscape that guided much of the urban development that occurred in the aftermath of World War II. As we saw in chapter 1, nature conservationists also operated on the outskirts of the city, defending even the smallest remains of unspoiled nature against unfettered expansion of urban settlements. By identifying "nature close to the city," they hoped both to rescue such remains of nature and to reconstitute the city as a more livable environment.

Nature and the city, however, were not preordained realms that just got separated or mixed differently in various historical phases and reform movements. Instead, they both have to be regarded as historical constructions that were constituted through the symbolic oppositions that contrasted them with one another, and through their association with specific attributes and paradigmatic spaces. Rather than reflecting any intrinsic features of the referential biophysical realities, nature and the city were the results of knowledge practices through which urban dwellers, designers and policy makers classified, explained, and practically engaged with real places and spaces. Concepts such as the park, the nature reserve, the allotment garden, and the city landscape functioned as elements of historically specific orderings that performed different versions of the natural and the urban. In chapter 4, I sketched the most striking differences between these orderings, both in programmatic discourse and in regard to their manifestations in tangible places.

Although the biotope-protection regime in Berlin shared many of the motives that had inspired earlier urban nature regimes, it established species—single, rare, and threatened species, as well as the overall diversity of species—as the main targets of its technical, political, and scientific concerns. The urban biotope thereby emerged as a new spatial entity that was constituted simultaneously as natural and as urban. Rather than treating nature as something opposite to the city, ecologists and their advocates promoted a political and cultural sensibility that asserted both nature's virtual omnipresence in the city and its vulnerability to conventional forms of urban development. At the same time, nature was reinterpreted through its connection with the urban. The species of the city were not simply considered remains of some pre-urban nature that had survived the process of urbanization. Rather, urban climate, land use, or the global transport networks that transported seeds of new species had, according to ecologists, created a new form of nature, one that was considered intrinsically urban. The emblematic example was, of course, urban wastelands, where the urban condition had created a new ruderal vegetation. The primary focus on the sheer presence and diversity of species distinguished urban nature protection from the aesthetic approach to the landscape associated with "homeland" nature. Cities, with their patchwork of biotopes, could even be evaluated more positively than the monotonous agricultural landscape of the surrounding, which indeed had a lower overall species diversity. Instead of always seeing human influence as a threat to nature, ecologists, and the regime communities with which they were allied, explored and cherished the new forms of conviviality between humans and non-human species.

Urban biotope protection, however, was not as coherent a political project as its programmatic rhetorics of an alignment of nature and city might suggest. As Maarten Hajer (1995) has suggested, public environmental discourses should be seen as assemblages of heterogeneous voices and motives whose intrinsic ambivalence persists under the umbrella of seemingly coherent story lines. This also goes for the nature regime through which the urban biotope has been constituted in Berlin. Beneath its guiding concepts and narratives, a number of tensions and ambiguities existed that gave the regime of urban biotope conservation a relatively ambivalent character.

First, the biotope-protection regime has been characterized by a fundamental ambivalence with regard to the resilience or the vulnerability of
nature in the city. One pillar of the program of nature conservation was
the claim that nature was much more resilient than earlier understandings
of nature conservation and urbanism had assumed. Since Sukopp's programmatic writings, it has been the mantra of Berlin's urban ecology that
there was no opposition between nature and the city, and that the full
spectrum of biotopes in the city should be valued. As many pathbreaking
studies sought to make clear, nature as wildlife was omnipresent in the
city, even in spaces that we might have considered as completely artificial
and urbanized. The city figured, not as a destructive force, but as a structuring principle that allowed nature to evolve in new spatial forms. Urban
ecology invited its audiences to look differently at the city, and thereby to
cherish the resilience of nature.

Such a broad notion of nature in ecological research coexisted with a
much more selective normative model of promoting or enhancing biotopes in the city. Contrary to the notion that the full spectrum of urban
biotopes should be valued, not all of them were considered equally worthy
of protection. Ecology drew on evaluative criteria in selecting areas that it
considered more worthy than others. Far from simply cherishing the de
facto resilience of nature, the program of urban nature conservation was
thus mainly about the extension of one form of urban nature at the
expense of the other.

Whereas traditional conservationists based their selection of valuable
areas on the imagined state of the pre-industrial landscape, urban nature
conservation had some more formal criteria for evaluating the ecological
value of biotopes, such as counting species, identifying rare species, or the
spectacularity of certain combinations of species that were deemed characteristic for the city. Such criteria were not only relevant for the selection
of protected areas, but also guided management programs, that, for example,
sought to direct the development of wasteland vegetations onto a certain
path. However, contrary to the programmatic embracing of the urban,
earlier criteria of nature conservation survived under the umbrella of an
urbanized understanding of nature. This should not be surprising, as the
project of urban nature conservation had developed successively from the
conservation ecology of the 1960s. For Sukopp and other advocates of
urban nature conservation, the fens and meadows at the fringe of the city
never lost their major fascination, even if they were not typical of the
urban nature that the advocates embraced elsewhere. As we have seen, the
Red Lists that guided the evaluation of biotopes took the state of the flora

in the middle of the nineteenth century as a normative yardstick against which the "loss" or "endangering" of species was estimated. By understanding the city as a patchwork of biotopes, or (as Kowarik has put it) as containing four different types of nature, conservation experts sought to integrate the more traditional definitions of nature and the new, urban-based definitions into a more comprehensive approach to the city.

Conventional distinctions between nature and culture survived in ecological schemes that classified spaces according to their different degrees of naturalness, and in conservation and management practices that treated some areas as more natural than others. Ecology's role thereby was ontologically ambivalent: it undermined the entrenched opposition of nature and culture, but at the same time it constantly re-constituted this opposition in its own spatial distinctions. The first aspect was visible in the promotion of ruderal areas as a specific form of urban nature: a nature that bears the imprint of human civilization. The other aspect was prevalent in the framing of such ruderal areas as a form of wildness that contrasted with the rest of the city. The first aspect was also articulated through the representation of the entire city as a patchwork of biotopes; the second survived in the classification of these biotopes according to different degrees of human influence.

In addition, the strategies of the new nature regime were torn between the explicit goal of making nature accessible to people where they actually lived and the actual tendency to protect such nature from the potentially damaging influence of people. In the first respect, urban ecology had a strong connection to motives that recently have come to the fore under the label of environmental justice. The natures that interested traditional conservationists were often located in the countryside and in suburban recreation landscapes, which were visited mostly by the well-to-do. Urban nature conservation, in contrast, sought to improve natural qualities in close proximity to the day-to-day environment of most citizens. For Sukopp and other planners this was a strong argument in favor of urban biotope protection, notably under the conditions provided by the wall. Also, the diagnosis of a "green deficit" in some inner-city districts, which had preoccupied earlier open-space planners, remained an important motive for the defense of biotope areas in Berlin. The reframing of these spaces as nature, however, often meant that regimes of order were established that were at odds with the spontaneous appropriation of these places as recreation spaces. In extreme cases, incursions of people into these areas were considered threatening, and thus, as in the first plans for the Südgelände, there was an attempt to restrict such incursions. Also, the management schemes according to which urban wasteland parks were reshaped followed criteria that clashed with the ways in which people made sense of them as recreation areas or appopriated them as anarachic counter-spaces of regulated urban life. As we have seen, the conflict between nature protection and recreation has been a major source of tension among communities of environmentalism that privileged either one of these aspects.

This theme is closely related to a more general ambivalence in urban ecology's planning discourse: a tension between an attempt to interpret nature in terms of human influences in the city and a relatively narrow concept of urban life. In ecologists' and planners' renderings of urban spaces, humans were not so much seen as active subjects who engaged cognitively and physically with the spaces of biotopes. In ecological studies, they were portrayed as material forces whose intentional or unintended "land use" had an impact on the composition of plant and animal wildlife in their surroundings. As we have seen, human land use served as a kind of black-box notion that allowed ecologists and planners to trace biological configurations back to the structuring impact of the city's human inhabitants. On this basis such activities could be judged as positive or negative. Ecological planning documents constituted urban dwellers as potential users who appropriated nature through the visual consumption of spaces. These ecological understandings of the role of humans in the city contrasted with the more spontaneous way in which nature was appropriated by activists in their informal activities on wastelands and other spaces. They also made it difficult for ecologists and planners to form links to, and thereby engage productively with, the new forms of political activism that have pursued urban nature conservation in civil society. Also, when in the late 1990s the global urban sustainablity initiative known as Agenda 21 led to various sustainability and participation projects in Berlin, issues of biotope protection found strikingly little resonance.[3]

Ecology as a Political Science

It has been a second argument of this book that urban nature conservation has been conducted through the generation and application of scientific knowledge. This is not unusual for environmental conflicts, in which both policy makers and activists often use scientific input to articulate or bolster their claims for action (Yearley 1991; Beck 1986; Bocking 2004). It also fits with a general trend toward a knowledge society where nearly all domains of action have become penetrated by the results and the procedures of science (Knorr Cetina 2007).

To some extent, earlier open-space policies and conservation efforts at the outskirts of cities also relied on specialized bodies of knowledge. With the open-space theory, the vegetation and geological mappings that were connected to the quest for an urban landscape, and the various, sporadic surveys that legitimized the designation of nature reserves, different kinds of experts were involved in traditional forms of green planning. Also, the technical skills and aesthetic sensibilities of horticulturalists were the knowledge base of such earlier policies.

With the biotope-protection regime, however, the knowledge generation of a specialized field of ecological expertise begun to occupy a central place in the very conduct the policy making for urban nature. The surveys that amateur naturalists and local academics had carried out in the city and those that were later undertaken for administrative reasons by university ecologists and landscape planning consultants contributed to this policy shift. By collecting evidence and representing the distribution of species and the mosaic of their biotopes, by underlying their various values and exposures to threats, and by proposing planning priorities, these surveys fundamentally redrew the cultural and political map of Berlin. What hitherto had been seen as a mass of houses, or a "sea of buildings," or as "steinernes Berlin" (Hegemann [1930] 1992) had been transformed into a variegated zone of cohabitation of human dwellers with varying assortments of plant and animal species. Far from being a purely descriptive interpretation of the city, these practices of knowledge making also embedded normative criteria that guided ecologists and naturalists to interesting fieldwork sites, or that underlay their planning-oriented evaluation of biotopes.

Ecological knowledge and related action claims were soon translated into political claims that also appealed to, and came to be supported by, other institutions and actors in the Berlin polity. Ecological knowledge became the core of a new environmental discourse that dovetailed with the concerns and routines of policy makers, administrators, and activist groups. A partiality for plants and ecological features of the city such as had been shared among naturalists and biologists became tied to the preoccupation with green space in the city, to professional agendas of the landscape planning profession, or to the broader political imageries of diverse strands of activism. Far from being a purely ecological concept, the biotope thereby became reconfigured as an object of administrative practices, a symbol of political resistance, or a physical site at which dominant planning policies were contested.

It would be misleading to see these interactions between science and politics simply as an application of existing ecological expertise to the city, as conventional models of the science-policy nexus have suggested. Rather than "speaking truth to power," ecologists and their knowledge practices and the political rationales of urban nature protection interacted and thereby co-produced each other. Three major mechanisms orchestrated the very process of generating ecological knowledge with the rationales of political claims making. First, spaces of fieldwork were never purely neutral objects about which factual knowledge could be gained. They were also cherished as worthy, and urban changes seen as threatening them were bemoaned. This implicit normativity, which partly had its roots in the early days of regional naturalism, guided the selection of fieldwork sites and the focus on species inventoried. Even if it was often restricted to a nostalgic undertone in statements about changes in the local flora and fauna, it bespoke the potential political significance of environmental knowledge. Second, political publics were the audiences that naturalists and ecologists had in mind when they selected research sites, conducted their observation work, and presented their results. From the first appeals of local naturalists, to the nature-conservation authorities, to Sukopp's programmatic statements about urban nature protection, to the first studies for the biotope mapping, the quest for environmental knowledge (aside from its relevance for academic research) has always been a means of communicating political claims to administrative and public audiences. Third, and most evidently, ecological research has been done in commission of, or in close institutional cooperation with, public institutions. As we have seen, such interrelations with the public had implications for what kind of knowledge was sought, how this knowledge was formatted and communicated, and under what conditions knowledge became—at least for public purposes—legitimized as credible and reliable. The interaction of research practices with the administrative structures of the city was significant here. The effect was an increasing alignment of ecological knowledge to the format of bureaucratic rationality that helped to make the biotope "governable" in terms of state-centered policy making. To an extent, activist groups also took an active role in the processing and generation of knowledge. Amateur naturalists had always fed their findings into the observation cycles of urban ecology. In the 1980s, citizen activists also engaged in the interpretation of ecological findings, often drawing alternative conclusions and relating data about wildlife to broader civic society in productive new ways.

The relationship between environmental experts and their publics was not free of tensions. On the one hand, administrators and policy makers sought to expand their control over the very terms and conditions under which environmental knowledge about urban nature was produced; on the other hand, ecologists, and to a lesser degree consultants, were eager to defend their own autonomy as academic experts. Ecologists seeking to expand their jurisdiction into public affairs experienced a need to make compromises with the rationales of administrators, which often threatened their academic standards. Nature conservation, in the perspective of the ecologists, should proceed mainly through the restructuring of the city under the guidance of their scientific expertise. This conflict became visible notably in Sukopp's skeptical evaluation of his own contribution to the Species Protection Program. This does not mean that Sukopp and other expert advisors considered their own knowledge incompatible with politics. In their view, however, political procedures had to be adapted more consistently to their standards. In this respect, they based their role on what can be characterized as a technocratic understanding of science-policy interaction. Accordingly, it was the exclusive role of ecologists to generate knowledge about biotopes and to draw conclusions as to how they should be protected or enhanced. Policy making and administration, in this view, became reduced to the general endorsement and accompanying promotional activities through which such policies were put into operation.

Other tensions existed between ecology and civic activism. Though activists often drew on (and thereby implicitly underwrote) the authority of science to speak legitimately about nature, in other instances they also held views that diverged from those of ecologists. We have seen that activists gave more emphasis to amenity issues, whereas ecologists focused more on species. In some instances—for example, when ecologists justified measures that were at odds with the place images that local activists had cherished—ecologists and consultants appeared to be rather "unreliable friends" (Yearley 1991) of these activists. The definition of common planning goals was often a fragile compromise between experts' views on biodiversity and the heterogeneous agendas of activists. On the other hand, the history of the nature regime shows that in many instances public knowledge and ecological knowledge were constructed in a more cooperative way. Many activist groups and campaigns can be seen as spontaneously emerging "agoras" (Nowotny, Scott, and Gibbons 2001) in which experts and activists exceeded, or at least minimized, the institutional boundaries that otherwise separated them as members of distinct social worlds. In Berlin's biotope regime, this more open model of civic science coexisted with the more technocratic advisory relationship that Sukopp and other planners developed to the administrative system.

The Limits of Urban Nature Conservation

The biotope-protection regime has changed Berlin's political landscape as well as its material spaces. Although it has lost much of the political significance that it enjoyed in the 1980s, it persists in the ongoing research of urban ecologists at the Technische Universität, and also in a persistent network of routinized negotiations among administrations and conservation representatives. The fundamental urban renewal under the guidance of ecology that Sukopp had dreamed of, however, did not take place. As we have seen, many of the projects of the original proposal of the Species Protection Program became watered down in the process of finalizing this planning scheme. Even the remaining goals lost much of their political backing. In the 1990s, urban ecology changed from a project of fundamentally reshaping the city to a more reactive policy of mitigating the harm of seemingly inevitable development policies. It is one of the most striking paradoxes of the history of ecological planning in Berlin that, contrary to the claim that nature conservation should be applied to the entire territory, it was eventually realized only in the most conventional form: the form of nature parks or reserves that set highly evaluated pieces of nature apart from the hostile environment. Another success for urban nature conservation was the policy of encroachment mitigation, which accepted development and sought only to remedy its most drastic impacts. Berlin certainly became greener through the efforts of ecologists, but at best it became "half-green," to adopt an expression used by Michael Bess (2003) in his study of ambivalent results of environmentalism in France. Much as the same glass of water can be seen as half-full or half-empty, half-green points to the mixed results of the compromises and the step-by-step progress through which environmentalist ambitions have materialized in political institutions or policies.[4]

The first and most obvious reason for the limitation of urban ecology's original program of urban renewal was the tension between urban ecology's far-reaching reform ambitions and the need to find compromises with the political and economic realities of a large city. Although in the 1980s the urban biotope-protection regime had achieved a rather stable position in the planning system, its representatives had to articulate their goals against the development orientation that remained greatly dominant within the administration and the government. The dynamics of legislation that followed the reform debates of the 1970s, or the short period of an SPD-Grünen government, paved the way for considerable steps toward a biotope-protection policy; however, they were too short-lived to bring about a more fundamental change, and they themselves were not free of contradictory forces.

The political changes after 1990 marked the end of the comprehensive preparatory planning policy of the 1980s. Since that time, biotope-protection policy has merely reacted to the ongoing efforts of urban renewal. This has been due in part to a weakening of the ties between ecologists and the civil-society actors who earlier had been active promoters of the regime. Additionally, urban development initiatives reflected the economic pressures that were unleashed by the unification and the need to make Berlin fit for the increasingly global competition with other cities for international investments. The realization of nature-conservation goals eventually followed a piecemeal pattern that was determined more by the economically motivated development imperatives than by the scientific rationalities of urban ecology. When nature parks were established or when less intensive forms of land use were stipulated, it was mainly in spaces that were only of marginal interest for development investments, and it happened only in conjunction with legally prescribed compensation requirements for development projects conducted elsewhere.

It would be wrong, however, to assume that ecological criteria were intrinsically at odds with development interests. As Desfor and Keil (2004) have argued, nature, has become an important asset through which even neoliberal policies of urban renewal seek to upgrade, and thereby increase, the economic value of certain areas. (Also see Evans 2007.) In Berlin, the integration of the Johannisthal airport into the science city can be read as an example of such a policy. Although such intentions were less explicit there, the Südgelände park and other green areas, once they had turned from metaphors of urban decay into highly appreciated pieces of nature, have certainly also upgraded their neighborhoods.

Ecological expertise played an ambivalent role. On the one hand, it remained a constantly warning yet ultimately disregarded voice against the destructive effects of development.[5] On the other hand, wherever it sought to remain in business, it had to adapt to institutional demands, to seek and justify compromises and thereby tended to become more an arbiter in ecology-development conflicts than a promoter of an alternative political vision. Much of the work of ecological consultants on environmental impacts and mitigation schemes was such a form of arbitrage. This need to seek for compromises had a rather paradoxical effect. Whereas urban ecology had started out with the aim to extend nature conservation to 100 percent of the country's surface, it saw itself forced to concentrate on a few areas it considered most worthy or in which win-win situations with dominant development policies existed or existing regulatory frameworks provided suffient leverage. Ecological expertise operated as a system of territorial triage that selectively secured some spaces, thereby implicitly legitimizing the neglect of others as unavoidable. As James Evans (2007) has argued in a study of nature-conservation conflicts in British cites, ecology thereby also tended to play a public role of a provider of an ideological legitimacy by flagging land-use decisions as scientifically motivated, thereby concealing their actual political rationales.

A second obstacle was the need to articulate ecological claims in terms of the institutional logics of the planning system. This meant that ecological claims could be articulated successfully only if they were expressed in the form of legal provisions for specific circumscribed spatial areas. A compartmentalization of space was as such not absent from ecology's practice. Ecology had always divided space into distinct parcels of space such as biotope types or ecochores. Ecologists now begun to use these units as a basis for defining planning zones and to redefine them to make them better suited to this purpose. That, however, was at odds with the holistic principles that guided the theoretical discourse of urban ecology. For ecologists, the mosaic of biotopes was not just a set of independent areas that could be considered separately from one another. Rather, biotopes were elements of a larger ecosystem, as was most plastically illustrated in Sukopp's concentric scheme of the city. Whereas biotope areas could be translated relatively easily into priority areas on planning maps, these systemic aspects had little effect on the actual planning practices. Even biotope networks were eventually realized through piecemeal efforts focusing on smaller spaces.

Another aspect of the planning logic was the need to legally fix certain states in planning areas. This dovetailed with a traditional approach of nature conservation, that sought to keep areas in their present shape. It was at odds, however, with the more dynamic character of urban biotopes that ecologists had described in much of their more scholarly work. It was also at odds with the ideals of wilderness and spontaneous nature that had inspired activism concerning urban wastelands. Berlin was interesting to ecologists and activists largely because of the changes in land use, the unexpected effects of the destruction that occurred during World War II, and the effects of new transportation practices. Promoting these fluid and capricious processes would have required other means than conventional regulatory planning such as earmarking areas permanently for preservation in a certain ecological state.

A third obstacle to realization of urban nature conservation in Berlin was the competition between biotope-protection policies and other forms of green-space politics and nature politics in the city. Ecologists and their allies were far from being the only people calling for nature in the city. The former traditions of green-space planning and horticulture were still alive, notably within the district administrations that were responsible for the implementation of many of the projected measures. Landscape planners, who in the 1970s had wholeheartedly embraced an ecological perspective, increasing sought to re-align their profession with a more artistically oriented ideal. It was in this vein that—parallel with the institutional success of ecological planning—a monument-care approach to urban gardening gained a foothold in Berlin's political administration. Similar tendencies could be observed in civil- society urban environmentalism. Under the umbrella of a loose political imagery, considerable differences continued to exist among academic ecologists, professional landscape planners, conservative conservationists, the left-wing counterculture, and middle-class neighborhood activists. These groups, however, embraced the discourse of urban nature conservation and the regime through which it was enacted only as long as this discourse resonated strategically or semantically with their own agendas. Such alignments had given the biotope-protection policy of the 1980s much of its momentum. In the long run, however, they turned out to be relatively short-lived, and often they were restricted to the dynamics of specific land-use conflicts.

The inability of ecology to forge a broad and sustained public consensus around the issue of biotope protection became visible when biotope protection began to clash with the more recreation-oriented motives of activists (as in conflicts over the design of the nature parks) or planners (as in the debate on mitigation planning schemes). Advocates of biotope protection therefore had increasing difficulty linking their concerns to the public discourses and practices in which matters about urban nature were articulated and politicized. Meanwhile, "green" activism has taken entirely new forms, such as appropriation of urban spaces through "guerilla gardening," community gardening projects, and urban agriculture.[6] The marginalization of the nature-park idea at the Gleisdreieck was as much a result of this new orientation as it was a result of the increasing development pressure in the unified capital. As examples from the 1980s have shown, there is no intrinsic tension between the ecological approach to biodiversity and civic forms of activism. The extent to which ecology will be able to play a role in this context in the future, however, will depend on its ability to integrate its conservationist goal with amenity and recreation issues, instead of playing them against one another.

Toward a Politically Reflective Urbanization of Nature

As the urban geographer David Harvey (1996: 186) put it in a widely quoted statement, "there is nothing *unnatural* about New York City." In this vein, recent work in environmental history and urban studies has begun to move beyond the entrenched opposition between city and nature. It has been argued that we understand cities only if we take them at the same time as human constructions and as natural structures and processes. As Matthey Gandhy (2002) showed in his environmental history of New York, far from being absent, nature has been "urbanized" or "reworked" through its incorporation into municipal water systems, parks, parkways, and industrialized neighborhoods.[7] In a similar way, also recalling Sukopp's ideas on urban vegetation, the geographers Steve Hinchliffe and Sarah Whatmore (2006) have pleaded for a symmetric attention to humans and animal species as the constituent components of cities. Whereas some studies in this vein have resorted to the natural-scientific concept of metabolism to develop a social scientific version of the ecosystem model of the city, others have invoked Bruno Latour's notion of "quasi-objects" and Donna Haraway's notion of "cyborgs" to do justice to the profound ontological hybridity of the city (Heynen, Kaika, and Swyngedouw 2006; Gandy 2002). Such studies have had both analytic goals (e.g., to yield a better understanding of how cities and their nature became what they are) and political goals (e.g., to provide vocabularies that help to address issues of urban quality of life, social justice, or the role that non-human entities should be accredited in urban worlds).

Our ability to shape new forms of coexistence of the social life of cities and the biophysical environments in which cities are embedded will depend on more than just a general shift in conceptual attention. It will depend on processes of conflict and social negotiation, and probably also ambivalences and dilemmas, such as those that have characterized the history of urban nature conservation in Berlin. It will depend on the generation of new environmental knowledges and the forging of coalitions that are able to mediate these concerns with the institutional practices of policy making. Even if they do not resort to essentialized notions of nature and culture, such attempts will also be bound to the making of some distinctions between spaces in which non-human species are given more place and spaces in which the needs of such species will be curtailed on behalf of more human-centered ambitions of land use.

One of the main conclusions to be drawn from the present study is that forging such new forms of coexistence will require not only the proliferation of knowledge about the city and nature but also public reflection on how we know what we know about the city and its natures. As we have seen, such knowledge is not self-evident, but is shaped and negotiated in situated regimes of nature. This raises a number of questions: Who is entitled to speak on behalf of urban natures. What are the traditions that shape our understanding of nature and society? What are the legitimate procedures through which such knowledge is processed and stabilized? How can we make such knowledge resonant in institutional policy and broader public debates? In other words, any attempt to forge more viable relations between the city and nature will be as much a politics of knowledge as a politics of urban space. No matter how such a future politics of nature will look, it seems to be clear from the Berlin experience that its creative power will depend on its ability to embrace both the subtlety and complexity of environmental expertise and the imagination and experimental attitude of a lively civil society.

Notes

Introduction

1. The term *nature* is also applied in a much broader sense to include the "inner nature" of humans or elements of a cultural order that are experienced as self-evident. In line with the quoted authors, my focus is on the materialities and spatialities of the non-human biophysical world. Although the term is charged with many semantic and philosophical ambivalences (Williams 1983), I prefer it to the possible alternative term *landscape*, which tends to put too much emphasis on the visual perception of nature. In this more specific sense, it will be treated as an actor's concept that is independent of and often in conflict with biotope protection.

2. Although in both respects nature has to be seen as fundamentally socialized, the term is indispensable when getting to grips with a specific set of human practices and concerns. In this respect I disagree with Bruno Latour (1993, 2004), who has suggested that the profound entanglement of human and non-human worlds has rendered the category of nature obsolete. As a consequence, however, nature can no longer be treated as a universal structure; rather, as Macnaghten and Urry (1998) have argued, multiple natures are locally embedded in historically specific social practices.

3. For example, environmental historians have employed the concept of *metabolism* to map the socially mediated flows of material and energy in cities. See Tarr 2002.

4. Escobar (1999) distinguishes the following three regimes: the regime of "organic" nature, which he identifies with non-Western cultures, the regimes of "capitalist" nature of the industrialized West, and the regime of "techno" nature that evolves from more recent approaches of bio-engineering.

5. The sense in which I use *regime* here is much broader than the sense in which it is used in "urban regime theory," where it refers merely to the social networks that dominate policy making in cities. (See, for example, Stone 1993.) Focusing on discourse and practices, my use of *regime* has more in common with recent uses of the

term in governmentality studies (Dean 1999) and in other work in STS—e.g., *tech-nopolitical regime* (Hecht 1998) or *regulatory regime* (Halffmann 2003).

6. On the notions of "place" and "sense of place," see Cresswell 2002 and Gieryn 2003.

7. The *performativity* of ecological classification has been analyzed by Waterton (2003). The term refers to the way in which classifications, through their repetitive and context-dependent enactment create the social and natural orders that they describe. Although Bowker and Star (1999) do not use the term *performance*, such consequences of classifications are also the theme of their sociology of classification infrastructures.

8. See, e.g., Latour 2004 and Hinchliffe 2007. Besides the strict application of actor-network theory, since the 1990s environmental sociology has witnessed a broad move from a social constructivism that focused primarily on the discursive repre-sentation of the environment toward a co-productionist or co-constructivist view that emphasizes the socio-material nature of practical engagements with the envi-ronment. See, e.g., Irwin 2002, Franklin 2002, and Macnaghten and Urry 1998. This book can be seen as a further contribution to that turn.

9. In this book I deal mainly with how the world—through the engagement of biophysical materialities and spatialities—has been made intelligible for the partici-pants in a set of culturally encoded practices. Although I attend to "non-human" agency—in the form of material constraints and affordances that practices are faced with—I see intentions and knowledge-ability as the exclusive realm of the human participants of these practices. In this respect I do not follow the suggestion by Latour and other actor-network theorists to treat human and non-human entities fully "symmetrically." For similar reservations about actor-network theory, see Pick-ering 1995.

10. On the division of Berlin and the political development of West Berlin, see Ribbe 1987.

11. The addendum *regierender* (governing) referred to the special status of West Berlin, which in terms of legislation was formally independent from West Berlin. The addendum was abolished with unification, when Berlin became a regular Land.

12. This also became visible in the different nominations that the FRG and the GDR used for this part of the city. In Western administrative terms it was called "Berlin (West)," emphasizing its provisional status and its character only as a part of a larger city. The GDR, by contrast, used the expression "Westberlin" and thereby disen-tangled this part of the city semantically from East Berlin, which was officially called "Berlin, capital of the GDR." Trying to keep some distance from both of these politi-cally invested expressions, I use the denominations "West Berlin" and "East Berlin" respectively. These have also become the common terms that historians, in hind-sight, have used to refer to the two political entities.

13. On the effects of the division of the city and the Cold War constellation on academia, technology, and culture in West and East Berlin, see Lemke 2008. Attempts to disentangle social life from the respective opposite side of the city (and to furnish the city parts as "showcases" for respective political systems) coexisted with technical and informal interdependences and pragmatic arrangements that transgressed the political boundary between the two parts of the city.

14. West Berlin's economy received financial aid from the United States under the Marshall Plan, and later received continuous subsidies from West Germany under the 1950 Aid for Berlin Act (Berlin Hilfe Gesetz) and the 1952 Act Concerning the Status of Berlin within the Fiscal System of the Federation (Gesetz über die Stellung Berlins im Finanzsystem des Bundes). See Ribbe 1987: 1076.

15. In the West German literature on social movements, West Berlin has been described as a "stronghold" of new protest movements. Although Rucht, Blattert, and Link (1997: 69–70) posit that the relative amount (in relation to the number of inhabitants) of activists or protest events was even higher in some smaller West German cities, they maintain that Berlin had functioned as a vanguard for the formation of new forms of protest and that these often developed in more radical forms.

16. In contrast to the federal structure of the FRG, the GDR was (from 1952 on) organized into 14 districts, which were purely administrative in function.

17. The landscape planner Sonja Pobloth (2008) has provided a historical inventory of landscape planning activities in Berlin. This work is purely descriptive and does not place these results in any broader interpretive context. In her intellectual history of urban ecology, the ecologist Monika Wächter (2003) has reconstructed the theoretical concepts of urban ecologists and conservationists in Germany during recent decades, and thereby also deals with the work of Sukopp and his co-workers in Berlin.

Chapter 1

1. The Hobrecht Plan of 1862, which determined the development of the agglomeration until World War I, only stipulated a basic street pattern of large regular blocks. See Bodenschatz 1987.

2. See Ladd 1990 and Lees 2002. Specifically on Berlin, see Peschken 1984 and Asmus 1984.

3. On the development of the concept of the garden city in Germany, see Hartmann 1976.

4. On the German life reform movement, see Hau 2003 and Williams 2007. Williams explicitly links strands of life reform activism, such as nudism and hiking, with longing for the protection of nature.

5. In 1906 the Staatliche Stelle für Naturdenkmalpflege in Preussen (Prussian State Office for the Care of Natural Monuments) was founded. In 1907 Provincial Commissions were established to implement this policy at a regional level, among them the Brandenburgische Provinzialkommission für Naturdenkmalpflege (Brandenburg Province Commission for the Care of Natural Monuments), which focused on the region around Berlin. In 1927 the Berliner Stelle für Naturdenkmalpflege (Berlin Body for the Care of Natural Monuments) was established as a Städtischen Kommission (City Commission) that was responsible only for this city. See Klose and Hilzheimer 1929.

6. The Prussian Acts against the Disfigurement of Places and Significant Landscape Areas of 1902 and 1907 introduced measures against a number of supposed visual annoyances, such as the excessive placing of billboards. A late success of the homeland-protection interest in landscape was the inclusion of the landscape reserve as a protective category into the German Nature Conservation Act in 1935. The same act also included a right for the commissions to formally comment on all measures that would interfere with the character of the landscape (Lekan 2004: 61).

7. For example, in the early 1920s the Arbeitsgemeinschaft für Wandern und Heimatpflege campaigned against the destruction of forests around Berlin (Stürmer 1991: 149).

8. Hugo Conventz, who died in the year of the foundation of the association, is often considered to have been one of its promoters. A leading founder and first president (until 1945) of the Volksbund Naturschutz (People's Nature Conservation Alliance) was the high school teacher Hans Klose, an early companion of Convenz and since 1923 director of the Brandenburg provincial commission. Under the National Socialist regime, which Klose supported wholeheartedly, he became the principal author of the German Nature Conservation Act of 1953. Another leading member was the zoologist Max Hilzheimer, who acted as Berlin's official Naturschutzbeauftrater (Nature Conservation Appointee) (since 1927). A co-founder and leading member of the VBN who did not play an official role in the nature-protection bureaucracy was Robert Potonié, a Privatdozent (private lecturer) and later a professor at the Technische Universität. (See Weiß 1997.)

9. Linse (1986) has explored this in more detail. For an analysis of the "proletarian" approach to nature in late Imperial and Weimar Germany, see Williams 2007: 67–99.

10. In contrast to the administration of green space, the administration of nature conservation in this period formed part of the police. In postwar years the central nature administration competence became relocated to the Senate Department for Construction and Housing (Senatsverwaltung für Bau- und Wohnungswesen), and was thereby brought closer to the administration of green space. The association of nature conservation with the police does not only show the extent to which it was

identified with matters of social control. This understanding was in line with the traditional concept of police as it had been developed by cameralist thinkers in the seventeenth and eighteenth centuries; it was much broader than the pursuing of criminal behavior that we now associate with the term.

11. On the use of slides to create awareness for "homeland-nature," see Strech 1931. On the Berlin Nature Protection Exhibition, see R.R. 1931. On earlier promotion activities related to homeland nature via the Naturkundemuseum (Museum of Natural History) in the German Kaiserreich, see Köstering 2003.

12. As the first bogs in the Grunewald, the Hundekehlefenn, the Langes Luch, the Pechsee, and the Barsee had been protected since 1920. See Ketelhut 1958.

13. During the National Socialist period, the VBN was absorbed by the Reichsbund Volkstum und Heimat (Imperial Alliance for Folkways and Homeland).

14. See also Witte 1957: 45.

15. The analogy between the city and the (human) organism was not new. It can already been found in Harvey's writing on blood circulation (Sennet 1996, chapter 4). On the use of such analogies in nineteenth-century public health writings, see Graeme and Davison 1983. In contrast to these earlier analogies, which had conceived of the actual complexity and functional integration of the city and its parts, the organism metaphor was now used as a normative ideal for a future city. See also Welter's 2002 study on the early-twentieth-century Edinburgh urban planner and biologist Patrick Geddes. Geddes' use of organic metaphors and his ideas on community building, as well as on the integration of city with the landscape, have similarities with the landscape idea as promoted after World War II by West German planning experts. These German urbanists, however, did not share the cultural and political regionalism that was the hallmark of Geddes' approach.

16. As the green-planning official Fritz Witte stated in 1952, the 40 million cubic meters of rubble material and the high number of unemployed people did not allow them to take much time "for theoretical ideal plans" and "quickly and energetically lead us to take the way of practical work and planning 'in the limits of the possible'" (Witte 1952b: 4)

17. According to the garden director of the district Neukölln (Lohrer 1958: 121), the creation of new parks in the district of Neukölln followed the suggestions given by the botanist Hueck.

18. Senator von Berlin, Senatsbeschluß Nr. 947/55 vom 19. Juli 1955. Bestellung des Landesbeauftragten für Naturschutz und Landschaftspflege in Berlin. Landesarchiv Berlin, Senatsbeschlüsse 2. Wahlperiode, Nr. 690-819.

19. Senator von Berlin, Senatsbeschluß Nr. 804/55 vom 18. Juli 1955. Bestellung des Landesbeauftragten für Naturschutz und Landschaftspflege in Berlin. Landesarchiv Berlin, Senatsbeschlüsse 2. Wahlperiode, Nr. 690-819.

20. "Bericht der 10. Sitzung der Landesstelle für Naturschutz und Landschaftspflege in Berlin am Mittwoch, dem 13. Februar 1957," p. 1, archive Senator für Stadtentwicklung und Umweltschutz, Berlin, Landesbeauftragter für Naturschutz und Landschaftspflege (hereafter archive Landesbeauftragter).

21. "Protokoll der ersten Sitzung der Naturschutzstelle am Freitag, dem 9. Dezember 1955," archive Landesbeauftragter.

22. For example at the so-called Alte Schanze in Ruhleben. See "Bericht über die 38. Sitzung der Landesstelle für Naturschutz und Landschaftspflege in Berlin," archive Landesbeauftragter.

23. "Bericht der 28. Sitzung der Landesstelle für Naturschutz und Landschaftspflege in Berlin am Mittwoch, dem 17. Oktober 1962"; "Bericht der 54. Sitzung der Landesstelle für Naturschutz und Landschaftspflege in Berlin am Mittwoch, dem 24. Mai 1972," archive Landesbeauftragter.

24. "Bericht über die 27. Sitzung der Landesstelle für Naturschutz und Landschaftspflege in Berlin am Freitag, dem 30. März 1962," archive Landesbeauftragter.

25. "Bericht über die 22. Sitzung der Landesstelle für Naturschutz und Landschaftspflege in Berlin am Donnerstag, dem 28. April 1960," archive Landesbeauftragter.

26. "Protokoll der 2. Sitzung der Landesstelle für Naturschutz und Landschaftspflege in Berlin am 9.3.1956 16 Uhr: 4," archive Landesbeauftragter.

27. "Bericht über die 12. Sitzung der Landesstelle für Naturschutz und Landschaftspflege in Berlin am Mittwoch, dem 22. Mai 1957, p. 2," archive Landesbeauftragter.

28. As Cheney (2008, chapter 2) has shown, the blaming of Allied troops for causing landscape destruction has been a widespread topic in West German nature-conservation debates.

29. Abgeordnetenhaus von Berlin, "Vorlage—zur Beschlußfassung—über den Flächennutzungsplan von Berlin für die Bezirke II, III, IV, VII, VIII, IX,X, XI, XII,XIII, XIV, XX vom 30. Juli 1965." Drucksache 5/502, October 4, 1968, p. 2.

30. Ibid., p. 3.

31. Ibid., p. 2.

32. Ibid., p. 3.

33. Ibid., p. 16.

34. "Bericht über die 58. Sitzung der Landesstelle für Naturschutz und Landespflege am Mittwoch, dem 17. Januar 1973," p. 2, archive Landesbeauftragter.

35. According to Hennicke (CDU) in Abgeordnetenhaus Berlin, Parlamentsprotokolle 5/71, April 23, 1970, p. 201 and Drucksache Nr. 5/502, September 4, 1968, p. 3.

36. As Harald Bodenschatz (1987: 171) argued, with the creation of new living space in the green area, policy makers also intended to attract new people to the walled city, which after the erection of the wall suffered from increasing depopulation, and scarcity of workforce.

37. "Bericht über die 55. Sitzung der Landesstelle für Naturschutz und Landespflege am Mittwoch, dem 16. August 1972, pp. 1–2"; "Bericht über die 50. Sitzung der Landesstelle für Naturschutz und Landespflege am Mittwoch, dem 1971"; "Bericht über die 57. Sitzung der Landesstelle für Naturschutz und Landespflege am Mittwoch, dem 16. Juni 1971" (all in archive Landesbeauftragter).

Chapter 2

1. Berlin ecologists also ran a field station in which they conducted experimental studies. Those studies, however, were only loosely connected with the emergence of the urban nature that forms the topic of this book. For descriptions of these sites and the experiments that were conducted there, see Bornkamm and Hennig 1982; Bornkamm 1987; Bornkamm and Köhler 1987.

2. In a 1986 book on Swedish ecology, Thomas Söderqvist uses the term *protoecology* to designate such pre-disciplinary forms of ecological research. The term, however, suggests a rather teleological understanding that takes unitary disciplines as the natural end result toward which such earlier activities of naturalists have evolved.

3. An early example in Berlin is the work of the zoologist Friedrich Dahl (1908), who coined the term *biotope*. In postwar Germany it was an animal ecologist, Wolfgang Tischler, who received the first chair that was formally designated for ecology. In contrast to vegetation ecology, the study of animal ecology was often based on a so-called *autecological* perspective—that is, it focused on the interaction of single species with their environment. In postwar Berlin such a perspective was represented by Fritz Peus, a professor of zoology at the Humboldt-Universität and later at the Freie Universität. While Sukopp developed his ecology of the city, Peus rejected the very concept of the biotope as an abstraction that could not be justified scientifically.

4. On the development of ecology in the United States and Britain, see Mitman 1992; McIntosh 2011; Bocking 1997; Kingsland 2005.

5. Although initially ecologists were much less significant in developing the new environmental policy programs that the German government launched in the early 1970s, they sought to involve themselves in various aspects of environmental planning (Chaney 2008, chapter 6; Küppers, Lundgreen, and Weingart 1978).

6. Compiled by the author on the basis of Stifterverband für die Deutsche Wissenschaft 1964, 1973, and 1968.

7. Scholz had studied biology before his military service in World War II. After completing a thesis on Berlin's ruderal vegetation, he took a position at the Botanisches

Museum in Dahlem. His main expertise was in botanical taxonomy, but as a member of the Botanischer Verein he also remained an active participant in Berlin's botanical inventory. Interview, Hildemar Scholz, Botanisches Museum, Berlin, December 6, 1999.

8. Naturalists studying the fauna formed smaller groups that specialized in different groups of animals. Ornithologists were often members of local chapters of the Bund für Vogelschutz (Alliance for Bird Protection) and the Deutsche Ornithologische Gesellschaft (German Ornithological Society). In the 1970s some West Berlin ornithologists organized themselves locally as the Ornithologische Arbeitsgruppe Berlin (West) (Ornithological Working Group Berlin (West)). A number of smaller societies had specialized in other groups of animals; one example was the Entomologische Gesellschaft Orion (Orion Entomological Association). An important role in the naturalist culture of Berlin was also played by the Volksbund Naturschutz Berlin (People's Alliance for Nature Protection Berlin), a nature-conservation organization founded in the Weimar period. Although mainly a political lobbying organization, it was also actively engaged in the coordination and publication of new insights to Berlin's flora and fauna. Some naturalists also maintained relations with the Gesellschaft Naturforschender Freunde (Association of Friends of Nature Research), which comprised a broad spectrum of scientific interests.

9. In day-to-day practice the political divisions did not seal West Berlin as completely as one might in hindsight suppose they did. At least until the erection of the wall, according to Lemke (2008: 8), the two Berlins remained a common "space of entanglement." On the continuing informal and illegal contact that naturalists maintained after 1961, see Behrends et al. 1993: 19.

10. "Liste der am 20.9.1972 in der Mittelbruchzeile in Berlin-Reinickendorf festgestellten wildwachsenden Pflanzenarten" (letter from Wieland Schneider), archive Technische Universität Berlin, Institut für Ökologie, Prof. H. Sukopp (henceforth cited as archive Sukopp).

11. Interview, Hildemar Scholz, Botanisches Museum, Berlin, December 6, 1999.

12. Herbert Sukopp, field notebooks "Luisenstadt," "Kreuzberg," "Schöneberg," Sukopp archive.

13. Röhrichtschutzgesetz, Gesetzes- und Verordnungsblatt Berlin. 1969: 2520.

14. For an ethnographic account of the skill-based nature of this practice, see Waterton 2002. On the use of "quadrats" in early-twentieth-century North American ecology and their role as a quantifying instrument, see Kohler 2002: 101–108.

15. Herbert Sukopp, field notebooks.

16. In the early 1900s, the American ecologist Frederic Clements formulated his well-known concept of succession. He assumed that succession would lead to a stable final state, the natural climax community. (See Kingsland 2005: 143–147.)

Among German plant sociologists, a more modest understanding of succession developed by Reinhold Tüxen (1956) dominated. According to the latter, succession would not level out man-made existing modifications of the environmental conditions and thus lead to a natural climax. The resulting "potential natural vegetation" would be shaped rather differently on the basis of humanly shaped starting conditions.

17. My description of the site is based on the information given by the Institute of Ecology's technical assistant Wilfried Tigges in an interview at the Insitut für Ökologie on January, 27, 2005.

18. Interview, Ullrich Asmus, formerly of Institute of Ecology, via telephone, January 12, 2001.

19. Interview, Wolfram Kunick, formerly of Institute of Ecology, Berlin, November 15, 2000.

20. Ibid.

21. Interview, Asmus.

22. A humorous comment on this conflict of perspectives appeared in the form of a cartoon (apparently taken from an English-language source) in the annual report that was printed in an excursion handout for the International Botanical Congress in 1987 that was organized by the Berlin ecologists. It depicted a woman brawling with another woman who had called a "weed" something that was, to her, a "flower." ("International Botanical Congress Berlin (West) 1987, Local Tours 28. July, Plant Ecology 14, Bornkamm Kehler Weg," archive Sukopp.)

23. This map had already been used by Frank Zacharias (1972) as a basis for his representation of "isochroms." As Kunick explained in an interview, he received a copy of this map from Zacharias more or less by accident, then used it as a basis for his own survey. (Interview, Wolfram Kunick, formerly of Institute of Ecology, Berlin, November 15, 2000.)

24. This suggests that Sukopp considered the ecochore map a more scientific representation of the city, whereas he considered the earlier biotope-type map to be mainly a pragmatic construction for planning purposes.

25. For an academic biography of Sukopp, see Kowarik, Starfinger, and Trepl 1995.

26. Sukopp cooperated with Berger-Landefeld only in a few minor projects (Berger-Landefeld 1962; Berger-Landefeld and Sukopp 1965).

27. "Protokoll Fachbereichsrat 14," Nr. 75, 16.1.1974: 3; Nr 76, 6.2.1974: 9; Nr 69., 1.10.1973. Archive Technische Universität Berlin, "Protokolle Fachbereichsrat 14," box 2.

28. Between 1982 and 1984, Sukopp also acted as a director of the entire Institute of Ecology.

29. Blume was a Privatdozent in 1970 and a professor in 1971.

30. Prof. Dr. H. Sukopp, Prof. Dr. H. P. Blume: "Gedanken zur den künftigen Aufgaben des heutigen Instituts für Angewandte Botanik im FB Landschaftsbau," 18.5.71. TUB, archive Herbert Sukopp. R. Bornkamm: Vorläufiger Antrag Institut für Ökologie (3 pages), Mai 6, 1971, archive Sukopp.

31. Interview, Barbara Markstein, Ökologie und Planung, Berlin, September 9, 2009.

32. On the Chicago school of sociology and its view of the city, see Pols 2003. On its links with biological ecology and the tensions between the two fields, see Cittadino 1993. Aspects of human community building also figured prominently in the work of the Scottish botanist Patrick Geddes, who around 1900 promoted an ecological concept of cities (Welter 2002).

33. Such analogies were widespread in system ecology. Even nuclear reactors—owing to their supposed self-regulating capacity—were compared to natural ecosystems (Bocking 1997: 84).

34. For a survey of these debates from an ecological point of view, see Wächter 2003: 78–90.

35. On the development of urban ecology in the United States from pioneering initiatives to recent approaches of human ecology, see Kingsland 2005: 257.

36. For an attempt to estimate the ecological "import-export" balance of a city, see Havelange, Duvigneaud, and Denaeyer-De Smet 1975: 20.

37. Today this view remains prominent in the discourse on sustainability under the metaphoric term "ecological footprint" (Girardet 1999).

38. An early example of the use of the term *ecology* was the title of the article on the "synecology" of dry meadows (Berger-Landefeld and Sukopp 1965).

39. In an article published in 1973 he referred to some "urban-ecological research projects" that had started at the institute (Sukopp 1973: 91–92). The term also appeared occasionally in other articles; see, e.g., Sukopp 1990: 179. The co-edited excursion guide for the Second European Ecological Symposium (Sukopp 1982) featured this term in its title. From that time on, it also appeared increasingly in the titles and introductions of Sukopp's publications—see, e.g., Sukopp 1990; Sukopp and Wittig 1993.

40. Rare exceptions were the studies of the Berlin neophyte *Chenopodium botrys* from different methodological angles (Sukopp et al. 1971).

41. As Sukopp told me in an interview, Duvigneaud had become aware of his scheme through his collaboration with Heinz Ellenberg in the International Biological Program. Although Sukopp and Duvigneaud never cooperated in common research endeavors, they met later on other occasions. (Interview, Herbert Sukopp, Institute of Ecologie, Berlin, September 9, 2009.)

42. According to Cittadino (1993), human ecology in the United States suffered from a constant inability to establish itself as a homogeneous discipline. Torn between biology, sociology, and geography it was too complex to link up more coherently with the biological (mostly botanical) core of ecology to form one intellectual and institutional whole.

43. For a more detailed account of system-theoretical models of urban ecology in Germany and elsewhere, see Wächter 2003: 81–90. Wächter also considers Sukopp's approach relatively independent from these theories.

44. See, for example, the skeptical remarks about Odum in Tischler's autobiography (1992: 117); Tischler defends the "qualitative" and "historical" dimension of ecology against reduction of research to "measuring, counting, weighing." Even Ellenberg, who had promoted the concept in Germany, claimed that the qualitative nature of fieldwork implied limits to the original optimism of quantitative system-modeling (Ellenberg, Mayer, and Schauermann 1986: 31).

45. Since the early 1960s, Sukopp had managed to maintain contacts to the other side of the "Iron Curtain," notably with botanists in Poland and Czechoslovakia. He took his concept of hemerobia from the Finnish botanist Jaako Jalas. In his programmatic 1973 paper he based many of his considerations on a case study of a Finnish town. Practically oriented work on nature conservation brought Sukopp in close intellectual contact with the Dutch biologist Victor Westhoff.

46. With the exception of the work by Wittig et al., these studies gave less attention to the detailed floristic inventory that was characteristic of the Sukopp approach.

47. For an overview of urban ecology in Britain, see Goode 1989. According to Goode, naturalist-based inventories of urban nature were, after about 1980, succeeded by more planning-oriented ecological surveys. In this context Goode also mentions the stimulating role of the Second European Ecological Symposium, held in Berlin in 1980. A comprehensive survey was carried out in London in 1984; later also other cities developed strategic nature-conservation plans.

Chapter 3

1. Abgeordnetenhaus von Berlin, Parlamentsprotokolle Wahlperiode 7/99, December 7, 1978: 4377.

2. Abgeordenetenhaus von Berlin, Parlamentsprotokolle vom 20.Oktober 1988: 4894; Drucksache 10/2582. The law came into force on January 1, 1979.

3. For critiques of such an instrumentalist view of the science-policy nexus, see Jasanoff 2003, Jasanoff 1990, Evans 2006, and Bocking 2004.

4. See, for example, Mäding 1951: 19–27. The publication also includes a proposal of the author for a new "Act on Land Care."

5. Bundesnaturschutzgesetz (Bundesgesetzblatt I, 3575, December 20, 1976).

6. On the background of the Mainau conference and the Grüne Charta that was approved here, see Engels 2006: 130–135.

7. The terms "biological diversity" or, even more so, "biodiversity" became commonly used in conservation discourse in the 1970s. It includes three dimensions: species diversity, the diversity of ecosystems (or biotopes), and genetic diversity. On the rise of the concept in an American context, see Hannigan 1995: 146–161; Tacacs 1996. Whereas these authors argue that the concept is an invention of the 1960 and the 1970s, Kohler (2006) traces the core of what we now call biodiversity further back, to nineteenth-century and early-twentieth-century naturalism in North America. Although German promoters of Heimatschutz generally cherished diversity in nature and landscape, they did refer to biodiversity or species diversity in a quantitative sense, as contemporary species-protection policies do.

8. Sukopp (1971) referred to studies in the Netherlands by the Dutch biologists Victor Westhoff and Chris van Leeuwen which characterized 33 percent of Dutch species as disappeared, endangered, or extinguished.

9. A first list, published in 1958, had established 34 endangered mammal species worldwide. In its 1960 Red Data book, the IUCN established a number of 135 (McCormick 1989: 39). Around 1970 it estimated that about 550 species had become rare or were directly threatened by extinction, a number that also included plants and smaller animals. (See Olsschowi 1972: 6.)

10. At the Species Protection Seminar, the ornithologist Gerhard Thielke (1982: 47) estimated that 16 of the former 238 breeding species in Germany had been extinguished.

11. The Commission of Ecology of the IUPN maintained in 1956 that priority should be given to habitat (biotope) protection. Two years later the International Zoological Congress in London called for the protection of "representative ecosystems." In line with this proposal, the 1980 Red Data Book estimated that 67 percent of all threatened species suffered from pollution or loss of their habitats (MacCormick 1989: 40–41).

12. On the ecological debate on diversity and stability, see Kingsland 2005: 211–213.

13. Erz explicitly criticized the so-called Stein draft for a Land Care Act for its supposed ignorance of species protection as a pillar of the household of nature.

14. On the tensions that existed between the land-care approach (as represented by the Grüne Charta von der Insel Mainau, the Rat für Landschaftspflege, and many scholars from land care or landscape care institutes) and nature-conservation organizationss, see Engels 2006: 76–77.

15. Abgeordnetenhaus von Berlin, "Antrag der Fraktion der SPD und der F.D.P. über Gesetz über Naturschutz und Landschaftspflege von Berlin," Drucksache 7/1024, November 10, 1977.

16. Abgeordnetenhaus von Berlin, "Sitzungsprotokolle des Unterausschuß Naturschutz" 7/13, November 7, 1978, Bibliothek des Berliner Abgeordnetenhauses.

17. Abgeordnetenhaus von Berlin, "Sitzungsprotokolle des Unterausschuß Naturschutz," February 27, 1978, Bibliothek des Berliner Abgeordnetenhauses.

18. Besides internal conflicts within the SPD, the fall of the Stobbe Senate was spurred by the "Garski scandal." Via its land-own bank, the Senate had given loans that supported corrupt business of the construction firm of the Berlin architect Dietrich Garski. After Stobbe's demise in January of 1981, his position was taken over by the social democrat Jochen Vogel. His social-liberal coalition lost its majority in the elections for the new Abgeordnetenhaus in May.

19. The first phase of the biotope mapping survey was financed by the Senate in 1981 and 1982. Senatsverwaltung für Stadtentwicklung und Umweltschutz, "Vermerk III a D 1, 7596," archive Sukopp.

20. Up to September of 1982, seventeen separate surveys had been carried out for the Landscape Program. Six of them were directly related to the Species Protection Program. "Vermerk III a D 1, 7596," archive Sukopp.

21. Senatsverwaltung für Stadtentwicklung und Umweltschutz, "Vermerk III a A 1 – 6235/6/2/3," archive Sukopp. The work for the Environmental Atlas begun in October of 1982 under the directorship of a steering committee (Steuerungsgruppe), of which also Sukopp was a member, Senatsverwaltung für Stadtentwicklung und Umweltschutz, "Vermerk III a A 12," October 20, 1982, archive Sukopp.

22. Interview, Klaus Ermer, formerly of Senatsverwaltung für Stadtentwicklung und Umweltschutz and Rita Mohrmann, formerly of Technische Universität Institute of Ecology, Berlin, September 7, 1999.

23. See also Sukopp in Abgeordnetenhaus von Berlin, Ausschuß für Bau- und Wohnungswesen, Unterausschuß "Naturschutz," "Inhaltsprotokolle" 7/3, 1978, Bibliothek des Berliner Abgeordnetenhauses.

24. The hearings took place on February 20 and 27, 1978. See Abgeordnetenhaus von Berlin, Ausschuß für Bau- und Wohnungswesen, Unterausschuß "Naturschutz," "Inhaltsprotokolle" 7/1, 1978, Bibliothek des Abgeordnetenhauses. The following citations refer to these records.

25. Abgeordnetenhaus, 7/2, p. 2.

26. Ibid., p. 8.

27. Ibid., p. 3.

28. Ibid., p. 8.

29. Interview, Herbert Sukopp, Technische Universität, Institute of Ecology, Berlin, June 7, 1999. Berlin, December 14, 1999; Interview, Friedrich Duhme (former member of AG Biotopkartierung im besiedelten Bereich), Technische Universität München Weihenstephan, September 20, 1998.

30. For an analysis of the AG Biotopkartierung im besiedelten Bereich and its internal debates, see Lachmund 2002.

31. The work of the specialists was partly paid with grants that Sukopp received in 1979 from the Senate for Construction and Housing. See letter Senatsverwaltung Bau und Wohnen, Preissler-Holl to Sukopp, December 12, 1979, archive Sukopp.

32. This is not self-evident. Nature-restoration projects in other countries have aimed at the reconstitution of postglacial (in the Netherlands) or pre-colonial (in the Chicago Savannah region) states of nature (Helford 1999; Keulhartz 1999). Projects of nature reconstruction have been based on different reference states in different cultural contexts. On landscape reconstruction in German mining areas, see Gross 2010.

33. In addition to these categories, which were also used in German Red Lists, the Berlin list included species that were considered rare but not endangered.

34. It has been suggested that the contemporary concern for indigenous plants reflected xenophobia or a even a Nazi-obsession with native species (Gröning and Wolschke-Buhlmann 1992). As Coates (2006) has argued in a study on the US context, attitudes against non-native species have never been exclusively determined by xenophobic motives. Not all alien species were combated, and if they were this was typically not simply due to their foreignness. For a similar differentiation concerning the German case, see Uekötter 2007. As the Berlin example shows, negative attitudes toward some species or species in some contexts could even coexist with calls for preservation of others.

35. Although they had a pioneering character for later biotope-mapping projects, these studies originated in a different context. They were commissioned by the district of Kreuzberg and helped in preparing an urban architectural exposition that was planned in these parts of the city. (See chapter 4.)

36. He hoped to reduce the financial expenses that were required for the biotope mapping from 1,720,000 to 431,250 DM (Schneider and Ökologie & Planung 1979: 16, 17).

37. In his 1979 proposal, Schneider estimated the expenses for biotope mapping at DM 431,250 (Schneider and Ökologie & Planung 1979: 17). According to Ermer, this significantly exceeded the usual dimensions of what hitherto had been paid for external expert reports, and kindled much reluctance among those responsible for the financial planning within the Senate. It was only year by year that Ermer

could convince the Senate to increase his budget. Eventually Ermer was able to mobilize a lump sum to pay Sukopp for the survey. Interview Klaus Ermer, formerly of Senatsverwaltung für Stadtentwicklung und Umweltschutz, Berlin, September 9, 1999.

38. Important sites that were investigated in such reports were the famous 1920s garden-city-like settlement areas Onkel-Toms Hütte (Drescher and Stöhr 1980) and Hufeisensiedlung Britz (Drescher, Mohrmann, and Stern 1981), the large wasteland areas along the railway tracks at the Potsdam- and Anhalt- railway stations (Asmus 1980), the "southern area" (Südgelände) in Schöneberg (Asmus 1981), and Berlin's central area (Arbeitsgruppe 1982). In addition to these official reports, students' qualification theses at the institute contributed further material to the study.

39. Letter, Senator für Stadtentwicklung und Umweltschutz Klaus Ermer to Prof. Dr. Sukopp, "Biotopkartierung zum Landschaftsprogramm," March 26, 1982, archive Sukopp.

40. Senator für Stadtentwicklung und Umweltschutz, "Vermerk III Landesbeauftragter 2/7095" (by Axel Auhagen), July 7, 1982, archive Sukopp.

41. Instead of the scale of 1:5,000 that had been envisaged in the first proposals, it was only carried out at a scale of 1:20,000. This implies that information had to be presented in a much more aggregated form, and that minor differences in spatial variations could not be as precisely represented as originally intended.

42. The conventional nature of such classificatory systems and the negotiations through which they are produced has been studied by Waterton in her work on the CORINE survey of the European Union (Waterton 2002).

43. See the conservation literature cited in Sukopp 1970.

44. Biotopes or biotope types were thereby given values with ordinal-scaled variables, which were aggregated stepwise to more synthetic indicators. Weight factors adjusted the relative emphasis that was given to such criteria. First, biotope types were grouped into larger categories, on which different criteria of evaluation were applied. A comparative evaluation in each of these groups resulted in the "biotope-type value" (ibid.: 124). On this basis, the survey group further estimated the "conservation claim" and "the urgency of conservation" of the biotope types, and included the most precarious ones on a Red List of Biotope Types. The survey also encompassed an evaluation of single biotopes within their own biotope-type category. The map Value of the Biotopes that was mentioned above, represented the results of this procedure.

45. The exhibition took place at the Ernst-Reuter-Haus, a public building in the central district of Charlottenburg, on June 5 and 6, 1984.

46. According to the minutes of the Sachverständigenbeirat of October 22, 1985, the department for landscape planning at the Senate had just declared that it was

no longer able to promote any provisions that differed from the FNP 84. It now intended to develop four coordinated sub-plans: one for recreation, one for landscape scenery, one for biotope and species protection, and one for the household of nature. "Protokoll der 34. Sitzung des Sachverständigenbeirats für Naturschutz und Landschaftspflege am Dienstag, dem 22.10.1985," p. 2, archive Landesbeauftragter.

47. According to the minutes of the Council for Nature Conservation and Landscape Care, the Landscape Program was revised in unit IIIaA (Referat IIIaA) by new staff members (including six full-time positions for three years). "Protokoll der 34. Sitzung des Sachverständigenbeirats für Naturschutz und Landschaftspflege in Berlin am 22.10.1985, archive Landesbeauftragter.

48. "Protokoll der 34. Sitzung des Sachverständigenbeirats für Naturschutz und Landschaftspflege am Dienstag, dem 22.10.1985," p. 2; "Protokoll der 36. Sitzung des Sachverständigenbeirats für Naturschutz und Landschaftspflege am Dienstag, dem 11.6.1986," p. 1; "Protokoll der 37. Sitzung des Sachverständigenbeirats für Naturschutz und Landschaftspflege am Dienstag, dem am 8.7.1986," p. 2–3; "Protokoll der 38. Sitzung des Sachverständigenbeirats für Naturschutz und Landschaftspflege am Dienstag, dem 23.9.1986," pp. 2–3. See also "Pressemitteilung des Landesbeauftragten für Naturschutz und Landschaftspflege 2/86," Berlin, 30.10.86, all in archive Landesbeauftragter.

49. For example, would the tolerance of spontaneous vegetation in public greeneries save a lot of money that otherwise would have been spent on gardening.

50. Abgeordenhaus von Berlin, "Große Anfrage der Fraktion der SPD über Landschaftsplanung in Berlin," Drucksache 10/264, 11 October 1985; Abgeordnetenhaus von Berlin, "Antrag der Fraktion der SPD über die Erfüllung der sich aus dem Berliner Naturschutzgesetz ergebenden gesetzlichen Pflichten," Drucksache 10/265, October 11, 1985.

51. Abgeordnetenhaus von Berlin, Parlamentsprotokolle, 10/11, October 17, 1985, pp. 535–536.

52. Abgeordnetenhaus von Berlin, "Anträge der AL-Fraktion zu Grün- und Verkehrsmaßnahmen im Zentralen Bereich Berlin (West)" Drucksache 9/1784–1790, Mai 9, 1984. Abgeordnetenhaus von Berlin, "Anträge der AL-Fraktion zu Sofortmaßnahmen zu Naturschutz Berlin," Drucksache 9/1844–1857, June 8, 1984.

53. Abgeordnetenhaus von Berlin, "Anträge der AL-Fraktion zu Sofortmaßnahmen zu Naturschutz Berlin," Drucksache 10/69–74, Mai 28, 1985; "Anträge der AL-Fraktion zu "Grün- und Verkehrsmaßnahmen im 'Zentralen Bereich," Drucksache 10/123–128, June 18, 1985.

54. The term "bearers of public interest" refers to public institutions, e.g., administration units, districts etc. which, according to German planning law have to be invited in such procedures to voice their opinion.

55. Opinions were rejected, for example, on grounds that according to the authoritative administration unit they lacked the spatial specificity required for the plan, they required a too detailed scale of regulation, or they were obviously based on a faulty interpretation of the map plans. Other opinions were rejected for more substantial reasons, such as being considered at odds with the legally defined aims of the Landscape Program with its stipulations, or being based on assumptions considered faulty according to scientific ecological knowledge claims (Senator für Stadtentwicklung and Umweltschutz, no year).

56. For the planning principles of the new FNP 84 and its difference to the former plan see Aust 2002: 39–42.

57. They were also directly voiced during the hearing of the "Special Committee Land-Use Plan." See "Wort-Protokoll Sonderausschuß Flächnnutzungsplan—FNP 84, 8. Sitzung," Mai 6, 1988, archive Bibliothek des Berliner Abgeordnetenhauses. Besides nature-conservation groups it included e.g., the Berlin Chamber of Commerce and Industry (Industrie- und Handelskammer zu Berlin), the Land Association of Free Housing Enterprise (Landesverband Freier Wohnungsunternehmen in Berlin), the Chamber of Crafts (Handwerkskammer), the Working Group of Private Associations of Traffic (Arbeitsgemeinschaft privater Verkehrsverbände), and the Interest Community Railway Berlin (Interessengemeinschaft Eisenbahn Berlin).

58. The initiative for a Green FNP was founded in early 1987 as a strategic alliance of 17 organizations. A year later it comprised 40 organizations, among them various citizen activist groups and the Alternative Liste. On the foundation, see the statement by its spokesman, Kühn, at the hearing of the special committee FNP 84. "Wort-Protokoll, Sonderausschuss 'Flächennutzungsplan—FNP 84, 8. Sitzung, 6. Mai 1988," Bibliothek des Abgeordnetenhauses Berlin. The call for a green FNP was also promoted by AL parliamentarian Kapek in the Abgeordnetenhaus. Abgeordnetenhaus von Berlin, Parlamentsprotokolle 10/77: 4625, 4628, June, 16, 1988.

59. "Protokoll der 37. Sitzung des Sachverständigenbeirats für Naturschutz und Landschaftspflege am Dienstag, dem 8.7.1986," p. 2, archive Landesbeauftragter.

60. In 1987 the SPD opposition requested that the FNP procedure should be abandoned until the results of the new census were available. (SPD parliamentarian Meisner in the Abgeordnetenhaus, Abgeordnetenhaus von Berlin, Parlamentsprotokolle 10/58, p. 3432, September 9, 1987).

61. SPD parliamentarian Meisner in the Abgeordnetenhaus, Abgeordnetenhaus von Berlin, Parlamentsprotokolle 10/77, p. 4622, June 16, 1988. Witt (president of the Berliner Landesarbeitsgemeinschaft Naturschutz) and Siederer (a legal expert) made that same point during a hearing on the FNP. "Wort-Protokoll, Sonderausschuss Flächennutzungsplan—FNP 84, 8. Sitzung, 6. Mai 1988," pp. 221, 225–226, Bibliothek des Berliner Abgeordnetenhauses.

62. Senator for Urban Development and Environment Jürgen Starnik in the Abgeordnetenhaus. Abgeordnetenhaus von Berlin, Parlamentsprotokolle 10/77, p. 4632, June 16, 1988.

63. Senator für Stadtentwicklung und Umweltschutz 1988: 6f. Also see Aust 2002: 39.

64. On the planning of the new capital and its manifestation in Berlin's urban space, see Strom 2001.

Chapter 4

1. This did not only include ecology, but also other environment-related fields, such as the so-called Environmental Studies (Umweltforschung) (Küppers, Lundgreen, and Weingart 1978). As Küppers, Lundgreen and Weingart argue, this area was mainly an artifact of the research policy in the early 1970s. It included a range of pre-existing disciplines that often had only strategically relabeled their agendas to make them fit into that program.

2. In an essay published in 1983, the Berlin ecologist Ludwig Trepl criticized what he considered a widespread misusage of ecology as a politicized "green key science."

3. This point was made convincingly by Callon and Rabeharisoa (2008), who advanced the concept of "emergent concerned groups" in their study of French patient activism.

4. The term *political entrepreneur* has been used widely in the political sciences to describe individuals who are significantly involved in changing policies. This does not mean that they are entrepreneurs in the economic sense of the term (Schneider and Teske 1992). Howard Becker (1963) used the term *moral entrepreneurship*.

5. In 1971, the Deutsche Rat für Landespflege suggested that the voluntary Landesbeauftragten and the related commissions of nature protection and landscape care should be replaced by so-called Sachverständigenräte für Landschafspflege (Councils for Landscape Care)—"boards which operate on an policy-expert base" (fachpolitisch wirkendes Gremium) (Deutscher Rat für Landespflege, 1971: 9), and not just a political representation of lobby groups. The replacement of the former Landesstelle by a Sachverständigenbeirat was provided by the new Berlin Nature Conservation Act in 1979. Basically, it implied a strengthening of the role of external experts in that committee.

6. See "Bericht über die 22. Sitzung der Landesstelle für Naturschutz und Landschaftspflege in Berlin am Donnerstag, dem 28. April 1960," p. 7; "Bericht über die 24. Sitzung der Landesstelle für Naturschutz und Landschaftspflege in Berlin am Donnerstag, dem 24. März 1961," p. 7; "Bericht über die 30. Sitzung der Landesstelle für Naturschutz und Landschaftspflege in Berlin am Mittwoch, dem 23. Oktober 1963," p. 2, all in archive Landesbeauftragter.

7. "Bericht über die 28. Sitzung der Landesstelle für Naturschutz und Landschaftspflege in Berlin am Mittwoch, dem 17. Oktober 1962," p. 4, archive Landesbeauftragter.

8. Before the Sachverständigenbeirat was founded, the membership of the Landesstelle of Nature Conservation had already expanded from five in the first meeting (1955) to ten. These members represented different disciplines, such as ecology, forestry, or urban planning, as well, as in one case the Berlin nature-conservation organizations (one member). The new Sachverständigenbeirat that was constituted in September of 1979 had twelve members (Elvers 1979/80): R. Böcker (botany); G. Bracht (environmental protection), A. Busch (urban- and regional planning), M. Großmann (landscape planning), J. Hopp (agriculture), M. Horbert (ecology), B. Kellermann (recreation planning), Klös (zoology), and I. Maas (open-space planning). This shows a clear strengthening of the planning disciplines as well as a broader dominance of academic researchers in contrast to administration personnel. The nature-conservation organizations were no longer represented in the Council. As "recognized" nature organizations, they were now represented in a different way via the Berliner Landesarbeitsgemeinschaft Naturschutz.

9. In 1980, the president of the VBN, Hans-Jürgen Mielke (formerly of the Land Office) noted that since the beginning of the year Berlin had not had a Landesbeauftragter (Mielke 1980). Mielke called for the immediate appointment of a new Landesbeauftragter. As Sukopp describes this event retrospectively, this was due to a formal interruption of his office due to the shift to the new regulatory conditions of the 1979 Nature Conservation Act. Interview, Herbert Sukopp, Technische Universität Berlin, Institut für Ökologie, September 9, 2009.

10. Röhrichtschutzgesetz Berlin (Gesetzes und Verordnungsblatt Berlin 1969: 2520).

11. Besides various contributions to *Naturschutzblätter*, articles by Sukopp appeared in *Umweltschutzforum*, in *Berliner Ärzteblatt* (Sukopp 1976), and in *Berliner Forum* (Schneider and Sukopp 1977).

12. This distinguished Sukopp from the public intellectuals of new environmentalism in West Germany, such as Bernhard Grizmek, Heinz Sielmann, or Horst Stern, who systematically made used of the TV and other mass media to promote their political agendas. For an analysis of these media authorities and on the difference between their public practices and the (publicly often invisible) lobbying style of earlier conservationists in the FRG, see Engels 2006: 214–256.

13. "Bericht über die 28. Sitzung der Landesstelle für Naturschutz und Landschaftspflege in Berlin am Mittwoch, dem 17. Oktober 1962," p. 4, archive Landesbeauftragter.

14. Interview, Herbert Sukopp, Technische Universität, Institut für Ökologie, Berlin, September 9, 2009.

15. For example, Sukopp rejected the offer by the VBN to follow the tradition, according to which the Landesbeauftragter had also been the president of this

association. As Sukopp put it in our interview, he had considered this a problematic mixing of political and scientific competences, which "in a democratic society" should be kept separate. Interview, Herbert Sukopp, Technische Universität, Institut für Ökologie, Berlin, September 9, 2009.

16. As Runge (1998: 111) notes, the term was coined by the garden architect Mielke in 1908 and popularized by his colleague Mäding in 1942.

17. For example, in his *Denkschrift über die derzeitige Organisation von Naturschutz und Landschaftspflege in der Bundesrepublik Deutschland und Vorschläge für eine künftige Entwicklung* Walter Mrass argued that the voluntary Landesbeauftragte (often teachers and administrative personnel) lacked the competence and the resources that were required to adequately fulfill their task (Mrass 1971: 15–16).

18. In 1973, land care (Landespflege) could be studied at the polytechnics (Fachhochschulen) in Berlin, Wiesbaden, Nürtingen, Osnabrück, Weihenstephan, and the Gesamthochschule Essen (Nimmann 1973).

19. Between 1934 and 1947 the chair was hold by Heinrich Friedrich Wiepking.

20. Many advisory institutions on landscape care were also involved in research. Besides the Bundesanstalt für Vegetationskunde, Naturschutz und Landschaftspflege (Federal Agency for Vegetation Studies, Nature Conservation, and Landscape Care) in Bonn-Bad Godesberg, Walter Mrass mentions institutions that dealt with specific issues such as waters, road construction, or agriculture (Mrass 1971).

21. The Association of German Landscape Architects (Bund Deutscher Landschaftsarchitekten, BDLA) and the German Society for Horticulture and Landscape Care (Deutsche Gesellschaft für Gartenkunst und Landschaftspflege, DGGL) included both aesthetically oriented horticulturalists and land-care specialists (Landespfleger). Among the other organizations landscape architects were also organized in the Federal Association of German Engineers (Bundesverband der Dipl. Ingenieure Gartenbau und Landespflege, BDGL) and the Association of Horticultural Engineers (Bund der Ingenieure des Gartenbaus, BIG). In 1970s, the professional organizations in this field formed an umbrella organization, the Working Group for Landscape Development (Arbeitsgemeinschaft für Landschaftsentwicklung, AGL). See Hübler, Kiemstedt, and Sittel 1981: 3, note 1.

22. Before 1960 it was only possible to stipulate compulsory landscape plans, if they were integrated into special planning procedures such as plot alignment, traffic planning or water resource management (Runge 1998: 97–99). According to Runge, the Federal Building Code (Bundesbaugesetz) in 1960 included a considerable extension of the legal scope of landscape planning. Instead of the punctiform interventions that had been typical for earlier landscape engineering projects it allowed local authorities to stipulate landscape plans that covered the entire area of a development plan. At the same time, it imposed restrictions on buildings in the external area of settlements. The result was an increase in attention to landscape planning

in cities and far-reaching expectations on the increasing need for professional land-scape planners in the administration (ibid., 104/5).

23. Anonymous 1963; Mrass 1971.

24. Nimmann 1978: 894, table 2.

25. This was one of the motivations for a social-scientific analysis of the situation of land-care professionals by Hübler, Kiemstedt, and Sittel (1981). As those authors state, however, detailed statistics about the amount of unemployed landscape plan-ners was lacking, because among other reasons, there was a high degree of uncer-tainty (ibid.: 7).

26. For an overview of the various definitions in this context, see Runge 1998.

27. The main trends in this field were the "emancipational open-space planning" of the Hannover planner Werner Nohl, the Kassel School around the plant sociolo-gist Karl Heinz Hülbusch, and the open space concept of the Munich landscape architect Günther Grzmek.

28. For responses to this critique, see Arnold et al. 1979: 364–366 and Mader 1979: 366.

29. For a detailed description of these conflicts in Berlin, see Eckebrecht 1991: 382–384.

30. For example, the holder of the Munich chair for landscape ecology described landscape planning as "applied ecology" (Haber 1992). For exemplary pleas for an integration of ecology and Landschaftsplanung, see Wedek 1973, Tomasek 1978, and Bierhals 1972. See also the plea for an application of geographical "landscape ecology" in Finke 1973.

31. Besides Nohl's "emancipatory open-space theory," this view has been promi-nently promoted by the Kassel School (e.g., Karl-Heinz Hülbusch) and the Osnabrück geographer Hard. For critiques of biotope protection from this perspective, see Hül-busch 1981a, 1981b; Hard 1992, 1994.

32. Besides this, students from other programs of the Technische Universität and the Freie Universität also had to take some selective courses at the institute: accord-ing to the institute's first decennial report in 1983, courses of Sukopp's department (Fachbereich Ökosystemforschung und Vegetationsforschung) were also offered in the Technische Universität programs in urban and regional planning, and environ-mental engineering. Other courses were also open for students from the Freie Uni-versität geography and biology programs (Institut für Ökologie 1983: 41).

33. For example, in 1980, students participating in the project on open-space plan-ning in railway areas, submitted a petition together with the AL, against the Senate's plans to built a new freight station at the so-called Südgelände. AL und "Projekt Stadtökologie und Freiraumplanung am Beispiel der Bahngelände in Schöneberg am

Fachbereich 14 der Technischen Universität Berlin," letter (no address included), with a statement for the press, 13.10.80, archive Landesbeauftragter.

34. For example, Rita Mohrmann, who joined the department of landscape planning at the Senate for Construction and Environment to develop the Species Protection Program, was also the co-founder and leading member of the activist group that campaigned for the protection of the Südgelände.

35. Abgeordnetenhaus von Berlin, "Antwort auf Kleine Anfrage Salomon (SPD) vom 23.3.1973 nach Besetzung der Dienststellen im Naturschutz," Drucksache 6/1020, pp. 32–33, June 28, 1973.

36. Ibid.

37. Abgeordnetenhaus von Berlin, "Antwort auf Kleine Anfrage Rainer Papenfuß (SPD) vom 12.11.1978 über neue Mitarbeiter in den Bezirken aufgrund des neuen Naturschutzgesetztes," Drucksache 8/331, p. 3, February 13, 1979.

38. The other term, *urban development*, referred to planning schemes that would steer the physical, social and economic evolution of the city according to predefined political priorities.

39. As the Senate put it in its proposal in 1980, the foundation should facilitate cooperation among "political decision takers, public administration, and nature-conservation associations and unions." In addition to giving advice to policy makers—in which respect it parallels the work of the Landesbeauftragter for nature conservation—the foundation was also commissioned to perform practical administrative responsibilities such as commissioning research, purchasing ground for nature reserves, as well as distributing awards for outstanding contributions to nature conservation, Abgeordnetenhaus von Berlin, "Gesetz über die Stiftung Naturschutz Berlin," Drucksache 8/575, November 7, 1980; Parlamentsprotokolle 8/50, December 3, p. 1981.

40. In April of 1979 he presented a co-authored report on the future landscape program (Ermer, Kellermann, and Schneider 1979). In December of 1979, his firm presented a more detailed methodological proposal for the future Species Protection Program (Schneider and Ökologie & Planung 1979).

41. For a self-presentation of the group, see Lamp 1983. In 1978 a similar group had been formed by landscape planning graduates in Hannover (Hübler, Kiemstedt, and Sittel 1981: 7).

42. On the development of environmentalism in West Germany, see Engels 2006; Chaney 2008; Brüggemeier and Engels 2005; Bergmeier 2002. Specifically on environmental organizations, see Markham 2008.

43. This development was not unique to Berlin. It could be observed all over Germany. The implications of the "institutionalization" of the environmental move-

ment have been an issue of controversial debate among movement scholars. Whereas some have posited that institutionalization would lead to a "sclerosis" and end of the movement, Dieter Rucht and Jochen Roose (2001) have used data from Berlin to argue that the number and size of environmental groups did not at all decline in the 1990s, even if their work may have become less visible to the public. On the basis of the *Stattbuch*, a local directory of alternative groupings, they estimated the number of environmental groups in Berlin to have developed from 139 in 1989 to 115 in 1995 to 120 in 1999.

44. After 2000, Berlin witnessed a debate on connections of the VBN to right-wing organizations that isolated the association from other parts of the Berlin environmentalist scene. Bernd Schütze (2005) accused members of the VBN of personal involvement of leading members in National Socialism and for having actively contributed to the ignorance of Jewish conservationists in the collective memory of the association. In additional, internal struggles led to a relative decline of public visibility of this organization.

45. This was due mainly to the engagement of Heinrich Weiß, who also noted the huge cultural differences between the conventional and often hierarchical organizational style of the VBN, and the informal, politicized activism of the new Bürgerinitiativen (Weiß 1978: 382). For the VBN, and other conventional nature conservation, he diagnosed a "crisis of identity" (ibid.: 382) that forced them either to develop a new position or to become politically irrelevant.

46. An example of the first is the Berliner Arbeitsgemeinschaft Igelschutz (Berlin Working Group for Hedgehog Protection), which split from the VBN in the early 1980s.

47. When the BLN was founded in 1979 it comprised ten associations: Agintha, Alliance of Bird Friends (Agintha, Bund der Vogelfreunde 1875 e.V.); the Berlin section of the DBV; the German Society for Herpetology and Terrariums (Deutsche Gesellschaft für Herpetologie und Terrarienkunde, Stadtgruppe Berlin e.V.); Entomological Association Orion (Entomologischer Verein Orion) (until 1986); Community for the Protection of Berlin's Trees (Gemeinschaft zum Schutz des Berliner Baumbestandes e.V.); the Berlin Society of Friends of Nature Research (Gesellschaft Naturforschender Freunde zu Berlin); the Society for Environmental Protection (Gesellschaft für Umweltschutz e.V.); the Kohlhas-Alliance for the Protection of Nature and Landscape in Albrechts Teerofen und Kohlhasenbrück (Kohlhasbund zum Schutz von Natur und Landschaft in Albrechts Teerofen und Kohlhasenbrück); Tourist Association Friends of Nature (Touristenverein Die Naturfreunde e.V.) (Source: "Protokoll der Gründungsversammlung der Berliner Landesarbeitsgemeinschaft Naturschutz am 12. Dezember 1979 im Konferenzraum des Berliner Buchhandelszentrums," Archive Berliner Landesarbeitsgemeinschaft Naturschutz, henceforth cited as archive BLN.) According to the 1981–82 annual report, the BLN already had twelve member organizations, which together amounted to 10,000 individual members. (Source: "Geschäftsbericht 1981/82," archive BLN.)

48. According to Cornelsen (1991: 55), the number of members doubled between 1983 and 1989 from 80,000 to 160,000. Only 10 percent of these members were active; others were just supporting the association with their membership fees.

49. For example, they blamed the Nature Conservation Ring (Naturschutzring), which hitherto had dominated the nature-conservation policy in Germany, for its positive stance toward nuclear energy and the ignorance against problems of species decline. Nature conservation, as the founders of the BUND maintained, had excessively been guided by the interests of nature users such as hunters, fishers and tourism associations. See Cornelsen 1991: 22 and Engels 2006: 311–322.

50. This was the impression of Johannes Schwarz, who at that time was a member of the DBV and later became the managing director of the BLN. Interview, Klaus Schwarz, Senatsverwaltung für Stadtentwicklung, Naturschutzamt, October 19, 2008.

51. This is noted in the annual report of the BLN 1990–91 (archive BLN).

52. For a self-presentation of Bürgerinitiativen in West Berlin see the compilation of articles that the magazine *Zitty* had published in 1977 and 1978. The very title "Bürgerinitiativen: Model Berlin" reflects the self-understanding of the alternative culture of the city as a center of gravitation for new political protest.

53. The origin of the term is unclear. Since the 1960s various groups that promoted specific political issues had been formed under the name Aktionsgruppen. One early use of the term was in the Sozialdemokratische Wählerinitiative of writers who supported the SPD in the federal elections. Andritzki and Terlinde also quote two 1968 articles that used the term Bürgerinitiative not to refer to specific political groupings but to refer to the public engagement of citizens in general. See Andritzki and Wahl-Terlinden 1978.

54. The work of the BBU suffered, on the one hand, from the lack of coordination among the often short-lived and loosely organized member associations. On the other, conflicts around the leader of the organization H.H. Wüstenhagen escalated in the late 1980s and further weakened the organization. On the development of the BBU, see Markham 2008: 101–103 and Engels 2006: 332–338.

55. Meyer-Tasch 1985; Andritzki and Wahl-Terlinden 1978.

56. In the 1980s and the 1990s, names of the initiatives and contact addresses were regularly published by the local environmentalist journal *Grünstift*.

57. Examples were AG Gleisdreieck für einen Naturpark auf dem Gleisdreieck and the various "fields" initiatives, such as BI Heiligensee "Rettet die Felder" and BI Mariendorf "Rettet die Mariendorfer Feldmark."

58. Examples include BI Energieplanung und Umweltschutz, BI gegen Fluglärm, Berliner Aktionsgemeinschaft gegen das Waldsterben, and BI gegen Müllexport in die DDR.

59. For a detailed political science analysis of this conflict, see Hager 1995. The plan was based on estimations of the Senate that a drastic expansion of the energy production capacity was needed in Berlin. The decision for this site (which was actually opposed by the FDP Senators within the Senate as well as the SPD governed district of Spandau) followed the dismissal of an alternative site next to the existing power plant Oberhavel. Since that site was closer to residential areas, the Senate feared that citizens would prevent the project by litigation on the basis of a new emission law (ibid.: 103–107).

60. Interview, Herbert Sukopp, Technische Universität, Institute of Ecology, Berlin, September 9, 2009.

61. *Spandauer Volksblatt*, quoted by Hager (1995: 119).

62. These examples are taken from the chronicle of the BIW of its former activities (Bürgerinitiative Westtangente 2010).

63. When members of the Maoist "K-groups" (communist groups), which increasingly embroiled themselves in this conflict, also advocated more violent forms of conflict, this led to tensions within the majority of the BI members, who tended to prefer non-violent forms of protest or at best "violence against things." As Hager (1995) has shown, this became a critical issue in the Action Alliance, which in the 1976 opposed the power plant in the Spandauer Forst.

64. The protest movement, which was also concerned with air pollution and noise, did not prevent construction of a power plant at an alternative site. Some local activists pursued litigation; see Hager 1995: 139–171.

65. Thus, when residents took legal action against a new hotel at the Dörnbergdreieck, the court turned down the procedure, because it argued that that the residents had no right to sue: one of the claimants was the owner of a house in the second row and therefore was not considered to be directly affected by the hotel. The Higher Administrative Court also denied a first row's resident's right to sue, because she had only rented her apartment and thus did not qualify as its owner (*Der Tagesspiegel* 1986).

66. On the resistance of the Higher Administrative Court Berlin in the first two court cases, see Bilzer, Ormond, and Riedle 1990: 79.

67. The main bone of contention was the so-called subsistence clause, which stipulated that that Nature Conservation Associations could only sue if no other rights for suing existed. The Federal Administrative Court interpreted this rule differently than the Berliner Higher Administrative Court. See Bilzer, Ormond, and Riedle 1990: 79.

68. According to Bilzer, Ormond, and Riedle (1990), the first successful collective litigation was against a sewage sludge dump site. In ensuing years, the BLN made regular use of this legal procedure.

69. The Federal Building Code of 1960 only provided that plans had to be made public and that on this basis could raise objections. In 1971 the Act for the Stimulation of City Building (Städtebauförderungsgesetz), which applied only for special urban renewal projects, introduced a form of public inquiry at which citizens could raise objection. In 1976 the revised version of the Federal Building Code introduced a two-level procedure of participation: an Early Citizen Procedure (frühzeitige Bürgerbeteiligung) in the early phase of the plan construction, and a second one, before the final draft was formally enacted.

70. Federal Nature Conservation Act (Bundesnaturschutzgesetz) § 29; Berlin Nature Conservation Act (Naturschutzgesetz Berlin) § 39. The details were regulated in a special decree. In 1982 and 1983 three organizations received funds: the Association for the Protection of the German Forest (Schutzgemeinschaft Deutscher Wald), the VBN, and the BLN. Four organizations were approved: the Association for the Protection of the German Forest (VBN), the German Association for Bird Protection (Deutscher Bund für Vogelschutz), the German Association for Herpetology and Terrariums (Deutsche Gesellschaft für Herpetologie und Terrarienkunde), and BUND. The BLN, which was an umbrella organization of four of these associations, was not itself approved, see: Abgeordnetenhaus von Berlin, "Antwort auf die Kleine Anfrage von Dr. Meisner (SPD) über Naturschutzogranisationen in Berlin," Drucksache 9/1397, pp. 25–26, October 23, 1983.

71. "BLN Jahresbericht 1981/82," p. 2, archive BLN.

72. As the ornithologist and BLN-member Klaus Witt stated in 1982, the relation between the Landesbeauftragter and the BLN was not free of tension. In this context, he referred particularly to conflicting views regarding the Jagen 86, the Teufelsfenn (both in the Grunewald) and the renovation of the Fritz-Schloß-Park. BLN, "Geschäftsbericht 1981/82," archive BLN.

73. Andreas Kalesse, an activist associated with the VBN and the former BI Save the Field of Gatow told me that in the late 1970s and the early 1980s some journalists in Berlin were always ready to take up his statements or suggestions. Interview, Andreas Kalesse, Berlin, July 1, 2001.

74. See the documentation of *Zitty* articles from 1978 and 1979 in Beer and Spielhagen 1978.

75. On the mobilization of expertise on energy policy during the conflict around the power plant in the Spandauer Forst, see Hager 1995: 87–89.

76. The paper by Knapp, an official at the Senate for Building and Housing, suggested that the Senate should use or even make up the need to search for remaining World War II munitions at the Südgelände to justify the beginning of clearing measures. By such clearing he wanted to render the argument obsolete that the site presented an ecologically valuable nature. "Knapp-papier," copy in possession of the author.

77. Abgeordnetenhaus von Berlin, "Kleine Anfrage von Peter Boroffka (CDU) über Teltowkanal und Naturschutz," Drucksache 8/75, July 9, 1979; "Kleine Anfrage von Christian Kayser (FDP) über Naturschutz- und Landschaftsplanung beim Ausbau des Teltow-Kanals," Drucksache 8/764, October 8, 1980; Swinne (FDP) in Berliner Abgeordnetenhaus, Ausschuß für Gesundheit und Umweltschutz, 1980/17, Mai 28, 1980.

78. The AL had already participated in the elections of 1979. At that time it had failed to pass the 5 percent quota that was required for getting into the Abgeordnetenhaus. Only at the district level did it become represented in four chambers, among them the district of Kreuzberg.

79. Abgeordnetenhaus von Berlin, "Anträge der AL-Fraktion zu Grün- und Verkehrsmaßnahmen im Zentralen Bereich Berlin (West)," Drucksache 9/1784–1790, Mai 9, 1984; "Anträge der AL-Fraktion zu Sofortmaßnahmen zu Naturschutz Berlin," Drucksache 9/1844–1857, June 8, 1984; "Anträge der AL-Fraktion zu Sofortmaßnahmen zu Naturschutz Berlin," Drucksache 10/69–74, Mai 28, 1985; "Anträge der AL-Fraktion zu Grün- und Verkehrsmaßnahmen im Zentralen Bereich," Drucksache 10/123–128, June 18, 1985.

80. The Culture League was founded in 1949. Next to nature-conservation and homeland-protection groups it comprised groups that specialized in literature, art, philosophy, history, philately, photography, and folk handicrafts. The organization was one of a number of mass organizations, similar to the state youth organization (Free German Youth, FDJ), the united unions (Free German Alliance of Unions, FDGB), association of the popular theaters (Volksbühnen), women organizations (Democratic Women Organization). According to Behrends et al. (1993: 32), many of the nature and homeland groups pre-dated the foundation of the Kulturbund and only reluctantly integrated into the new institutional framework.

81. On the role of Eingaben (petitions) as a major accepted form of environmental protest in the GDR, see Nölting 2002: 68.

82. On this policy shift, see Nölting 2002. One of the main ways in which the GDR sought to prevent the public discussion of environmental problems was a decision by the Council of Ministers (Ministerrat) in 1982, which forbade the publication of any data on the environmental situation. Collecting and publishing such data, was thereby criminalized (ibid.: 70–71).

83. Nölting (2002: 218) describes the work of the urban ecology group in Berlin-Köpenik, which organized campaigns for courtyard-greening, composting, bicycle traffic. It even organized an annual "Eco-Fair," which developed into a de facto demonstration against the SED regime.

84. Thus, in 1987 two members of the Environmental Library were arrested. This event which received much national and international attention became one of the landmark events for the burgeoning opposition movement in East Berlin (Nölting 2002: 78).

85. The GNU itself was dissolved shortly afterwards. Its remaining members joined other nature-conservation organizations, notably the Nature Protection League (Naturschutzbund), which had formerly been created as a merger of the West German League for Bird Protection (Bund für Vogelschutz) and the East German Nature Protection League. On the end of the GNU and the formation of the Green League see (Behrends et al. 1993).

86. On the reorganization of the East Berlin nature administration, see Pobloth 2008: 271–281.

87. The Grüne Liga was one of the early proponents of a park at the site of the former wall. (See Grüne Liga 1992.) A much smaller version of that park was later realized under the auspices of a landscape gardening office. Because of a mistake, protected areas in the GDR were not included in the new Berlin in 1990. In 1992 some of these areas were provisionally secured by means of a Sicherstellungsprogramm, and later some achieved formal status. This process was critically accompanied by East Berlin nature conservationists (see, for example, Anonymous 1992). On the tensions and convergences of East German and West German environmentalism, see Rink 2002.

Chapter 5

1. Since 1956 West Berlin witnessed the development of a large system of highway tangents. The further extension of the system was a central goal of the Flächenutzungsplan 1965 (Land Use Plan 1965) that was enacted in 1970. Until 1969, 22.3 kilometers of highways had already been constructed. At the base of this policy, was the attempt to make the car the dominant form of mobility in the city. As Ciesla (2008) has maintained, this policy was due in part to the fact that the car was seen as a symbol of American lifestyle.

2. In February of 1981 the Senate presented a proposal for a decision to cancel the "planning and construction of the so-called highway Westtangente": Abgeordnetenhaus von Berlin, "Vorlage—zur Beschlußfassung—über Billigung der Richtlinien der Regierungspolitik für den Rest der Legislaturperiode," Drucksache 8/704, February 20, 1981. A few days later, the Abgeordnetenhaus supported this decision: Abgeordnetenhaus von Berlin, Parlamentsprotokolle 8/48, February 26, 1981, p. 2008.

3. Abgeordnetenhaus von Berlin, "Vorlage zur Beschlußfassung über die Veränderung von Eisenbahnanlagen im Südbereich von Berlin." Drucksache 8/36, Mai 11, 1979.

4. At January 24,1980 the Senate and the East German Reichsbahn signed a contract concerning the future use of the area. The Reichsbahn relinquished its rights to use the Postdamer und Anhalter Bahnhof. At the same time, the administration authority on the site was transferred from the Administration of the former Fortune of the German Imperial Railway (Verwaltung des ehemaligen Reichsbahnvermögens) to

West Berlin. The Senate committed itself to construct a new freight station, which would be maintained by the Reichsbahn.

5. As the architect Jaap Engel (1983) stated, IBA-members did not participate in the core group that developed the plans.

6. The closeness and even corruptness of these networks became visible through a number of building scandals throughout the 1980s. In 1974 the Berlin Senator for Finance had to resign because he had agreed to a risky bailout for the construction of the Steglitzer Kreisel, a high-rise office building with shopping mall and underground station. When it became public in 1981 that the Senate had given a bail-out to the construction firm Garski, which had already been almost bankrupt by this time, this led to the demise of the social-liberal Senate of Dietrich Stobbe.

7. The term was introduced during the debate on the IBA (Maas 1982). The term was also used by the BI Westtangente (BI Westtangente 1983). In 1984 the Alternative Liste published a booklet in which it presented its concept of the Green Center (Alternative Liste Berlin 1984). In the ensuing debate in the Abgeordnetenhaus, members of FDP and CDU also embraced the concept of the Green Center, even if they disagreed with the specific claims of the motion of the AL. Abgeordnetenhaus von Berlin, Parlamentsprotokolle 9/69, June 14, 1984.

8. Institute of Ecology, "IDÖ-Beschluß im Umlaufverfahren vom 23.1.1986," archive Landesbeauftragter); letter Prof. Dr. Blume to Landesstelle für Naturschutz, March 3.1980 (archive Landesbeauftragter).

9. Abgeordnetenhaus von Berlin, "Antrag der AL-Fraktion über Zentraler Bereich - NSG Lützowplatz. Ruderralfläche östlich des Lützowplatzes," Drucksache 9/1789, Mai 9, 1984.

10. Stadtökologisches Symposium "Erklärung zum Schöneberger Südgelände (als Naturpark erhalten)," September 30, 1980 (archive Landesbeauftragter).

11. For example in spring 1979 the landscape planners Andreas Brand and Hanna Köstler and the action artist Ben Wargin initiated a "wildlife plant garden" at the Straße des 17. Juni. The garden existed until it was removed, presumably accidentally by garden maintenance workers in the fall of the same year (Auhagen 1979b).

12. Such an aesthetic can be found in various publications of the proponents of wasteland conservation. For an example, see Bürgerinitiative Südgelände 1985. The new aesthetic of spontaneous vegetation and wastelands was not restricted to West Berlin. See, for example, the essays in Andritzky and Spitzner 1981.

13. Alternative Liste and Projekt "Stadtökologie und Freiraumplanung am Beispiel der Bahngelände in Schöneberg am Fachbereich 14 der Technischen Universität Berlin," October 13. 1980, Lbftr. 6224–58–369, archive Landesbeauftragter.

14. Interview, Norbert Rheinländer, Bürgerinitiative Westtangente, Berlin, July 7, 2007.

15. This discursive motive was not entirely new. It can be found in earlier forms of romantic celebration of ruins (see Simmel 1965). For a more recent approach, which relates similar ideas to industrial ruins, see Edensor 2005.

16. The number was mentioned in a letter of the BI Schönberger Südgelände to the Deutsche Gesellschaft für Herpetologie und Terrarienkunde, Berlin, November, 28, 1982, archive BLN.

17. See Auhagen 1980; letter H. P. Blume to Landesstelle für Naturschutz und Landespflege, Abt. III., March 4.1980, archive Landesbeauftragter; letter Sukopp to Gartenamt Tiergarten und Senator for Bau und Wohnungswesen III a C. "Unterstützung des Antrags auf Unterschutzstellung" (no date), GA-Tiergarten file Dörnbergdreieck, March 24, 1980); letter Gartenamt Tiergarten to Senator für Bau und Wohnungswesen, Kubla (III a C 31), archive Bezirk Berlin Tiergarten, Gartenamt.

18. "Protokoll der 5. Sitzung des Sachverständigenbeirates für Naturschutz und Landschaftspflege am Dienstag, dem 8. Juli 1980 im Hause der Senatsverwatltung für Bau- und Wohnungswesen," archive Sukopp.

19. Letter (II a B 1–6133-II-5–222-Wuthe an IIIa), archive Gartenamt-Tiergarten. "Antrag der AL-Fraktion zu Zentraler Bereich - NSG Lützowplatz. Ruderalfläche östlich des Lützowplatzes," Abgeordnetenhaus von Berlin, Drucksache 9/1789, May 9, 1984.

20. Personal communication, Bernd Latzel (former member of BI Gleisdreieck), November 20, 2009.

21. The significance of the "biotope" as a symbolic marker of left-wing protest is also exemplified by students, who squatted a building of the Freie Universität (during their protest against a study reform) and who, alluding to the Lennédreieck, called the building a "second autonomous biotope" (*Die Tageszeitung* 1988).

22. Müller, Stühler, und Muhs (Senate for Construction and Housing) at meeting of the Sachverständigenbeirat "Protokoll der 33. Sitzung des Sachverständigenbeirates für Naturschutz und Landschaftspflege am Dienstag, dem 25.06.1985," p. 5.

23. Abgeordnetenhaus von Berlin, "Vorlage—Konzept der Bundesgartenschau Berlin," Drucksache 10/2039, May, 5, 1988.

24. Abgeordnetenhaus von Berlin, "Vorlage—zur Kenntnisnahme—über die neue Konzeption der Bundesgartenschau Berlin 1995," Drucksache 11/453, November, 16, 1989.

25. On these grounds ecologists had already critized the BUGA that had taken place in 1985 in the Southern district of Neukölln. Although this exhibition embraced a broader approach of landscape design that moved beyond the purely horticultural character of previous IBA's, ecologists would have preferred to preserve the agricultural areas that had existed at that site (Sukopp and Launhardt 1985). When first

plans for the BUGA were discussed, members of the Sachverständigenbeirat maintained, that the design was based on a romantic imagery, which emphasized the opposition of nature and city, and thereby did not do justice to the real problems of the big city. "Niederschrift über die 75. Sitzung der Landesstelle für Naturschutz und Landschaftspflege in Berlin am Dienstag, dem 27. September 1977," p. 6.

26. "Protokoll der 48. Sitzung des Sachverständigenbeirats für Naturschutz und Landschaftspflege am Dienstag, dem 19.01.1988," p. 3.

27. Abgeordnetenhaus von Berlin, Parlamentsprotokolle 11/77, p. 4627, June 16, 1988.

28. On the architectural debates on how to develop the new center, see Hertweck 2010. This debate developed mainly among architects who wanted to develop the new Berlin on the basis of the historical structure of block building, and others who opted for new high-rise development. Another issue that accompanied the development of the inner center concerned the place of collective memory of the history of National Socialism and the holocaust in the newly developing cityscape (Ladd 2000; Till 2005).

29. This process met the resistance by ecologists and landscape planning experts. In March of 1990 the German Council for Land Care published a Berlin Declaration (Berliner Erklärung) which was meant to prevent possible development projects at the Südgelände and Gleisdreieck. Deutscher Rat für Landespflege, "Berliner Erklärung, Erhaltung naturnaher Biotope," archive Landesbeauftragter. The same request was made again in September of 1991 by the German Society for Ecology and the AG Biotopemapping in Developed Areas. "Pressemitteilung. Ökologen rufen zur Erhaltung städtischer und außerstädtischer Ökosysteme im Großraum Berlin auf"; "Petition an den Berliner Senat," by Kongreß zur Biotopkartierung im besiedelten Bereich, both in possession of the author.

30. Verordnung zum Schutz der Landschaft des Schöneberger Südgeländes und über das Naturschutzgebiet Schöneberger Südgelände im Bezirk Schöneberg von Berlin, March 3, 1999 (available at http://www.stadtentwicklung.berlin.de).

31. The Citizens' Committee Johannisthal (Bürgerkommitee Johannisthal) and the Interest Community Airport Johannisthal (Interessengemeinschaft Flugplatz Johannisthal) were founded in the aftermath of the fall of the GDR. They aimed at representing local residents, and, with the Interessengemeinschaft, formed part of the broader of culture of citizens' protest, which existed in East Germany during and after the fall of the GDR. In 1991 the Interest Community Airport Johannisthal submitted a first proposal for a park on the area, which had been elaborated in cooperation with local landscape-planning offices (Interessengemeinschaft 1991). As with many East German citizens' groups, it dissolved very soon after unification, and only few people remained active in the Citizens' Committee. Interview, Andreas Nestke, former Interessengemeinschaft Flugplatz Johannisthal, Berlin, February 2, 2001.

32. In 2002 the inner area of the park became designated a nature reserve. Verordnung zum Schutz der Landschaft des ehemaligen Flugfeldes Johannisthal und über das Naturschutzgebiet ehemaliges Flugfeld Johannisthal im Bezirk Treptow-Köpenick von Berlin vom 4. September 2002 (available at http://www.stadtentwicklung.berlin.de).

33. Gutachterverfahren ehemaliges Flugfeld Johannisthal-Adlershof, archive Landesbeauftragter.

34. The "exchange of letters" (Notenwechsel), through which the Berlin and the Eisenbahn-Bundesamt had agreed upon this procedure raised much public attention in that year. Later a leaked copy of that letter was published by the activist group AG Gleisdreieck; it is available at http://www.berlin-gleisdreieck.de. The author has attended a public meeting at which these issues were debated among different actors: Stadtteilausschuss Kreuzberg e.V. (Organizer), "Wird das Gleisdreieck zugebaut?" Berlin, Gemeindesaal der Christuskriche, Juli 16, 2001.

35. This was the case with Schaumann's plan (Schaumann, no year) for the Gleisdreieck and the student project Gleisdreieck (Behrens, Hühn, and Karbowski 1982).

36. Interviews, Rita Mohrmann, BI Südgelände, Postdam, November 1, 2001; Hans Göhler, Grün Berlin, January 8, 2001, Lutz Spandau, Allianz-Stiftung (supporting foundation), via telephone, January 17, 2001. The following history is partly based on these interviews.

37. "Workshop IV, Protocol of the meeting on September 9, 1993," archive Landesbeauftragter.

38. Of 30 landscape planners who had been invited for participation about 27 submitted a proposal. Six of these landscape planning offices were then short-listed for participation in the further process of selection and discussion. The procedure was organized by the Berlin Adlershof Reconstruction Society (Berlin Adlershof Aufbaugesellschaft) and by the Berlin landscape planning office Becker, Gieseke, Mohren, Richards. Hearings during which the proposals were discussed took place on January 17,18, February 16, April 3, and May 13, 1996. Kolloquium Gutachterverfahren Flugfeld Johannisthal-Adlershof, 1.-4, Protocols, all in archive Landesbeauftragter. See also Becker 1996.

39. For example, a French architecture group called Agence TER suggested encircling the central nature reserve with a ring of wood. This was not only at odds with the idea, widely shared among the hearing's participants, that "wideness and openness" should be made a central feature of the park. It was also criticized by Kowarik, who participated as an expert advisor in the hearings. From the perspective of urban ecology, the creation of a wood would have diminished the ecological value of the area. Moreover, as Kowarik pointed out, various woods already existed in the neighborhood of the site. 2. Kolloquium Gutachterverfahren Flugfeld Johannisthal-Adlershof, February 16, 1996, minutes from February 22, archive Landesbeauftragter.

40. The concept only provided general guidelines that had and partly still have to be concretized in further steps.

41. Interview, Holger Brand and Karin Heintze, Senatsverwaltung für Stadtentwicklung und Umweltschutz, Berlin, January, 29, 2001.

42. Interviews Rita Mohrmann, BI Südgelände, Postdam, November 1, 2001; Göhler, Grün Berlin, January 8, 2001, Lutz Spandau, Allianz-Stiftung (supporting foundation). telephone interview, January 17, 2001. The following history is partly based on these interviews.

43. Interview, Klaus Hartmann, Gruppe Odius, Berlin, January 19, 2001.

44. What follows is partly based on interviews with participants of the process of negotiating the management scheme: Interviews, Holger Brandt, Berlin, January 1, 2001; Karin Heinze, Berlin, January 1, 2001; Langer, Berlin, January 23, 2001; Ingo Kowarik, formerly of ÖkoCon, Berlin, February 2001; Christoph Saure, freelance biologists consultant, Berlin, January 15, 2001; Hanna Köstler, vegetation studies consultant, January 22, 2001.

45. Interview, Hannah Köstler, Berlin, January 22, 2001.

46. On parks in Vienna, see Rotenberg 1995. On nature and urbanization in Toronto, see Keil 1995. On the symbolism of ecologically designed landscapes in French science cities, see Wakeman 2000.

Chapter 6

1. Bundesnaturschutzgesetz § 8; Naturschutzgesetz Berlin § 14. A first version of what later became known as the impact regulation had been introduced in 1966 with the Land Care Act (Landespflegegesetz) of the Land Baden-Württemberg (see Deiwick 2002: 11).

2. Grundsätze für die Prüfung der Umweltsverträglichkeit öffentlicher Maßnahmen des Bundes, Bekanntmachungen des BMI vom 12.9.1975 (U 11–500 110/9).

3. Council Directive 85/337/EEC of 27 June 189 on the Assessment of the Effects of Certain Public and Private Projects on the Environment.

4. Following European regulation in 2001 (Directive on Strategic Environmental Assessment, 2001/42/EG), the German construction law in 2004 introduced the so-called strategic environmental assessment that also applies for development plans in the city: Europaanpassungsgesetz Bau (Bundesgesetzblatt 1, 2004, p. 1359); revision Baugesetzbuch § 13a (Bundesgesetzblatt 1, 2004, p. 2414).

5. Naturschutzgesetz Berlin, § 39.

6. Interview, Louise Preissler-Holl, former landscape planner at the Senatsverwaltung für Stadtentwicklung und Umweltschutz, November 25, 2003.

7. In August of 1985 the garden authority of the district of Tiergarten had claimed that the investors had to pay a compensation fee of 360,000 DM. This was based on a calculation of the replanting of a similar vegetation area at another site. But since Berlin had not passed a decree on the determination of a compensation fee, the legal basis was missing (letter Gartenamt Tiergarten, Schaaf to Senatsverwaltung für Stadtentwicklung und Umweltschutz, Abt. III, August 22, 1985, archive Bezirksamt Berlin Tiergarten, Grünflachenamt). Finally, the creation of a new park in Moabit (Bremer Strasse/Wiglebstrasse) was stipulated in the construction permit, letter SenBauWohn-Wittwe an BLN Witt, Mai 21, 1986, archive Bezirk Berlin Tiergarten, Grünflächenamt.

8. During the conflict around the new highway in Tegel, Auhagen (1979a) complained that the representatives of nature conservation had been deliberately excluded by the choice of a simple planning procedure instead of the usual Planfeststellungsverfahren, because the latter would have implied higher demands for participation.

9. Senatsbeschluss, December 18, 1991.

10. An underground U-bahn, S-bahn, and regional train station at the Postdamer Platz (SW of the Tiergarten) and an intercity railway station at the site of the former S-Bahn station Lehrter Strasse (in the north of the park).

11. For more detailed analyses, see Blais and Spars 1990 and Schwambach 2006. I base my account here partly on interviews with planners and consultants who were involved in this process.

12. Eisenbahnbundesamt, Deutsche Bahn, Senat von Berlin, "Verkehrsanlagen im Zentralen Bereich". Unterlagen zur Planfeststellung. Neumann + Hoffmann, Bd. 19, Landschaftspflegerischer Begleitplan, März 1994. In Archive of the Senat für Bau- und Wohnungswesen Berlin.

13. Potsdamer/Leipziger Platz. "Begründungsentwurf für Bebauungspläne," April 21, 1992. Archive Bauamt Tiergarten.

14. Investitionserleichterungs- und Wohnbaulandgesetz, April 22, 1993. Bundesgesetzblatt I 446. The main intention behind this law was to facilitate infrastructural renewal in the former East, by for example, restricting the involvement of many different agencies and citizens in planning procedures.

15. Eisenbahnbundesamt, Deutsche Bahn, Senat von Berlin—Senatsverwaltung für Bau- und Wohnungswesen, "Verkehrsanlagen im Zentralen Bereich. Unterlagen zur Planfeststellung," Berlin, 1994, volumes 19, 22, 34–39. Archive Senatsverwaltung Berlin für Bau- und Wohnungswesen, now part of Senatsverwaltung für Stadtentwicklung und Umwelt.

16. Ibid., vol. 1, 529. The EIA also established environmental effects of the project that could not easily be prevented by such measures, among them the considerable

encroachment on "flora, fauna and biotopes." In order to compensate for these inflictions, the Landschaftspflegerischer Begleitplan opted for the unsealing of surrounding areas and the planting of new trees. It was with the creation of a nature park at the Südgelände, which at that time was still formally designated as a railway area, that the railway company was meant to compensate for the loss of nature at the Tiergarten.

17. Interviews, Christian Rau, Amt für Umwelt und Naturschutz Tiergarten, Berlin, September 13, 2003; Lydia Ebermann, Beratungsgesellschaft für Stadterneuerung und Modernisierung (BSM), Berlin, September 15, 2003.

18. For similar criticism of the expertise on climate, see Berliner Landesgemeinschaft Naturschutz and Anti-Tunnel AG 1994: 49.

19. In the "wider exploration area," the EIS drew mainly on the results of the biotope mapping for the new Artenschutzprogramm in 1993, which was of a much larger scale than surveys usually conducted for EIS (Berliner Landesgemeinschaft Naturschutz and Anti-Tunnel AG 1994: 52).

20. These methods either seek to define a functional-equivalent on the basis of the value of biotopes (Biotopwertverfahren) (or other aspects of the ecosystem or the landscape image) or directly by reference to the financial cost necessary to remedy the encroachment (Wiederherstellungsansatz). For a discussion of these methods, see Bruns 2007.

21. Examples of standardization efforts that preceded the development in Berlin were the approaches taken in Hesse (in 1991), in Hamburg (in 1991), North-Rhine Westphalia (in 1987), and of the Landschaftsverband Rheinland (in 1991). When Auhagen (1994) developed his approach for Berlin he based his own work on a critical discussion of this earlier work. As Bruns (2006) established, the continuous debate on the impact regulation has not led to any consensus over method. Approaches differ considerably between different Länder or even smaller political bodies. Also within the planning community no consensus has been reached about a standard method for encroachment evaluation.

22. See the Senatsverwaltung für Stadtentwicklung und Umweltschutz, Gruppe F, 2001: 2 for a summary of these demands given by Auhagen in retrospect (Senatsverwaltung für Stadtentwicklung und Umweltschutz and Gruppe F 2001).

23. For a discussion of these methods, see Bruns 2007.

24. For critical evaluations of the Berlin situation, see Rink, Jung, and Marks 1995 and Schröder and Barsig 2004. For a critical evaluation of the implementation of compensation and substitutions measures, see Hube, Runge, and Sukopp 1998: 32–34.

25. For the German Working Group of the Länder on Nature Protection and Land Care (Länderarbeitsgemeinschaft Naturschutz und Landschaftspflege) (LANA).

26. On the political and expert debates that followed the report, see Bruns 2007: 235–269. On how the critique by Kiemstedt and his co-workers critique was received in Berlin, see Brandl 2000: 23.

27. Baugesetzbuch 1998, § 1a.

28. Abgeordnetenhaus von Berlin, "Vorlage—zur Beschlußfassung—über Ergänzungen des Landschafts-/Artenschutzprogramms Berlin (LaPro)," Drucksache 14/1472, 2001. The concept was endorsed by the Abgeordnetenhaus in February 19, 2004, Abgeordnetenhaus von Berlin, Parlamentsprotokolle 14/13, p. 1840. For the published and textually elaborated version of the concept, see Senatverwaltung für Stadtentwicklung 2004.

29. Berliner Landesarbeitsgemeinschaft Naturschutz, "Stellungnahme zur Ergänzung des Landschaftsprogramms/Artenschutzprogramms gemäß § 13 NatSchGBln i.V. mit § 7 NatSchGBln (Ausgleichsflächenkonzeption/FFH-Gebiete)," Berlin, December 6, 1999: 5; Landesbeauftragter Naturschutz und Landschaftspflege, "Stellungnahme zur Berliner Ausgleichskonzeption" December 8, 1999: 1, both in archive BLN.

30. Berliner Landesarbeitsgemeinschaft Naturschutz, "Stellungnahme zur Ergänzung des Landschaftsprogramms/Artenschutzprogramms gemäß § 13 NatSchGBln i.V. mit § 7 NatSchGBln (Ausgleichsflächenkonzeption/FFH-Gebiete)," Berlin, December 6, 1999, p. 5, archive BLN.

31. Landesbeauftragter Naturschutz und Landschaftspflege, "Stellungnahme zur Berliner Ausgleichskonzeption," 8.12.1999, p. 1, archive BLN.

32. Berliner Landesarbeitsgemeinschaft Naturschutz, "Stellungnahme zur Ergänzung des Landschaftsprogramms/Artenschutzprogramms gemäß § 13 NatSchGBln i.V. mit § 7 NatSchGBln (Ausgleichsflächenkonzeption/FFH-Gebiete)," Berlin, December 6, 1999: 5; Landesbeauftragter Naturschutz und Landschaftspflege, December 8, 1999, "Stellungnahme zur Berliner Ausgleichskonzeption," p. 1, archive BLN.

33. The legal basis for such a levy scheme was given by the Federal Construction Code (Baugesetzbuch § 135c) as well as § 14a Abs 3 of the Berlin Nature Conservation Act.

34. In 2007 an "accelerated procedure" was introduced in the Federal Construction Code (§ 13 a). For small development plans (smaller than 20,000 square meters) environmental assessments and impact compensation are no longer required. This might in the future reduce the extent of nature compensation policy in the city. Since March of 2010 a new Federal Nature Conservation Act is in power. It diminishes the leeway that the Länder hitherto had in regulating nature-conservation issues. This has some implications for the impact regulation (§ 15, 16), whose federal definition now includes rather detailed provisions (e.g., for definition of impacts,

financial compensation, eco-account) from which individual Länder may no longer divert.

Conclusion

1. See Gieryn 2006 and Kohler 2002.

2. See, for example, the "Thematic Strategy on the Urban Environment" of the European Union, EU COM (2004/60). For an overview and a critique of global initiatives of "urban environmentalism," see Brand and Thomas 2005.

3. The local "agenda process" developed relatively independent from the institutionalized structure of urban nature conservation that had evolved earlier in Berlin. The project included new forms of citizen participation and thereby covered issues such as mobility, information economy, future of work, climate change, and gender equality. On the Berlin agenda process and its main fields of activity, see Schophaus 2001.

4. In her study of nature conservation in West Germany, Chaney (2008) also uses Bess' metaphor of "half-green," and on this basis comes to similar conclusions about the mixed results of the conservation efforts in this time.

5. For a pessimistic evaluation of past conservation efforts in Berlin, see Hube, Runge, and Sukopp 1998.

6. In recent years, "guerilla gardening" has become a widespread and visible phenomenon in various parts of Berlin. Although based on informal grass-roots activities, it is a form of active reshaping of land that differs from earlier activists' enthusiasm for ruderal vegetation. The exploration of the relation between these two attempts toward greening the city is beyond the scope of this book.

7. For similar case studies of the mutual implication of cities and natural environmental processes, see Kaika 2005, Hinchliffe and Whatmore 2006, and Heynen, Kaika, and Swyngedouw 2006.

References

Unpublished Sources

Abgeordnetenhaus von Berlin, Bibliothek: minutes of special committees on Natur-schutzgesetz Berlin (NSchGBln) (Berlin Nature Conservation Act) Berlin, FNP 1988 (Land-Use Plan 1984) and Landscape Program

Berliner Arbeitsgemeinschaft Naturschutz: administrative documents, newsletters, pamphlets, flyers, policy papers

Bezirksamt Tiergarten, Grünflächenamt: administrative documents on development of Dörnbergdreieck

Bezirksamt Wedding, Bauamt: administrative documents on development of Potsdamer/Leipziger Platz

Bürgerinitiative Westtangente, Berlin: newsletters, pamphlets, flyers, policy papers

Landesarchiv Berlin: decisions of Berlin Senate

Personal files of Axel Auhagen, Andreas Kalesse, and Rita Mohrmann

Senator für Stadtentwicklung, Berlin: Landesbeauftragter für Naturschutz und Landes-pflege, Berlin; Sitzungsberichte der Landesstelle für Naturschutz und Landschaftspflege; Sitzungsberichte des Sachverständigenbeirats für Naturschutz und Landschaftspflege; administrative documents of Abteilung für Naturschutz

Technische Universität Berlin: Universitätsarchiv and archive of Institut für Ökologie

Published Parliamentary Materials

Abgeordnetenhaus von Berlin, Plenarprotokolle (minutes, Berlin House of Representatives)

Abgeordnetenhaus von Berlin, Drucksachen (printed materials, Berlin House of Representatives)

Published Sources

Abbott, Andrew. 1988. *The System of Professions: An Essay on the Division of Expert Labor.* University of Chicago Press.

Adam, Doug, Sidney Tarrow, and Charles Tilly. 2001. *Dynamics of Contention.* Cambridge University Press.

Adams, Lowell W. 2005. Urban Wildlife Ecology and Conservation: A Brief History of the Discipline. *Urban Ecosystems* 8: 139–156.

AG Gleisdreieck. 1991. Anforderungen an ein Planungsverfahren für das Gleisdreieck. In *Gleisdreieck morgen.* Bundesgartenschau Berlin 1995 GmbH and Bezirksamt Kreuzberg.

Agrawal, Arun. 2005. *Environmentality: Technologies of Government and the Making of Subjects.* Duke University Press.

AGU Arbeitsgemeinschaft Umweltplanung Berlin. 1993. Ermittlung der Kompensationsmaßnahmen im Rahmen der Bebauungsplanung Potsdamer/Leipziger Platz in Berlin: Auftraggeber: Senatsverwaltung für Bau- und Wohnungswesen Berlin.

AGU Arbeitsgemeinschaft Umweltplanung Berlin and Büro für Kommunal- und Raumplanung. No year. Umweltverträglichkeitsuntersuchung zur Bebauungsplanung Potsdamer/Leipziger Platz: Auftraggeber: Senatsverwaltung für Bau- und Wohnungswesen Berlin.

Akademie für Raumforschung und Landesplanung. 1960. *Deutscher Planungsatlas.*

Allinger, Gustav. 1950. *Der deutsche Garten.* Bruckmann.

Allinger, Gustav. 1956. Die Panke-Landschaft in Berlin-Wedding. *Hilfe durch Grün* 4: 6–7.

Alternative Liste Berlin. 1984. *Zum Thema: Stadtentwicklung. Die 'Grüne Mitte'. Das Konzept der Alternativen Liste zum Zentralen Bereich.*

Anders, Klaus. 1979. Zur Vogelwelt des Tiergartens. *Ornithologische Berichte für Berlin (West)* 4 (1): 3–62.

Andritzki, Michael, and Klaus Spitzner, eds. 1981. *Grün in der Stadt. Von oben von selbst für alle von allen.* Rowohlt.

Andritzki, Walter, and Ulla Wahl-Terlinden. 1978. *Mitwirkung von Bürgerinitiativen an der Umweltpolitik,* volume 6. Schmidt.

Andritzki, Walter, and Ulla Wahl-Terlinden. 1982. Rodung in Tegel. *Bürgerinitiative Westtangente Rundbrief* 1: 5–7.

Andritzki, Walter, and Ulla Wahl-Terlinden. 1990. Unsere Meinung. *Berliner Naturschutzblätter* 34 (2): 3–5.

Andritzki, Walter, and Ulla Wahl-Terlinden. 1992. Einstweilige Lebensräume—zukünftige Lebensräume; Ungeschützt und doch geschützt. *Der Rabe Ralf,* June: 10.

Anonymous. 1955. Probleme der städtebaulichen Grünplanung in Berlin. *Das Gartenamt* 2: 27–28.

Anonymous (probably Reinhard Grebe). 1963. Großer Bedarf an Landschaftsplanern im Städtebau. *Garten und Landschaft* 193.

Arbeitsgruppe Artenschutzprogramm Berlin. 1984. *Grundlagen für das Artenschutzprogramm Berlin.* Technische Universität Berlin.

Arbeitsgruppe Eingriffsregelung. 1988. *Empfehlungen zum Vollzug der Eingriffsregelung.* Bonn–Bad Godesberg: Bundesforschungsanstalt für Naturschutz und Landschaftsökologie.

Arbeitskreis Eingriffsregelung und Umweltverträglichkeitsprüfung an der TU Berlin. 2000. Flexibilisierung der Eingriffsregelung—Modetrend oder Notwendigkeit? Landschaftsentwicklung und Umweltforschung. Technische Universität Berlin.

Arndt, A. 1941. Das Vogelleben des Schöneberger Stadtparkes. Naturdenkmalpflege und Naturschutz in Berlin und Brandenburg 45: 362–367.

Arnold, Falk, Georg Fritz, Hanno Heßke, Christian L. Krause, and Arnd Winkelbrandt. 1979. Verzichtet der BDLA auf die Landschaftsplanung? *Natur und Landschaft* 54 (10): 364–366.

Ascherson, Paul. 1859. *Verzeichnis der Phanerogamen und Gefäßkryptogamen, welche im Umkreise von sieben Meilen um Berlin vorkommen.* Hirschwald.

Ascherson, Paul. 1864. *Flora der Provinz Brandenburg, der Altmark und des Herzogthums Magdeburg.* Berlin.

Ascherson, Paul. 1954. Die verwilderten Pflanzen in der Mark Brandenburg. *Zeitschrift für gesamte Naturwissenschaften* 3: 435–463.

Asmus, Gesine. 1984. Wohnungselend 1001–1920. In *Exerzierfeld der Moderne,* ed. J. Boberg. Beck.

Asmus, Ullrich. 1980. *Vegetationskundliches Gutachten über den Potsdamer und Anhalter Güterbahnhof in Berlin.* Berlin: Senator für Bau- und Wohnungswesen.

Asmus, Ulrich. 1981. Vegetationskundliches Gutachten über das Südgelände des Schöneberger Güterbahnhofs. Berlin: im Auftrag des Senators für Bau- und Wohnungswesen.

Asmus, Ulrich, Claudia Martens, and Elmar Scharfenberg. 1983. *Biotopkartierung Berlin (West). II. Kreuzberg-Süd.* Technische Universität Berlin.

Atelier Loidl. 2007. Park auf dem Gleisdreieck Erläuterungsbericht. Vorplanung 1.Realisierungsstufe. Stand 24.06.2007 (available at http://www.berlin-gleisdreieck .de).

Auhagen, Axel. 1978. Nachrichten aus der Berliner Landschaft. *Berliner Naturschutzblätter* 3: 476–478.

Auhagen, Axel. 1979a. Autobahn durch den Tegeler Forst. *Berliner Naturschutzblätter* 23 (67): 515.

Auhagen, Axel. 1979b. Wildpflanzengarten am Berlin Pavillion (Nachrichten aus der Berliner Landschaft). *Berliner Naturschutzblätter* 23 (68): 550–551.

Auhagen, Axel. 1980. Nachrichten aus der Berliner Landschaft. *Berliner Naturschutzblätter* 3: 727–740.

Auhagen, Axel. 1994. *Wissenschaftliche Grundlagen zur Berechnung einer Ausgleichsabgabe.* Berlin Senatsverwaltung für Stadtentwicklung und Umweltschutz Berlin.

Auhagen, Axel. 1995. Einige Gedanken zur Bewertung von Eingriffen in Natur und Landschaft und zur Erhebung einer Ausgleichsabgabe. In Dynamik und Konstanz. Zum 65. Geburtstag von Herbert Sukopp, ed. I. Kowarik, U. Starfinger, and L. Trepl. *Schriftenreihe für Vegetationskunde* 3.

Auhagen, Axel, and Herbert Sukopp. 1982. Auswertung der wildwachsenden Farn- und Blütenpflanzen von Berlin (West) für den Arten- und Biotopschutz. In *Rote Listen der gefährdeten Pflanzen und Tiere in Berlin (West)*, ed. H. Sukopp and H. Elvers. Technische Universität Berlin.

Auhagen, Axel, and Herbert Sukopp. 1983. Ziel, Begründungen und Methoden des Naturschutzes der Stadtentwicklungspolitik von Berlin. *Natur und Landschaft* 58 (1): 9–15.

Aust, Bruno. 2002. *Berliner Pläne 1862–1994.* Berlin: Senator für Stadtentwicklung und Umweltschutz.

Auster, Regine. 2006. Schutz den Wäldern und Seen! Die Anfänge des sozialpolitischen Naturschutzes in Berlin und Brandenburg. In *Naturschutz und Demokratie!?* ed. G. Gröning and J. Wolschke-Bulmahn. Meidenbauer.

Balée, William. 1994. *Footprints of the Forest: Ka'apor Ethnobotany—the Historical Ecology of Plant Utilization by an Amazonian People.* Columbia University Press.

Barry, Andrew. 2001. *Political Machines: Governing a Technological Society.* Athlone.

Barsig, Michael, Nicols A. Bisom, Katja Bisom, Axel Wichmann, and Pia Paust-Lassen. 1995. Baumvitalität, Wasserstatus und Altlastenproblematik im Bereich des Tiergarten. Im Auftrag der Berliner Landesarbeitsgemeinschaft Naturschutz e.V.

Barsig, Michael, and Nicolas A. Klöhn. 1997. *Baumvitalität im Berliner Tiergarten währen der Großbaumaßnahmen.* Im Auftrag der Berliner Landesarbeitsgemeinschaft Naturschutz e.V.

Beck, Ulrich. 1986. *Risikogesellschaft.* Suhrkamp.

Becker, Carlo. 1996. Landschaftspark Adlershof: Kooperatives Gutachterverfahren. *Garten + Landschaft* 7: 35–39.

Becker, Carlo, and Undine Giseke. 1993. Behandlung des Eingriffstatbestandes im Rahmen der verbindlichen Bauleitplanung. Neue Anforderungen aufgrund der Änderungen durch das Gesetz zur Erleichterung von Investitionen und der Ausweisung und Bereitstellung von Wohnbauland (Investitionserleichterungs- und Wohnbaulandgesetz) vom 22. April 1993. Berlin.

Becker Gieseke Mohren Richard. 1995. *Landschaftsplanerisches kooperatives Gutachterverfahren für den naturnahen Freizeit- und Erholungspark auf dem ehemaligen Flugfeld Johannisthal/Adlershof.* Berlin: Adlershof Aufbaugesellschaft.

Becker Gieseke Mohren Richard. 1999. *Johannisthal/Adlershof: Pflege- und Entwicklungskonzept zur Biotopsicherung und -entwicklung im Landschaftspark Berlin-Adlershof.* Berlin: Adlershof Aufbaugesellschaft.

Becker, Howard. 1963. *Outsiders.* Free Press.

Becker, Kuno. 1932. *Die Naturschutzgebiete von Groß Berlin. Ein Heimatbuch für Naturfreunde in Wort und Bild und ein Führer für die Groß Berliner Schulen.* Bergmann.

Beer, Wolfgang, and Wolfgang Spielhagen, eds. 1978. *Bürgerinitiativen: Modell Berlin.* Zitty-Verlag.

Behrends, Hermann, Ulrike Benkert, Jürgen Hopfmann, and Uwe Maechler. 1993. *Wurzeln der Umweltbewegung: Die "Gesellschaft für Natur und Umwelt" (GNU) im Kulturbund der DDR.* Bund demokratischer Wissenschaftler.

Behrendsen, W. 1896. Zur Kenntnis der Berliner Adventivflora. *Verhandlungen des Botanischen Vereins der Provinz Brandenburg* 38: 76–100.

Behrens, M., B. Hühn, and E. Karbowski. 1982. Naturschutz in der Stadt. Berlin-Gleisdreieck. *Landschaftsentwicklung und Umweltforschung* 14: 163–228.

Beratungsgesellschaft für Flächen-Informations-Systeme mbH. 1991. *Biotopwertverfahren nach Aicher und Leyser.* Hessisches Ministerium für Landwirtschaft, Forsten und Naturschutz.

Berger-Landefeld, Ulrich, and Herbert Sukopp. 1962. *Gutachten über die Auswirkungen der beim Bau der Westtangente, Teilstück Tiergarten, erforderlichen Grundwasserabsenkungen auf den Bewuchs im Ostteil des Großen Tiergartens.* Technische Universität Berlin.

Berger-Landefeld, Ulrich, and Herbert Sukopp. 1965. Zur Synökologie der Trockenrasen, insbesondere der Silbergrasflur. *Verhandlungen des Botanischen Vereins der Provinz Brandenburg* 102: 41–98.

Bergmann, Klaus. 1970. *Agrarromantik und Grossstadtfeindschaft.* Marburger Abhandlungen zur Politischen Wissenschaft.

Bergmeier, Monika. 2002. *Umweltgeschichte der Boomjahre 1949–1973*. Waxman.

Berliner Landesgemeinschaft Naturschutz and Anti-Tunnel AG. 1994. "Eigentlich spricht alles gegen die Tiergartentunnel. . . ."

Berliner Landesgemeinschaft Naturschutz and Anti-Tunnel AG. 1995. Stellungnahme zu den Planänderungen zu den "Verkehrsanlagen im Zentralen Bereich."

Berliner Morgenpost. 1981. Aus dem Gutachten über das Schöneberger Südgelände. Sensationeller Fund: Drei Tiere die niemand kannte. November 22.

Berliner Naturschutzgesetz (Gesetz zum Schutz der Natur und Landschaft in Berlin vom 30. Januar 1997). Gesetz- und Verordnungsblatt für Berlin: 183.

Berliner Zeitung. 1988. Tümpel, Bäche, Gräben—Schutz der Natur ist billiger als viele glauben. April 9.

Bess, Michael. 2003. *The Light-Green Society: Ecology and Technological Modernity in France, 1900–2000*. University of Chicago Press.

Bierhals, Erich. 1972. Gedanken zur Weiterentwicklung der Landespflege. *Natur und Landschaft* 47 (10): 281–285.

Bijker, Wiebe E. 1995. *Of Bicycles, Bakelits, and Bulbs: Toward a Theory of Sociotechnical Change*. MIT Press.

Bilzer, Johann Ormond, and Thomas Riedle. 1990. *Die Verbandsklage im Naturschutzrecht*. Blottner.

Blackbourn, David. 2006. *The Conquest of Nature: Water, Landscape, and the Making of Modern Germany*. Norton.

Blais, Rudolf, and Guido Spars. 1990. *Großbaumaßnahmen in Berlin und das Konzept nachhaltiger Entwicklung*. Verlag für Wissenschaft und Forschung.

Blume, H. P., K. Cleve, W. Kunick, F. Peus, and H. Sukopp. 1975. Ökologisches Gutachten über die Auswirkungen des Erweiterungsbaues des Kraftwerkes Oberhavel auf das umgebende Natur- und Landschaftsschutzgebiet. Berlin: Senators für Bau- und Wohnungswesen.

Blume, Hans-Peter, and Marlies Runge. 1978. Genese und Ökologie innerstädtischer Böden aus Bauschutt. *Zeitschrift für Pflanzenernährung und Bodenkunde* 141: 727–740.

Bocking, Stephen. 1997. *Ecologists and Environmental Politics*. Yale University Press.

Bocking, Stephen. 2004. *Nature's Experts: Science, Politics, and the Environment*. Rutgers University Press.

Bodenschatz, Harald. 1987. *Platz frei für das neue Berlin! Geschichte der Stadterneuerung seit 1871*. Transit.

Böhme, Gernot. 1989. *Für eine ökologische Naturästhetik*. Suhrkamp.

Bornkamm, Reinhard. 1987. Veränderungen der Phytomasse in den ersten zwei Jahren einer Sukzession auf unterschiedlichen Böden. *Flora* 179: 179–192.

Bornkamm, Reinhard, and Axel Auhagen. 1977. Versuche zur Begrünung von Dächern mit Vegetationsplatten. *Das Gartenamt* 77: 143–147.

Bornkamm, Reinhard, and Ursula Hennig. 1982. Experimentell-ökologische Untersuchungen zur Sukzession von ruderalen Pflanzengesellschaften. I. Zusammensetzung der Vegetation. *Flora* 172: 267–316.

Bornkamm, Reinhard, and Manfred Köhler. 1987. *Ein Naturgarten für Lehre und Forschung. Der Garten des Instituts für Ökologie der Technischen Universität Berlin. Landschaftsforschung und Umweltforschung*. Technische Universität Berlin.

Bournot, Helmut. 1970. Das Grüne Territorium West Berlin. *Das Gartenamt* 10: 482–484.

Bowker, Geoffrey C. 2004. Time, Money, Biodiversity. In *Global Assemblages: Technology, Politics, and Ethics as Anthropological Problems*, ed. A. Ong and S. J. Collier. Blackwell.

Bowker, Geoffrey C., and Susan Leigh Star. 1999. *Sorting Things Out: Classification and Its Consequences*. MIT Press.

Bramwell, Anna. 1989. *Ecology in the 20th Century*. Yale University Press.

Brand, Karl-Werner. 1999. Dialectics of Institutionalisation: The Transformation of the Environmental Movement in Germany. *Environmental Politics* 8 (1): 35–58.

Brand, Peter, and Michael J. Thomas. 2005. *Urban Environmentalism: Global Change and the Mediation of Local Conflict*. Blackwell.

Brandl, Heinz. 2000. Praxis der Eingriffsregelung in Berlin—ein Beispiel. In *Flexibilisierung der Eingriffsregelung—Modetrend oder Notwendigkeit?* Technische Universität Berlin.

Breitenreuter, Gerda, Axel Dehn, Burghard Nickel, Hinrich Elvers, and Johannes Schwarz. 1978. Die Sommervögel des Märkischen Viertels 1977. Projektbericht einer Arbeitsgruppe des Instituts für Allgemeine Zoologie der FU Berlin. *Ornithologische Berichte für Berlin (West)* 3 (2): 147–170.

Brickman, Ronald, Sheila Jasanoff, and Thomas Ilgen. 1985. *Controlling Chemicals: The politics of regulation in Europe and the United States*. Cornell University Press.

Bringmann, Matthias. 1992. Ehrenamtliche Helfer nicht mehr gefragt? *Der Rabe Ralf*, June: 10.

Brück, Michael. 1999. *Pflege- und Entwicklungskonzept für das "Wäldchen" auf dem Gleisdreieck in Berlin*. Berlin: Naturschutz- und Grünflächenamt.

Brüggemeier, Franz-Joseph, Mark Cioc, and Thomas Zeller, eds. 2005. *How Green Were the Nazis? Nature, Environment, and Nation in the Third Reich.* Ohio University Press.

Brüggemeier, Franz-Joseph, and Jens Ivo Engels, eds. 2005. *Natur- und Umweltschutz nach 1945. Konzepte, Konflikte, Kompetenzen.* Campus.

Brunner, Manfred, Friedrich Duhme, Hermann Mück, Johann Patsch, and Elmar Wenisch. 1979. Kartierung erhaltenswerter Lebensräume in der Stadt. *Das Gartenamt* 28:72–78.

Bruns, Elke. 2007. Bewertungs- und Bilanzierungsverfahren in der Eingriffsregelung. Analyse und Systematisierung von Verfahren und Vorgehensweisen des Bundes und der Länder, Technische Universität, Berlin (available at http://www.baufachinformation.de).

Bruns, Elke, Johann Köppel, and Wolfgang Wende. 2000. Einführung. In *Flexibilisierung der Eingriffsregelung—Modetrend oder Notwendigkeit?* Technische Universität Berlin.

Buchwald, Konrad, W. Lendholt, and E. Preising. 1964. Was ist Landespflege? *Garten und Landschaft* 7: 229–231.

Buhnemann, Michael. 1984. *Die Alternative Liste Berlin: Entstehung, Entwicklung, Positionen.* LitPol Verlagsgesellschaft.

Bund Deutscher Landschaftsarchitekten (BDLA). 1979. Ungesetzliche Hochschulausbildung für Landschaftsarchitekten? *Natur und Landschaft* 7/8: 262–263.

Bundesgartenschau Berlin 1995 GmbH and Bezirksamt Kreuzberg. 1991. *Gleisdreieck morgen. Sechs Ideen für einen Park.*

Burchell, G., C. Gordon, and P. Miller. 1991. *The Foucault Effect.* Harvester Wheatsheaf.

Bürgerinitiative Schöneberger Südgelände, no year (1980?). *Das Schönberger Südgelände.*

Bürgerinitiative Schöneberger Südgelände. 1982. *Das Südgelände—eine Landschaft in der Stadt.*

Bürgerinitiative Südgelände. 1985. *Das verborgene Grün von Schöneberg—Naturpark Südgelände.* Freihold.

Bürgerinitiative Westgelände. 1983. *Grüne Mitte. Ein Konzept der Bürgerinitiative Westtangente.*

Bürgerinitiative Westtangente. *Chronik 2010* (available at http://www.bi -westtangente.de).

Callon, Michel, and Vololona Rabeharisoa. 2008. The Growing Engagement of Emergent Concerned Groups in Political and Economic Life: Lessons from the French Association of Neuromuscular Disease Patients. *Science, Technology & Human Values* 33 (2): 230–261.

Chaney, Sandra. 2008. *Nature and the Miracle Years: Conservation in West Germany, 1945–75.* Berghahn.

Chappuis, Ulrich von. 1930–31. Das bisherige Ergebnis der Durchforschung der Berliner Naturschutzgebiete. *Naturdenkmalpflege und Naturschutz in Berlin* 4: 100–104, 7: 218–120.

Chappuis, Ulrich von. 1931. Aus den bisherigen Ergebnissen der Durchforschung der Berliner Naturschutzgebiete—Die Pfaueninsel. *Heimat und Ferne: Beilage zum Teltower Kreisblatt.*

Ciesla, Burghard. 2008. Konkurrierende Stadttechnik im Kalten Krieg. Die Deutsche Reichsbahn und der Straße-Schiene-Konflikt in den sechziger Jahren. In *Konfrontation und Wettbewerb. Wissenschaft, Technik und Kultur im geteilten Berliner Alltag (1948–1973)*, ed. M. Lemke. Metropol.

Cioc, Mark. 2002. *The Rhine: An Eco-Biography 1815–2000.* University of Washington Press.

Cittadino, Eugene. 1990. *Nature as the Laboratory: Darwinian Plant Ecology in the German Empire, 1880–1900.* Cambridge University Press.

Cittadino, Eugene. 1993. The Failed Promise of Human Ecology. In *Science and Nature: Essays in the History of the Environmental Sciences*, ed. M. Shortland. British Society for the History of Science.

Cloos, Ingrid. 1999. Das Landschafts-einschließlich Artenschutzprogramm in Berlin und seine Umsetzung. *TU International* 46/47: 21–24.

Coates, Peter. 2006. *American Perceptions of Immigrant and Invasive Species.* University of California Press.

Cornelsen, Dirk. 1991. *Anwälte der Natur. Umweltschutzverbände in Deutschland.* Beck.

Cramer, Jacqueline. 1987. *Mission-Orientation in Ecology: The Case of Dutch Fresh-Water Ecology.* Rodopi.

Cresswell, Tim. 2002. Theorizing Place. *Thamyris/Intersecting* 9: 11–32.

Cronon, William, ed. 1995. *Uncommon Ground.* Norton.

Czepluch, Anneliese. 1966. Botanische Beobachtungen im Volkspark Hasenheide. *Berliner Naturschutzblätter* 10 (29/30): 91–105, 121–136.

Dahl, F. 1908. Grundsätze und Grundbegriffe der biozönotischen Forschung. *Zoologischer Anzeiger* 33: 349–353.

Dahl, Friedrich. 1902. *Das Tierleben im Grunewald.* Fischer.

Darier, Éric. 1996. Environmental Governmentality: The Case of Canada's Green Plan. *Environmental Politics* 5 (4): 585–606.

Davison, Graeme. 1983. The City as a Natural System: Theories of Urban Society in Early Nineteenth Century Britain. In *The Pursuit of Urban History*, ed. D. Fraser and A. Sutcliffe. Arnold.

Dean, Mitchell. 1999. *Governmentality: Power and Rule in Modern Society*. Sage.

Dear, Michael. 2000. *The Postmodern Urban Condition*. Blackwell.

Deiwick, Britta. 2002. *Entwicklungstendenzen der Eingriffsregelung, Landschaftsentwicklung und Umweltforschung*. Technische Universität Berlin.

Der Tagesspiegel. 1977. Grundsatz: "Energie ist gut, Natur ist besser." May 3.

Der Tagesspiegel. 1982. Kritik an Hassemers Plänen für Nord-Süd-Straßenverbindung. April 21.

Der Tagesspiegel. 1986. Baustop für Hotel am Lützowplatz aufgehoben. April 19.

Desfor, Gene, and Roger Keil. 2004. *Nature and the City: Making Environmental Policy in Toronto and Los Angeles*. University of Arizona Press.

Deutscher Rat für Landespflege. 1971. *Organisation der Landespflege*.

Deutsche Sektion des Internationalen Rats für Vogelschutz. 1971. Die in der Bundesrepublik Deutschland gefährdeten Vogelarten und der Erfolg von Schutzmaßnahmen. *Die Vogelwelt* 92 (2):75–80.

Die Tageszeitung. 1988. Kein Ausverkauf. Dokumentation einer Soli-Erkärung der DDR-Christen "von unten" zum Kubat-Dreieck. June 28.

Die Tageszeitung. 1988. Letzte Meldung: autonomes Biotop. June 23.

Die Tageszeitung. 1988. Plastikband als Mauerersatz. May 31.

Die Tageszeitung. 1988. Vor der Mauer liegt der Strand. May 30.

Die Tageszeitung. 1989. Grünklau? January 26.

Die Tageszeitung. 1990. Geschäftshäuser bedrohen Gartenschau. April 3.

Die Tageszeitung. 1991. Streit um Naturpark. Bausenator will Buga-Gelder für Hellersdorf. July 8.

Dittmann, R. 1958. Öffentliche Grünanlagen, ein Faktor zur Lärmbekämpfung. *Das Gartenamt* 8: 175–180.

Drescher, Barbara, Rita Mohrmann, and Stefan Stern. 1981. *Untersuchung des Gehölzbestandes der Hufeisen-Siedlung in Berlin-Britz*. Technische Universität Berlin.

Drescher, Barbara, and Manfred Stöhr. 1980. *Untersuchung des Gehölzbestandes der Onkel-Tom-Siedlung im Bezirk Zehlendorf von Berlin*. Technische Universität Berlin.

Duhme, Friedrich, et al. 1983. *Kartierung schutzwürdiger Lebensräume in München*. TU München-Weihenstephan.

Düll, Ruprecht, and Herbert Werner. 1955–56. Pflanzensoziologische Studien im Stadtgebiet von Berlin. *Wissenschaftliche Zeitschrift der Humboldt-Universität zu Berlin* 5 (4): 321–331.

Durth, Werner, and Niels Gutschow. 1988. *Träume in Trümmern. Planungen zum Wiederaufbau zerstörter Städte im Westen Deutschlands 1940–1950.* Vieweg.

Duvigneaud, P. 1974. L'ècosysteme "urbs." *Mémoire de la Societé Botanique Belgique* 6:5–35.

Eade, John. 2000. *Placing London.* Berghahn.

Eckebrecht, Berthold. 1991. Die Entwicklung der Landschaftsplanung an der TU Berlin—Aspekte der Institutionalisierung seit dem 19. Jahrhundert im Verhältnis von Wissenschaftsentwicklung und traditionellem Berufsfeld. In *Geschichte und Struktur der Landschaftsplanung,* ed. U. Eisel and S. Schultz. Technische Universität Berlin.

Edensor, Tim. 2005. *Industrial Ruins: Space, Aesthetics and Materiality.* Berg.

Ellenberg, Heinz. 1973. Die Ökosysteme der Erde, Versuch einer Klassifikation nach funktionalen Gesichtspunkten. In *Ökosystemforschung,* ed. H. Ellenberg. Springer.

Ellenberg, Heinz, Robert Mayer, and Jürgen Schauermann. 1986. *Ökosystemforschung. Ergebnisse des Sollingprojekts 1966–1986.* Ulmer.

Elvers, Hinrich, Ingo Kowarik, Barbara Markstein, Bernhard Palluch, Hans-Peter Flechner, and Martina Nath. 1982. Landschaftspflegerischer Begleitplan für den Bau des Südgüterbahnhofs. Berlin: Ökologie und Planung and Büro Hans-Peter Flechner and Senatsverwaltung für Bau- und Wohnungswesen.

Emery, Malcolm. 1986. *Promoting Nature in Cities and Towns: A Practical Guide.* Croom Helm.

Engel, Horst. 1949. Die Trümmerpflanzen von Münster. *Natur und Heimat* 9 (2): 1–15.

Engel, Jaap. 1983. Einige Bemerkungen zum Verfahren, um die formalen Entscheidungen zum Zentralen Bereich, Berlin, herbeizuführen: Adaption des Flächennutzungsplan 1982. In *Dokumentation zum Planungsverfahren Zentraler Bereich Mai 1982–Mai 1983.* Berlin: Senator für Stadtentwicklung und Umweltschutz.

Engels, Jens Ivo. 2005. "In Stadt und Land': Differences and Convergences between Urban and Local Environmentalisms in West Germany, 1950–1980. In *Resources of the City: Contributions to an Environmental History of Modern Europe,* ed. D. Schott, B. Luckin, and Massard-Guilbaud. Ashgate.

Engels, Jens Ivo. 2006. *Naturpolitik in der Bundesrepublik. Ideenwelt und politische Verhaltensstile in Naturschutz und Umweltbewegung 1950–1980.* Schöningh.

Ermer, Klaus. 1977. NC-Fach Landschaftsplanung/ Landschaftspflege—Was heisst das? *Das Gartenamt* 1: 41–42.

Ermer, Klaus, Britta Kellermann, and Christian Schneider. 1979. *Materialien zur Umweltsituation in Berlin.* Berlin: Senator für Bau und Wohnungswesen.

Ermer, Klaus, Britta Kellermann, and Christian Schneider. 1979–80. *Wissenschaftlich-methodische Grundlagen für ein Landschaftsprogramm Berlin.* Berlin: Senator für Bau- und Wohnungswesen.

Erz, Wolfgang. 1972. Grundlegendes zum Schutz von Pflanzen- und Tierarten in einem Bundesnaturschutzgesetz. In *Artenschutz,* ed. G. Olschowy. Bonn–Bad Godesberg: Bundesanstalt für Vegetationskunde, Naturschutz und Landschaftspflege.

Erz, Wolfgang. 1978. Zur Aufstellung von Artenschutzprogrammen. In *Natur- und Umweltschutz in der Bundesrepublik Deutschland,* ed. G. Olschowi. Parey.

Escobar, Arturo. 1999. After Nature: Steps to an Antiessentialist Political Ecology. *Current Ecology* 40 (1): 1–30.

Eser, Uta. 1999. *Der Naturschutz und das Fremde.* Campus.

Essholtz, Johann Sigismund. 1663. *Flora marchia, sive catalogus plantarum, quae partim in Hortis Electoralibus Marchiae Brandenburgicae primariis, Beronlenensi, Aurangiburgico & Potstamensi excoluntur: partim sua sponte passim proveniununt (Märkische Flora, oder Katalog der Pflanzen, die teils in den ansehnlichen Kurfürstlichen Gärten der Mark Brandenburg in Berlin, Oranienburg udn Potsdam sorgsam angebaut werden, teils überall wild vorkommen).* Berlin.

Evans, James P. 2006. Lost in Translation? Exploring the Interface between Local and Environmental Reseach and Policymaking. *Environment and Planning* 38: 517–531.

Evans, James P. 2007. Wildlife Corridors: An Urban Political Ecology. *Local Environment* 12 (2): 129–152.

Eyermann, Ron, and Andrew Jamison. 1991. *Social Movements: A Cognitive Approach.* Polity Press.

Falck, Zachary. 2010. *Weeds: An Environmental History of Urban America.* University of Pittsburgh Press.

Fichtner, Volkmar. 1977. *Die anthropogen bedingte Umwandlung des Reliefs durch Trümmeraufschüttungen in Berlin (West) seit 1945.* Geographisches Institut der Freie Universität Berlin.

Finke, Lothar. 1973. Zur Bedeutung neuerer geographischer Forschungen für die Landespflege. *Natur und Landschaft* 48 (2): 44–48.

Fischer, M. 1955. Grünflächen auf Trümmerschutt und Ödland. Neue Wohnsiedlungen in Mannheim. *Hilfe durch Grün* 3: 23.

Foucault, Michel. 1991. Governmentality. In *The Foucault Effect: Studies in Governmentality*, ed. G. Burchell, C. Gordon, and P. Miller. Harvester Wheatsheaf.

Foucault, Michel. 1997. The Birth of Biopolitics. In *Ethics: Subjectivity and Truth*, ed. P. Rabinow. New Press.

Frank, Sybille. 2000. Der Potsdamer Platz: Das 'Herzstück' der Metropole und das Unbehagen in der Geschichte In *Steinbruch. Deutsche Erinnerungsorte*, ed. C. Carcenac-Lecomte, K. Czarnowski, S. Frank, S. Frey, and T. Lüdtke. Lang.

Franklin, Adrian. 2002. *Nature and Social Theory*. Sage.

Franklin, Adrian. 2010. *City Life*. Sage.

Frickel, Scott. 2004. *Chemical Consequences: Environmnetal Mutagens, Scientist Activism, and the Rise of Genetic Toxicology*. Rutgers University Press.

Fritsche, Peter. 1996. *Reading Berlin 1900*. Harvard University Press.

Fugmann, Harald, Martin Janotta, and Christian Schneider. 1986. *Gutachten über ein mittelfristiges Forschungskonzept "Naturschutz—Berlin."* Berlin: Senator für Stadtentwicklung und Umweltschutz.

Fujimura, Joan H. 1996. *Crafting Science: A Sociohistory of the Quest for the Genetics of Cancer*. Harvard University Press.

Gandy, Matthew. 2002. *Concrete and Clay: Reworking Nature in New York City*. MIT Press.

Gaßner, Hartmut, and Wolfgang Siederer. 1987. *Eingriffe in Natur und Landschaft. Erläuterungen zum Berliner Naturschutzgesetz Landschaftsentwicklung und Umweltforschung*. Technische Universität Berlin.

Gembitzki, Jürgen. 1980. Eine grüne Brücke nach Schöneberg. *Stadtgespräch. Wahlkreiszeitung der SPD Kreuzberg* 4.

Gerstberger, Manfred. 1992. *Die Schmetterlingsfauna auf dem Gelände des ehemaligen Flugplatzes Johannisthal in Berlin-Treptow*. Berlin: Landesbeauftragter für Naturschutz und Landschaftspflege.

Gieryn, Thomas F. 1999. *Cultural Boundaries of Science: Credibility on the Line*. University of Chicago Press.

Gieryn, Thomas F. 2003. A Space for Place in Sociology. *Annual Review of Sociology* 26: 463–496.

Gieryn, Thomas F. 2006. City as Truth-Spot: Laboratories and Field-Sites in Urban Studies. *Social Studies of Science* 36 (1): 5–38.

Gille, Zsuzsa. 2006. Detached Flows or Grounded Place-Making? In *Governing Environmental Flows*, ed. G. Spaargarten, A. Mol, and F. Buttel. MIT Press.

Girardet, H. 1999. *Creating Sustainable Cities*. Green Books.

Gleditsch, Johann Gottlieb. 1751. Verzeichnis der Gewächse, die sich in der Mark Brandenburg finden. In *Historische Beschreibung der Chur und Mark Brandenburg*, ed. J. C. Beckmann. Berlin.

Gobster, Paul H., and R. Bruce Hull. 2000. *Restoring Nature: Perspectives from the Social Sciences and Humanities*. Island.

Göderitz, Johannes, Rainer Roland, and Hubert Hoffmann. 1957. *Die gegliederte und aufgelockerte Stadt*. Vieweg.

Golley, Frank Benjamin. 1993. *A History of the Ecosystem Concept in Ecology: More Than the Sum of the Parts*. Yale University Press.

Goode, D. A. 1989. Urban Nature Conservation in Britain. *Journal of Applied Ecology* 26 (3): 859–873.

Gottdiener, Mark. 1985. *The Social Production of Urban Space*. University of Texas Press.

Gottlieb, Robert. 2007. *Reinventing Los Angeles*. MIT Press.

Gröning, Gert, and Joachim Wolschke-Buhlmahn. 1992. Some Notes on the Mania for Native Plants in Germany. *Landscape Journal* 11 (2): 116–126.

Gross, Matthias. 2010. *Ignorance and Surprise: Science, Society, and Ecological Design*. MIT Press.

Gross, Matthias, and Holger Hoffmann-Riem. 2005. Ecological Restauration as a Real-World Experiment: Designing Robust Implementation Strategies in an Urban Environment. *Public Understanding of Science* 14: 269–284.

Grün Berlin. 1995. *Natur-Park Südgelände*.

Grüne Charta von der Insel Mainau. 1961. *Natur und Landschaft* 36: 151–152.

Grüne Liga, Projektgruppe Mauerpark. 1992. Der Mauerpark—eine drei Jahre alte Forderung. *Der Rabe Ralf*, November: 1.

Grüntangente, Studentische Projektgruppe. No year [1981?]. Projekt Grüntangente. Technische Universität Berlin.

Gusfield, Joseph. 1981. *The Culture of Public Problems: Drinking-Driving and the Symbolic Order*. University of Chicago Press.

Haas, Peter. 1990. *Saving the Mediterranean*. Columbia University Press.

Haber, Wolfgang. 1992. Erfahrungen und Erkenntnisse aus 25 Jahren der Lehre und Forschung in Landschaftsökologie: Kann man ökologisch planen?" In *25 Jahre Landschaftsökologie in Weihenstephan mit Prof. Dr. Dr. h.c. W. Haber*, ed. F. Duhme and L. Spandau. Weihenstephan: Lehrstuhl für Landschaftsökologie.

Hager, Carol J. 1995. *Technological Democracy*. University of Michigan Press.

Hajer, Marten. 1995. *The Politics of Environmental Discourse: Ecological Modernization and the Policy Process*. Clarendon.

Halffman, Willem. 2003. Boundaries of Regulatory Science: Eco/Toxicology and Aquatic Hazards of Chemicals in the US, England, and the Netherlands. Dissertation, University of Amsterdam.

Hall, Marcus. 2005. *Earth Repair: A Transatlantic History of Restoration*. University of Virginia Press.

Hall, Marcus. 2010. Editorial: The Native, Naturalized and Exotic. Plants and Animals in Human History. *Landscape Research* 28 (1): 5–9.

Hannigan, John A. 1995. *Environmental Sociology*. Routledge.

Hard, Gerhard. 1992. Konfusionen und Paradoxien. Natur- und Biotopschutz in Stadt- und Industrieregionen. *Garten und Landschaft* 1: 13–18.

Hard, Gerhard. 1994. Die Natur, die Stadt und die Ökologie. Reflexionen über "Stadtnatur" und "Stadtökologie." In *Pathways to Human Ecology*, ed. H. Ernste. Lang.

Hard, Gerhard. 2011. Geography as Ecology. In *Revisiting Ecology*, ed. A. Schwarz and K. Jax. Springer.

Harrich. 1931. Praktischer Vogelschutz in der Grossstadt. *Naturdenkmalpflege und Naturschutz in Berlin und Brandenburg* 8: 266–269.

Hartmann, Kristiana. 1976. *Deutsche Gartenstadtbewegung. Kulturpolitik und Gesellschaftsreform*. Heinz Moos.

Harvey, David. 1996. *Justice, Nature and the Georgaphy of Difference*. Blackwell.

Harvey, David. 2009. *Social Justice and the City*. University of Georgia Press.

Hassemer, Volker. 1995. Politikberatung zum Naturschutz in Berlin. In *Dynamik und Konstanz. Festschrift für Herbert Sukopp*, ed. I. Kowarik, I. Starfinger, and L. Trepl. Bonn–Bad Godesberg: Bundesamt für Naturschutz.

Hau, Michael. 2003. *The Cult of Health and Beauty in Germany: A Social History*. University of Chicago Press.

Häussermann, Hartmut, and Andreas Kapphan. 2000. *Berlin: von der geteilten zur gespaltenen Stadt? Sozialräumlicher Wandel seit 1990*. Leske + Budrich.

Havelange, P., P. Havelange, and S. Denaeyer-De Smet. 1975. L'écosystèm Bruxelles. Edition Agglomeration Bruxelles.

Hecht, Gabrielle. 1998. *The Radiance of France: Nuclear Power and National Identity after World War II*. MIT Press.

Hegemann, Werner. [1930] 1992. *Das steinerne Berlin. Geschichte der größten Miets-kasernenstadt der Welt.* Vieweg.

Heitmann, Gunter, and Christian Muhs. 1972. Freiraumplanung Berlin. *Das Gartenamt* 1: 12–14.

Helford, Reid M. 1999. Rediscovering the Presettlement Landscape: Making the Oak Savanna Ecosystem "Real." *Science, Technology & Human Values* 24 (1): 55–79.

Henke, Cristopher R. 2000. Making a Place for Science: The Field Trial. *Social Studies of Science* 30 (4): 483–511.

Henneke, Stefanie. 2011. German Ideologies of City and Nature: The Creation and Reception of Schiller Park in Berlin. In *Greening the City: Urban Landscapes in the Twentieth Century*, ed. D. Brantz and S. Dümpelmann. University of California Press.

Hentzen, K. 1955. Großstadt und Landschaft. Untersuchungen an Beispielen aus Berlin und Köln. In *Festschrift zum 25jährigen Bestehen des gärtnerischen Hochschul-studiums in Deutschland*, ed. Bund deutscher Diplomgärtner. Parey.

Hentzen, Kurt. 1950. Über die Landschaft Gross Berlins vor den Zerstörungen des letzten Weltkrieges. Unpublished manuscript.

Hertel, Reiner. 2000. *Rot-grüne Politik und die Regulation gesellschaftlicher Naturverhält-nisse in Frankfurt am Main.* Westfälische Dampfboot.

Hertweck, Florian. 2010. *Der Berliner Architekturstreit. Stadtbau, Architektur, Geschichte und Identität in der Berliner Republik.* Mann.

Heydemann, Bernd. 1985. *Folgen des Ausfalls von Arten—am Beispiel der Fauna. Schriftenreihe des Deutschen Rates für Landespflege* 46: 581–594.

Heynen, Nik, Maria Kaika, and Eric Swyngedouw, eds. 2006. *In the Nature of Cities: Urban Political Ecology and the Politics of Urban Metabolism.* Routledge.

Hilzheimer, Max. 1929. Naturschutzprobleme in der Grossstadt. *Naturdenkmalpflege und Naturschutz in Berlin und Brandenburg* 1 (1): 6–10.

Hilzheimer, Max. 1931. Zweck der Vogelschutzgebiete in der Grossstadt. *Naturdenk-malpflege und Naturschutz in Berlin und Brandenburg* 8: 260–263.

Hinchliffe, Steve. 2007. *Geographies of Nature: Societies, Environments, Ecologies.* Sage.

Hinchliffe, Steve, and Sarah Whatmore. 2006. Living Cities: Towards a Politics of Conviviality. *Science as Culture* 15 (2): 123–138.

Hoffmann-Axthelm, Dieter. 1980. Gegeneinander IBA und Bausenat.

Hoffmann-Axthelm, Dieter. 1996. Der Tunnel ist Selbstlähmung. *Die Tageszeitung*, March 15.

Hommels, Anique. 2005. *Unbuilding Cities: Obduracy in Urban Sociotechnical Change.* MIT Press.

Howard, Ebenezer. 1970. *Garden Cities of To-morrow.* Faber and Faber.

Hube, Elke, Marlies Runge, and Herbert Sukopp. 1998. Anspruch und Wirklichkeit des kommunalen Naturschutzes. *Archiv für Kommunalwissenschaften* 37 (1): 19–37.

Hübler, Karl-Hermann, Hans Kiemstedt, and Wolfgang Sittel. 1981. *Berufsfeldanalyse für Absolventen der Fachrichtung Landespflege an den Hoch- und Fachhochschulen der Bundesrepublik Deutschland : Forschungsauftrag des Bundesministers für Ernährung, Landwirtschaft und Forsten.* Münster-Hiltrup: Landwirtschaftsverlag.

Hueck, Kurt. (no year) Vorschläge für die Wiederbepflanzung der Grünanlagen auf pflanzensoziologischer Grundlage im Landschaftsraum von Gross Berlin.

Hülbusch, Karl Heinrich. 1981a. Das wilde Grün der Städte. In *Grün in der Stadt. Von oben, von selbst, für alle, von allen,* ed. M. Andritzki and K. Spitzner. Rowohlt.

Hülbusch, Karl Heinrich. 1981b. Zur Ideologie der öffentlichen Grünplanung. In *Grün in der Stadt. Von oben, von selbst, für alle, von allen,* ed. M. Andritzki and K. Spitzner. Rowohlt.

Hülbusch, Karl Heinrich, Heidbert Bäuerle, Frank Hesse, and Dieter Kienast. 1979. *Freiraum- und landschaftsplanerische Analyse des Stadtgebietes von Schleswig.* Kassel: Gesamthochschulbibliothek.

Hünemörder, Kai F. 2004. *Die Frühgeschichte der globalen Umweltkrise und die Formierung der deutschen Umweltpolitik.* Franz Steiner.

Institut für Ökologie, Technische Universität Berlin. 1974. *Jahresbericht für 1974.*

Institut für Ökologie, Technische Universität Berlin. 1976. *Jahresbericht des Instituts für Ökologie der TUB 1976.*

Institut für Ökologie, Technische Universität Berlin. 1982. *Landschaftsökologisch differenzierte Nutzungstypen.*

Institut für Ökologie, Technische Universität Berlin. 1983. *10 Jahre Institut für Ökologie.*

Interessengemeinschaft Flugplatz Johannithal. 1991. Der ehemalige Flugplatz in Johannisthal. Vorstellungen zur Entwicklung einer Stadtregion.

Irwin, Alan. 2002. *Sociology and the Environment.* Polity.

Jacobshagen, A., R. Bornkamm, and W. Heinze. 1977. Untersuchungen zur kostensparenden Begrünung von Dachflächen. *Das Gartenamt* 3: 148–151.

Jahn, Edvard. 1983. *Räumliche Ordnung im Zentralen Bereich. 1. Bericht, März 1983.* Berlin: Senator für Stadtentwicklung und Umweltschutz.

Jalas, Jaakko. 1955. Hemerobe und hemerochore Pflanzenarten. Ein terminologischer Reformversuch. *Acta Societas pro Fauna et Flora Fennica* 72 (11): 1–15.

Jamison, Andrew. 2001. *The Making of Green Knowledge.* Cambridge University Press.

Jasanoff, Sheila. 1990. *The Fifth Branch: Science Advisors as Policy-Makers.* Harvard University Press.

Jasanoff, Sheila, ed. 2003. *States of Knowledge: The Co-Production of Science and Social Order.* Routledge.

Kaika, Maria. 2005. *City of Flows: Modernity, Nature and the City.* Routledge.

Kalesse, Andreas. 1979. Was soll nur aus Gatow werden? *Berliner Naturschutzblätter* 23 (68): 505–511, 541–548.

Köhler, Henning. 1987. Berlin in der Weimarer Republik. In *Geschichte Berlins, Zweiter Band: Von der Märzrevolution bis zur Gegenwart,* ed. W. Ribbe. Beck.

Kaule, Giselher, Jörg Schaller, and Hans-Michael Schober. 1979. *Schutzwürdige Biotope in Bayern. Auswertung der Kartierung. Außeralpine Naturräume.* München: Bayerisches Landesamt für Umweltschutz.

Keil, Roger. 1995. The Environmental Problematic in World Cities. In *World Cities in a World System,* ed. P. Knox and P. Taylor. Cambridge University Press.

Ketelhut, Otto. 1957. Aus unseren Naturschutzgebieten. *Berliner Naturschutzblätter* 1 (2): 10–11.

Ketelhut, Otto. 1958. Aus der Arbeit der Landesstelle für Naturschutz und Landespflege und der höheren Naturschutzbehörde in Berlin. *Berliner Naturschutzblätter* 2 (5): 33–35.

Ketelhut, Otto. 1958. Zahlen und Paragraphen. Eine Übersicht über den geschützten Bestand und die rechtlichen Grundlagen für die Naturschutzarbeit in Berlin. *Berliner Naturschutzblätter* 2 (6): 61–67.

Ketelhut, Otto. 1961. Fünf Jahre Landesstelle für Naturschutz und Landschaftspflege in Berlin. *Berliner Naturschutzblätter* 5 (13): 268–270.

Keulhartz, J. 1999. Engineering the Environment: The Politics of 'Nature Development. In *Living with Nature: Environmental Politics as Cultural Discourse,* ed. F. Fischer and M. Hajer. Oxford University Press.

Kielhorn, U., and K. H. Kielhorn. 1993. *Ökologisch-faunistisches Gutachten der Spinnen- und Laufkäfer des ehemaligen Flugplatzes Johannisthal (Berlin-Treptow).* Berliner Landesarbeitsgemeinschaft Naturschutz.

Kiemstedt, H. 1969. Bewertungsverfahren als Planungsgrundlage in der Landschaftsplanung. *Landschaft und Stadt* 4 (69): 154–158.

Kiemstedt, Hans. 1995. Eingriffsregelung im Abseits? In *Dynamik und Konstanz.* *Festschrift für Herbert Sukopp,* ed. I. Kowarik, I. Starfinger, and L. Trepl. Bonn–Bad Godesberg: Bundesamt für Naturschutz.

Kiemstedt, Hans, M. Mönnecke, and Stefan Ott. 1996a. *Methodik der Eingriffsregelung.* *Teil II: Analyse.* Stuttgart: Umweltministerium Baden-Württemberg.

Kiemstedt, Hans. 1996b. *Methodik der Eingriffsregelung. Teil III: Vorschläge zur bundeseinheitlichen Anwendung der Eingriffsregelung nach §8 Bundesnaturschutzgesetz.* Stuttgart: Umweltminitsterium Baden-Würtemberg.

Kiemstedt, Hans, and Stefan Ott. 1993. Methodik der Eingriffsregelung. In *Auftrag der Länderarbeitsgemeinschaft Naturschutz, Landschaftspflege und Erholung.* Institut für Landschaftspflege und Naturschutz der Universität Hannover.

Kiemstedt, Hans, and Stefan Ott. 1994. *Methodik der Eingriffsregelung. Teil I: Synopse.* Stuttgart: Umweltminitsterium Baden-Würtemberg.

Kienast, Dieter. 1978. Die spontane Vegetation der Stadt Kassel in Abhängigkeit von bau- und stadtstrukturellen Quartierstypen. Kassel: Gesamthochschulbibliothek.

King, Anthony D., ed. 1996. *Re-Presenting the City: Ethnicity, Capital and Culture in the 21st Century Metropolis.* St. Martins.

Kingsland, Sharon E. 2005. *The Evolution of American Ecology, 1890–2000.* Johns Hopkins University Press.

Kleihues, Joseph Paul. 1981. Sieben Essentials zum Rahmenplan für die Neubaugebiete der Internationalen Bauausstellung Berlin. *Stadtbauwelt* 71: 1589.285–1595.291.

Klose, Hans, and Max Hilzheimer. 1929. Zur Einführung. *Naturdenkmalpflege und Naturschutz in Berlin und Brandenburg* 1: 1–6.

Knorr Cetina, Karin. 1981. *The Manufacture of Knowledge.* Pergamon.

Knorr Cetina, Karin. 1999. *Epistemic Cultures.* Harvard University Press.

Knorr Cetina, Karin. 2007. Culture in Global Knowledge Societies: Knowledge Cultures and Epistemic Cultures. *Interdisciplinary Science Reviews* 32 (4): 361–375.

Kohler, Alexander, and Herbert Sukopp. 1964. Über die Gehölzentwicklung auf Berliner Trümmerstandorten. *Berichte der Deutschen Botanischen Gesellschaft* 76: 389–406.

Kohler, Alexander, and Herbert Sukopp. 1964. Über die soziologische Struktur einiger Robinienbestände im Stadtgebiet von Berlin. *Sitzungsberichte der Gesellschaft natur-forschender Freunde zu Berlin* 4 (2): 74–88.

Kohler, Robert E. 1982. *From Medical Chemistry to Biochemistry: The Making of a Bio-medical Discipline.* Cambridge University Press.

Kohler, Robert E. 2002. *Landscapes and Labscapes: Exploring the Lab-Field Border in Biology*. University of Chicago Press.

Kohler, Robert E. 2006. *All Creatures: Naturalists, Collectors, and Biodiversity, 1850–1950*. Princeton University Press.

Köppel, Johann, and Britta Deiwick. 2004. *Verfahren zur Bewertung und Bilanzierung von Eingriffen im Land Berlin*. Berlin: Senatsverwaltung für Stadtentwicklung (available at http://www.stadtentwicklung.berlin.de).

Korneck, D., and H. Sukopp. 1988. Rote Liste der in Deutschland ausgestorbenen, verschollenen und gefährdeten Farn- und Blütenpflanzen. *Schriftenreihe für Vegetationskunde* 19: 1–120.

Köstering, Susanne. 2003. *Natur zum Anschauen. Das Naturkundemuseum des deutschen Kaiserreichs 1871–1914*. Böhlau.

Köstler, Hanna, Thomas Müller, Christoph Saure, and Biella Vossen. 1999. Monitoring im Landschaftspark Berlin-Adlershof. Berlin: Im Auftrag der Berlin Adlershof Aufbaugesellschaft mbH.

Köstler, Hannah, and M. Stöhr. 1993. *Floristisch-vegetationskundliches Gutachten über den ehemaligen Flugplatz Johannisthal*. Berlin: Landesbeauftragter für Naturschutz und Landschaftspflege.

Kowarik, Ingo. 1980. Naturschutz in der Innenstadt. Das Gelände des ehemaligen Potsdamer und Anhalter Güterbahnhofs. *Berliner Naturschutzblätter* 24 (72): 631–636.

Kowarik, Ingo. 1986. Vegetationsentwicklung auf innerstädtischen Brachflächen. Beipiele aus Berlin (West). *Tuexenia* 6 (75–98): 75–99.

Kowarik, Ingo. 1991. Berücksichtigung anthropogener Standort- und Florenveränderungen bei der Aufstellung Roter Listen. In *Rote Listen der gefährdeten Pflanzen und Tiere in Berlin, Schwerpunkt Berlin (West)*, ed. A. Auhagen, R. Platen, and H. Sukopp. Technische Universität Berlin.

Kowarik, Ingo. 1991. Unkraut oder Urwald? Natur der vierten Art auf dem Gleisdreieck. In *Gleisdreieck morgen. Sechs Ideen für einen Park*. Bundesgartenschau Berlin GmbH and Berzirksamt Kreuzberg.

Kowarik, Ingo. 1992. Stadtnatur—Annäherung an die "wahre" Natur der Stadt. In *Ansprüche an Freiflächen im urbanen Raum*, ed. J. Gill. Stadt Mainz.

Kowarik, Ingo, Stephan Körner, and L. Poggendorf. 2004. Südgelände: Vom Natur- zum Erlebnispark. *Garten und Landschaft* 2: 24–27.

Kowarik, Ingo, and Andreas Langer. 1994. Vegetation einer Berliner Eisenbahnfläche (Schöneberger Südgelände) im vierten Jahrzehnt der Sukzession. *Verhandlungen des Botanischen Vereins Berlin Brandenburg* 127: 5–43.

Kowarik, Ingo, Uwe Starfinger, and Ludwig Trepl. 1995. Dynamik und Konstanz. Zum 65. Geburtstag von Herbert Sukopp. In *Festschrift für Herbert Sukopp*, ed. I. Kowarik, I. Starfinger, and L. Trepl. Bonn–Bad Godesberg: Bundesamt für Naturschutz.

Krämer, Steffen. 2007. Urbanität durch Dichte. Die neue Maxime im deutschen Städtebau der 1960er Jahre. In *Denkmal! Moderne. Architektur der 60er Jahre*, ed. A. von Buttler and C. Heuter. Jovis.

Krauke, Günter, and Wolfgang A. Schnell. 1982. *Das Südgelände—eine Landschaft in der Stadt*. Berlin: Bürgerinitiative Schöneberger Südgelände.

kubus. 1995. *Dokumentation der Fachkolloquien "Tiergartentunnel—das Ende des Tiergartens?" und "Tiergartentunnel—ein Weg aus dem Verkehrschaos?"*.

Kühnelt, Wilhelm. 1955. Gesichtspunkte zur Beurteilung der Großstadtfauna (mit besonderer Berücksichtigung der Wiener Verhältnisse). *Österreichische Zoologische Zeitschrift* 6: 30–54.

Kunick, Wolfram. 1974. *Veränderungen von Flora und Vegetation einer Großstadt, dargestellt am Beispiel von Berlin (West)*. Technische Universität Berlin.

Kunick, Wolfram. 1978. Schutzvegetation in Siedlungen. *Garten + Landschaft* 7: 451–456.

Kunick, Wolfram. 1979. *Stadtbiotopkartierung Berlin*. Technische Universität Berlin.

Kunick, Wolfram. 1982. *Zonierung des Stadtgebietes von Berlin West—Ergebnisse floristischer Untersuchungen*. Technische Universität Berlin.

Kunick, Wolfram. 1984. Verbreitungskarten von Wildpflanzen als Bestandteil der Stadtbiotopkartierung, dargestellt am Beispiel Köln. *Verhandlungen der Gesellschaft für Ökologie* 12: 268–275.

Küppers, Günter, Peter Lundgreen, and Peter Weingart. 1978. *Umweltforschung—die gesteuerte Wissenschaft?* Suhrkamp.

Kwa, Chung Lin. 1989. Mimicking Nature: The Development of Systems Ecology in the United States, 1950–1975. Dissertation, University of Amsterdam.

Ladd, Brian. 1990. *Urban Planning and Civic Order in Germany: 1860–1914*. Harvard University Press.

Ladd, Brian. 2000. *The Ghosts of Berlin: Confronting German History in the Urban Landscape*. University of Chicago Press.

Lamp, Jochen. 1983. Arbeitslosigkeit und Landschaftsplanung. *Natur und Landschaft* 58 (7/8): 310–311.

Latour, Bruno. 1987. *Science in Action*. Harvard University Press.

Latour, Bruno. 1988. *The Pasteurization of France*. Harvard University Press.

Latour, Bruno. 1993. *We Have Never Been Modern*. Harvard University Press.

Latour, Bruno. 1995. The "Pedofil" of Boa Vista: A Photo-Philosophical Montage. *Common Knowledge* 4 (1): 144–187.

Latour, Bruno. 1999. *Pandora's Hope*. Harvard University Press.

Latour, Bruno. 2004. *Politics of Nature: How to Bring the Sciences into Democracy*. Harvard University Press.

Latour, Bruno, and Steve Woolgar. 1979. *Laboratory Life: The Social Construction of Scientific Facts*. Sage.

Lees, Andrew. 2002. *Cities, Sin, and Social Reform in Imperial Germany*. University of Michigan Press.

Lefèbvre, Henry. 1991. *The Production of Space*. Blackwell.

Lekan, Thomas. 2004. *Imagining the Nation in Nature: Landscape Preservation and German Identity, 1885–1945*. Harvard University Press.

Lekan, Thomas M. 2005. *Germany's Nature: Cultural Landscapes and Environmental History*. Rutgers University Press.

Lemke, Michael. 2008. Einleitung. In *Konfrontation und Wettbewerb. Wissenschaft, Alltag und Kultur im geteilten Berliner Alltag (1948–1973)*, ed. M. Lemke. Metropol.

Lemke, Michael, ed. 2008. *Konfrontation und Wettbewerb. Wissenschaft, Alltag und Kultur im geteilten Berliner Alltag (1948–1973)*. Metropol.

Lenoir, Timothy. 1997. *Instituting Science: The Cultural Production of Scientfic Disciplines*. Stanford University Press.

Lenz, Michael. 1971. Zum Problem der Erfassung von Brutvogelbeständen in Stadtbiotopen. *Die Vogelwelt* 92 (2): 41–52.

Leps, G. 1985. Zur Geschichte der Ökologie in Berlin. *Wissenschaftliche Zeitschrift der Humboldt-Universität Berlin, Mathematisch-Naturwissenschaftliche Reihe* 34: 342–359.

Le Roy, Louis. 1973. *Natuur uitschakelen, natuur inschakelen*. AnkhHermes.

Linse, U. 1986. *Ökopax und Anarchie: Eine Geschichte ökologischer Bewegungen*. Deutscher Taschenbuch Verlag.

Lohrer. 1958. Grünzug Britz im Verwaltungsbezirk Neukölln. *Das Gartenamt* 58: 119–200.

Londo, Ger. 1977. *Natuurtuinen en -parken. Aanleg en onderhoud. Zuthpen*. Thieme.

Löschau, Martin, and Michael Lenz. 1967. Zur Verbreitung der Türkentaube (*Streptopelia decaocto*) in Groß Berlin. *Journal für Ornithologie* 108 (1): 51–64.

Ludwig, D., and H. Meining. 1991. *Methode zur ökologischen Bewertung der Biotopfunktion von Biotoptypen.* Fröhlich und Sporbeck.

Luke, Timothy W. 1995. On Environmentality: Geo-Power and Eco-Knowledge in the Discourse of Contemporary Environmentalism. *Cultural Critique,* fall: 57–81.

Lynch, Michael. 1985. *Art and Artifact in Laboratory Life.* Routledge & Kegan Paul.

Lynch, Michael, and Steve Woolgar, eds. 1990. *Representation in Scientific Practice.* Harvard University Press.

Maas, Inge. 1982. Zusammenfassung der Anforderungen des Gutachtens "Zentraler Bereich" und "Grünkonzept Südliche Friedrichstadt," erstellt von I. Maas, P. Krusche, J. Schmalz. In *Stellungnahmen zum IBA-Neubaugebiet,* ed. Senator für Stadtentwicklung und Umweltschutz Berlin.

MacCormick, John. 1989. *Reclaiming Paradise: The Global Environmental Movement.* Belhaven.

Macnaghten, Phil, and John Urry. 1998. *Contested Natures.* Sage.

Mader, H. J. 1979. Zur Problematik der Lehrinhalte im Studium der Landespflege. *Natur und Landschaft* 54: 366.

Mädig, Erhard. 1951. *Verwaltungsaufbau und Organisation der Landespflege in der Bundesrepublik Deutschland,* second edition. Bonn–Bad Godesberg: Institut für Raumordnung.

Mahler, Erhard. 1991. Grußwort an die Teilnehmer des Kolloquiums "Rote Liste 1991" am 11.–12. Oktober 1989. In *Rote Listen der gefährdeten Pflanzen und Tiere in Berlin, Schwerpunkt Berlin (West),* ed. A. Auhagen, R. Platen, and H. Sukopp. Technische Universität Berlin.

Markham, William T. 2008. *Environmental Organizations in Modern Germany.* Berghan.

Markstein, Barbara, and Bernhard Palluch. 1981. *Sytematisierung von ökologischen Grundlagenuntersuchungen zur Bewertung von Eingriffen in Natur und Landschaft.* Berlin: Ökologie und Planung.

Martens, Claudia, and Elmar Scharfenberg. 1982–83. *Stadtbiotopkartierung Berlin (West). Vegetationskundliche Untersuchungen der Parkanlagen Volkspark Hasenheide, Park am Buschkrug.* Technische Universität Berlin.

Masson, Dominique. 2006. Constructing Scale/Contesting Scale. *Social Politics* 14 (4): 462–486.

Matless, David. 1997. Moral Geographies of English Landscape. *Landscape Research* 22 (2): 141–155.

Mayer, Margit. 2000. Social Movements in European Cities: Transitions from the 1970s to the 1990s. In *Cities in Contemporary Europe,* ed. A. Bagnasco and P. Le Gales. Cambridge University Press.

Mbg (Mecklenburg). 1957. Natur in Not. *Berliner Naturschutzblätter* 2: 10.

McDonnel, Mark J., and Steward T. Picket. 1993. *Humans as Components of Ecosystems: The Subtle Human Effects and Populated Areas.* Springer.

McIntosh, Robert. 2011. The History of Early British and US-American Ecology to 1950. In *Revisiting Ecology*, ed. A. Schwarz and K. Jax. Springer.

Melosi, Martin V. 2001. *Effluent America: Cities, Industry, Energy and the Environment.* University of Pittsburgh Press.

Metzler, Gabriele. 2005. *Konzeptionen politischen Handelns von Adenauer bis Brandt. Politische Planung in der pluralistischen Gesellschaft.* Schöningh.

Meyer-Tasch, Peter. 1985. *Die Bürgerinitiativbewegung: Der aktive Bürger als rechts- und politikwissenschaftliches Problem,* fifth edition. Rowohlt.

Meynen, Emil, and Josef Schmithüsen, eds. 1953–1962. *Handbuch der naturräumlichen Gliederung.* Bad Godesberg: Bundesamt für Landeskunde und Raumforschung.

Mielke, Hans-Jürgen. 1980. Aus der Arbeit des Volksbundes Naturschutz. *Berliner Naturschutzblätter* 24 (69): 574–576.

Migge, Hans. 1953. Westberliner Kinderspielplätze. *Das Gartenamt* 3: 46–47.

Miller, Peter, and Nicolas Rose. 1990. Governing Economic Life. *Economy and Society* 19 (1): 1–31.

Mitman, Gregg. 1992. *The State of Nature: Ecology, Community, and American Social Thought, 1900–1950.* University of Chicago Press.

Mrass, W. 1971. Denkschrift über die derzeitige Organisation von Naturschutz und Landschaftspflege. In *Organisation der Landespflege. Schriftenreihe des Deutschen Rates für Landespflege* 15.

Muhs, Christian. 1979. Das neue Berliner Naturschutzgesetz. *Berliner Naturschutzblätter* 23 (66): 451–453.

Murdoch, Jonathan. 1997. Governmentality and Territoriality: The Statistical Manufacture of Britain's "National Farm." *Political Geography* 16 (4): 307–324.

Nicolson, Malcolm. 1989. National Styles, Divergent Classifications: A Comparative Case Study from the History of French and American Plant Ecology. *Knowledge and Society: Studies in the Sociology of Science. Past & Present* 8: 139–186.

Nimmann, Hans. 1973. Wie ist die Ausbildung und was umfasst sie. In *Das Berufsbild des Landschaftsarchitekten,* ed. Bund Deutscher Landschaftsarchitekten. Callwey.

Nimmann, Hans. 1978. Ausbildung in der Landespflege. In *Natur- und Umweltschutz in der Bundesrepublik Deutschland,* ed. G. Olschowi. Parey.

Nolte, Paul, and Dieter Gosewinkel, eds. 2008. *Planung im 20. Jahrhundert*. Vanden-hoeck & Ruprecht.

Nölting, Benjamin. 2002. *Strategien und Handlungsspielräume lokaler Umweltgruppen in Brandenburg und Ostberlin, 1980–2000*. Lang.

Nowotny, Helga, Peter Scott, and Michael Gibbons. 2001. *Re-Thinking Science: Knowledge in an Age of Uncertainty*. Polity.

Nyhard, Lynn K. 2009. *Modern Nature: The Rise of the Biological Perspective in Germany*. University of Chicago Press.

Odum, Howard T. 1971. *Environment, Power, and Society*. Wiley.

Oels, A. 2005. Rendering Climate Change Governable: From Biopower to Advanced Liberal Government. *Journal of Environmental Policy and Planning* 7 (3): 185–207.

ÖkoCon. 1991. *Ökologisch-Landschaftsplanerisches Gutachten Natur-Park Südgelände, 1. Zwischenbericht*. Bundesgartenschau Berlin 1995 GmbH.

ÖkoCon. 1992. *Naturpark Südgelände. Bestand, Bewertung, Planung*. Bundesgar-tenschau Berlin 1995 GmbH.

ÖkoCon and Planland. 1998. *Natur-Park Südgelände—Bio-Monitoring. Zielsetzung, Konzept, Aufnahmeflächen—Lage und Erstaufnahme*. Grün Berlin und Gartenschau GmbH.

Olschowi, Gerhard. 1972. Vorwort. In *Artenschutz*, ed. G. Olschowy. Bonn–Bad Godesberg: Bundesanstalt für Vegetationskunde, Naturschutz und Landschaftspflege.

Peschken, Goerd. 1984. Spielwiesen für die arbeitende Bevölkerung. In *Exerzierfeld der Moderne. Industriekultur in Berlin im 19. Jahrhundert*, ed. J. Boberg. Beck.

Peters, H. 1956. Aufgaben der Kommunal-Biologie während Krieg, Zerstörung und Wiederaufbau. In *Festschrift zur Jahreshauptversammlung Hamburg 1956*, ed. Verband Deutscher Biologen. Wissenschaftliche Verlagsgesellschaft.

Peus, Fritz. 1952. Steppenvögel mitten in Berlin. *Die Vogelwelt* 73 (1): 1–6.

Pflug, Wolfram. 1973. Erfassung und Bewertung des Naturhaushaltes als Aufgabe der Landespflege. *Natur und Landschaft* 48 (1): 20–23.

Pflug, Wolfram. 1980. Aachen: Stadtentwicklung auf landschaftsökologischer Grundlage. *Garten + Landschaft* 9: 740–743.

Phillipps, Denise. 2003. Friends of Nature: Urban Sociability and Regional Natural History in Dresden, 1800–1850. In *Science and the City (Osiris* 18), ed. S. Dierig, J. Lachmund, and J. A. Mendelsohn. University of Chicago Press.

Pickering, Andrew. 1995. *The Mangle of Practice*. University of Chicago Press.

Pinch, Trevor J., and Wiebe E. Bijker. 1987. The Social Construction of Facts and Artifacts: Or How the Sociology of Science and the Sociology of Technology Might Benefit Each Other. In *The Social Construction of Technological Systems*, ed. W. Bijker, T. Hughes, and T. Pinch. MIT Press.

Planland, unter Mitarbeit von Prof. Dr. Ingo Kowarik. 2000. *Schöneberger Südgelände. Pflege- und Entwicklungsplan*. Berlin: Senatsverwaltung für Stadtentwicklung.

Plantage. 1993. *Wertvolle Flächen für Flora und Fauna*. Berlin: Senatsverwaltung für Stadtentwicklung und Umweltschutz.

Pobloth, Sonja. 2008. *Die Entwicklung der Landschaftsplanung in Berlin im Zeitraum 1979 bis 2004 unter besonderer Berücksichtigung der Stadtökologie*. MBV.

Pols, Hans. 2003. Anomie in the Metropolis: The City in American Sociology and Psychiatry. In *Science and the City* (*Osiris* 18), ed. S. Dierig, J. Lachmund, and J. A. Mendelsohn. University of Chicago Press.

Porter, Theodore. 1995. *Trust in Numbers*. Princeton University Press.

Potonié, Robert. 1922. *Wanderbuch für den Berliner Naturfreund*. Reimer.

Rada, Ulrich. 1997. *Hauptstadt der Verdrängung*. Schwarze Risse.

Radkau, Joachim, and Frank Uekötter, eds. 2003. *Naturschutz und Nationalsozialismus*. Campus.

Rating, Katrin, Norfried Pohl, Ulrike Tschörner, and Wilhelm Meier. 1970. Das Märkische Viertel in Berlin. Ausdruck einer undemokratischen Planung. *Das Gartenamt* 70: 152–154.

Reichow, Hans Bernhard. 1948. *Organische Stadtbaukunst. Von der Großstadt zur Stadtlandschaft*. Westermann.

Reuss, Martin, and Stephen H. Cutcliffe, eds. 2010. *The Illusory Boundary: Environment and Technology in History*. University of Virginia Press.

Rheinländer, Norbert. 2000. Laudatio auf Prof. Sukopp anlässlich der Verleihung des Dr. Victor-Wendland-Ehrenringes 2000. Unpublished manuscript.

Ribbe, Wolfgang. 1987. Berlin zwischen Ost und West (1945 bis zur Gegenwart). In *Geschichte Berlins*, zweiter Band: *Von der Märzrevolution bis zur Gegenwart*, ed. W. Ribbe. Beck.

Richter, Günther. 1987. Zwischen Revolution und Reichsgründung (1948–1870). In *Geschichte Berlins*, zweiter Band: *Von der Märzrevolution bis zur Gegenwart*, ed. W. Ribbe. Beck.

Rieck, Günter. 1954. Der Volkspark Rehberge in Berlin. *Das Gartenamt* 6: 102–104.

Rink, Dieter. 2002. Environmental Policy and the Environmental Movement in East Germany. *Capitalism, Nature, Socialism* 13 (3): 73–90.

Rink, Uwe, Elisabeth Jung, and Brigitte Marks. 1995. *Untersuchung über die Umsetzung von Ausgleichs- und Ersatzmaßnahmen in Berlin.* Berlin: Im Auftrag der Berliner Landesarbeitsgemeinschaft Naturschutz e.V.

Rodenstein, Marianne. 1988. *Mehr Licht, mehr Luft: Gesundheitskonzepte im Städtebau seit 1750.* Campus.

Rollins, William H. 1997. *A Greener Vision of Home: Cultural Politics and Environmental Reform in the German Heimatschutz Movement 1904–1918.* University of Michigan Press.

Rosenberg, Charles. 1979. Toward an Ecology of Knowledge. In *The Organization of Knowledge in Modern America,* ed. A. Oleson and J. Voss. Johns Hopkins University Press.

Rotenberg, Robert. 1995. *Landscape and Power in Vienna.* Johns Hopkins University Press.

R.R. 1931. Die Berliner Naturschutzausstellung 1931. *Naturdenkmalpflege und Naturschutz in Berlin und Brandenburg* 8: 272–278.

Rucht, Dieter, Barbara Blattert, and Dieter Rink. 1997. *Soziale Bewegungen auf dem Weg zur Institutionalisierung? Zum Strukturwandel 'alternativer' Gruppen in beiden Teilen Deutschlands.* Campus.

Rucht, Dieter, and Jochen Roose. 2001. Neither Decline nor Sclerosis: The Organisational Structure of the German Environmental Movement. *West European Politics* 24: 55–81.

Runge, Karsten. 1998. *Entwicklungstendenzen der Landschaftsplanung.* Springer.

Runge, Marlies. 1974. Ruderalstandorte. *Mitteilungen der Deutschen Bodenkundlichen Gesellschaft* 18: 386–387.

Runge, Marlies. 1975. West Berliner Böden anthropogener Litho- und Pedogenese. Doctoral thesis, Technische Universität Berlin.

Saure, C. 1992. *Zur Stechimmenfauna des Südgeländes in Berlin-Schöneberg.* Berliner Landesarbeitsgemeinschaft Naturschutz.

Saure, Christoph, and Peter Steinlein. 1994. *Pflege- und Entwicklungsplan für den ehemaligen Flugplatz Johannisthal (Berlin-Treptow).* Berlin: Landesstelle für Naturschutz und Landschaftspflege.

Schaumann, Martin. 1992. Vergleichende Untersuchung von Unteren Naturschutzbehörden im ländlichen- und großstädtischen Bereich, dargestellt am Beispiel des Landkreises Göttingen und des Berliner Bezirks Kreuzberg. Doctoral dissertation, Technische Universität Berlin.

Schaumann, Martin. 1986. *Landschaftsplan VI-1a*. Berlin: Bezirksamt Kreuzberg, Abteilung Bau- und Wohnungswesen, Gartenbauamt/Landschaftsplanung.

Schindler, Norbert. 1971. Auftragsforschung für das öffentliche Grün. Berliner Forschungsaufträge. *Das Gartenamt* 2: 74–78.

Schindler, Norbert. 1972. Das Berliner Grün der Nachkriegszeit. *Das Gartenamt* 9: 483–596, 10: 564–570.

Schmidt, Erika. 1970. Vom Deutschen Naturschutztag 1970 in Berlin. *Das Gartenamt* 381: 381–386.

Schmoll, Friedemann. 2004. *Erinnerung an die Natur*. Campus.

Schneider, Bernhard. 1983. Verwaltungsintern zum Verfahren Zentraler Bereich/ Definition von Vorgaben. In *Dokumentation zum Planungsverfahren Zentraler Bereich Mai 1982–Mai 198*. Berlin: Senator für Stadtentwicklung und Umweltschutz.

Schneider, Christian, and Ökologie & Planung. 1979. *Zur Durchführung und Aufarbeitung wissenschaftlich-methodischer Grundlagen für ein Landschaftsprogramm Berlin*. Berlin: Teilbereich Naturschutz (Arten- und Biotopschutz).

Schneider, Mark, and Paul Teske. 1992. Toward a Theory of the Political Entrepreneur: Evidence from Local Government. *American Political Science Review* 86 (3): 737–747.

Schneider-Wilkes, Rainer. 2001. *Engagement und Misserfolg in Bürgerinitiativen: Politische Lernprozesse von Berliner Verkehrsinitiativen. Otto-Suhr-Institut für Politikwissenschaften*. Freie Universität Berlin.

Scholz, H., and H. Sukopp. 1960. Zweites Verzeichnis von Neufunden höherer Pflanzen aus der Mark Brandenburg und angrenzenden Gebieten. *Verhandlungen des Botanischen Vereins der Provinz Brandenburg* 98–100: 23–47.

Scholz, H., and H. Sukopp. 1965. Drittes Verzeichnis von Neufunden höherer Pflanzen aus der Mark Brandenburg und angrenzenden Gebieten. *Verhandlungen des Botanischen Vereins der Provinz Brandenburg* 102: 3–40.

Scholz, H., and H. Sukopp. 1967. Viertes Verzeichnis von Neufunden höherer Pflanzen aus der Mark Brandenburg und angrenzenden Gebieten. *Verhandlungen des Botanischen Vereins der Provinz Brandenburg* 104: 27–47.

Scholz, Hildemar. 1956. *Die Ruderalvegetation Berlins*. Dissertation, Freie Universität Berlin.

Schophaus, Malte. 2001. Bürgerbeteiligung in der Lokalen Agenda 21 in Berlin. Discussion paper FS II 01-306, Wissenschaftszentrum Berlin für Sozialforschung.

Schreier, K. 1955. Die Vegetation auf Trümmerschutt. *Schriftenreihe der Naturschutzstelle Darmstadt* 3: 1–49.

Schröder, R., and Michael Barsig. 2004. *Recherchen zur Umsetzung der Ausgleichs- und Ersatzmausnahmen zu den planfestgestellten Vorhaben im 'Zentralen Bereich' von Berlin*. Kubus.

Schulte, W., H. Sukopp, V. Voggenreiter, and P. Werner, eds. 1986. Flächendeckende Biotopkartierung im besiedelten Bereich als Grundlage einer ökologisch bzw. am Naturschutz orientierten Planung. Grundprogramm für die Bestandsaufnahme und Gliederung des besiedelten Bereichs und dessen Randzonen. *Natur und Landschaft* 61 (10): 371–389.

Schulz, Johann Heinrich. 1845. *Fauna marchica: Die Wirbelthiere der Mark Brandenburg*. Eyssenhardt.

Schulz, Jürgen. 1991. Raumwissenschaft und Raumplanung als Rahmen der Entwicklung der Profession und der Hochschuldisziplin Landschaftsplanung. In *Geschichte und Struktur der Landschaftsplanung. 1991*, ed. U. Eisel and S. Schultz. Technische Universität Berlin.

Schütze, Bernd. 2005. Verdrängte Geschichte: Juden im Naturschutz. Der Zivilisationsbruch für den Naturschutz in Deutschland und Berlin ab 1933 und das aktive Vergessen nach 1945. *trend onlinezeitung* 03/05.

Schwambach, Gabriele. 2006. *Das unsichtbare Geschlecht der Stadtplanung*. Mosnenstein und Vanderdat.

Schwarz, Astrid. 2003. *Wasserwüste—Mikrokosmos—Ökosystem: eine Geschichte der Eroberung des Wasserraumes*. Rombach.

Schwarz, Astrid, and Kurt Jax. 2011. Early Ecology in the German-Speaking World Through WWII. In *Revisiting Ecology*, ed. A. Schwarz and K. Jax. Springer.

Schwarzer, Udo. 1985. "Naturschützer gehören zu denen, die folgsam sind." *Grünstift* 5: 13.

Scott, James. 1998. *Seeing Like a State*. Yale University Press.

Senator für Bau- und Wohnungswesen. 1983. *Erläuterungsbericht zum Planfeststellungsverfahren*. Berlin: Senator für Bau- und Wohnungswesen.

Senator für Stadtentwicklung und Umweltschutz. No year. Stellungnahme zu den im Rahmen der öffentlichen Auslegung eingegangenen Bedenken und Anregungen zum Landschaftsprogramm Berlin.

Senator für Stadtentwicklung und Umweltschutz. 1984. *Landschaftsprogramm—Artenschutzprogramm*. Berlin: Senator für Stadtentwicklung und Umweltschutz.

Senator für Stadtentwicklung und Umweltschutz. 1988. *Landschaftsprogramm—Artenschutzprogramm*. Berlin: Senator für Stadtentwicklung und Umweltschutz.

Senator für Stadtentwicklung und Umweltschutz. 1994. *Landschaftsprogramm/Artenschutzprogram Berlin*. Berlin: Senator für Stadtentwicklung und Umweltschutz.

Senatsverwaltung für Stadtentwicklung. 2004. *Artenschutzprogramm, Landschaftsprogramm—Ergänzung 2004.* Berlin: Senatsverwaltung für Stadtentwicklung (available at http://www.stadtentwicklung.berlin.de).

Senatsverwaltung für Stadtentwicklung. 2006. Umweltprüfungen. Berliner Leitfaden für die Stadt- und Landschaftsplanung (available at http://www.stadtentwicklung.berlin.de).

Senatsverwaltung für Stadtentwicklung, Umweltschutz und Technologie. 1999. *Leitfaden: Umweltverträglichkeitsprüfung und Eingriffsregelung in der Stadt- und Landschaftsplanung.* Kulturbuchverlag (available at http://www.stadtentwicklung.berlin.de).

Senatsverwaltung für Stadtentwicklung und Umweltschutz and Gruppe F. 2001. Protokoll des Workshops zur Evaluierung der 'Auhagen-Methode' am 08.12.2001. Berlin.

Senat von Berlin. 1978. Rundschreiben zur Durchführung der Umweltverträglichkeitsprüfung vom 5. September 1978. *Dienstblatt des Senats von Berlin* 17 (1): 211–216.

Sennet, Richard. 1996. *Flesh and Stone.* Norton.

Shapin, Steve, and Simon Schaffer. 1985. *Leviathan and the Air-Pump.* Princeton University Press.

Shields, Rob. 1991. *Places on the Margin: Alternative Geographies of Modernity.* Routledge.

Simmel, Georg. 1965. The Ruin. In *Essays on Sociology, Philosophy and Aesthetics*, ed. K. Wolff. Harper.

Sitte, Camillo. [1909] 2001. *Der Städtebau nach seinen künstlerischen Grundsätzen. Vermehrt um "Großstadtgrün."* Birkhäuser.

Söderqvist, Thomas. 1986. *The Ecologists: From Merry Naturalists to Saviours of the Nation.* Almqvist & Wiksell.

Soja, Edward. 1996. *Thirdspace: Journeys to Los Angeles and Other Real and Imagined Places.* Blackwell.

Spandauer Volksblatt. 1986. Nur ein Sofortprogramm kann noch helfen.

Spindler. 1958. Anlagen im Bezirk Reinickendorf. *Das Gartenamt* 6: 122–124.

Star, Susan Leigh, and James R. Griesemer. 1989. Institutional Ecology, "Translations" and Boundary Objects: Amateurs and Professionals in Berkeley's Museum of Vertebrate Zoology, 1907–39. *Social Studies of Science* 19: 387–420.

Sterns, Forest, and Tom Montag, eds. 1974. *The Urban Ecosystem: A Holistic Approach.* Dowden, Hutchinson & Ross.

Stifterverband für die Deutsche Wissenschaft. 1964. *Vademecum Deutscher Lehr und Forschungsstätten.*

Stifterverband für die Deutsche Wissenschaft. 1968. *Vademecum Deutscher Lehr und Forschungsstätten.*

Stifterverband für die Deutsche Wissenschaft. 1973. *Vademecum Deutscher Lehr und Forschungsstätten.*

Stimmann, Hans, and Dittmar Machule. 1980. Wofür braucht Berlin eine neue Mitte? Zwei Stadtplaner der TU formulieren Forderungen für ein Konzept des "Zentralen Bereiches." *Berliner Stimme,* December 5.

Stone, Clarence N. 1993. Urban Regimes and the Capacity to Govern: A Political Economy Approach. *Journal of Urban Affairs* 15 (1): 1–28.

Strech, A. 1931. Im Lichtbild durch die Heimat-Natur. *Naturdenkmalpflege und Naturschutz in Berlin und Brandenburg* 8: 263–265.

Stricker, W. 1974. Botanisches Neuland: Potsdamer Bahnhof und Eiskeller. *Berliner Naturschutzblätter* 18 (52): 36–41.

Stricker, W. 1974–75. Die Wildpflanzen der Altstadt von Berlin. *Berliner Naturschutzblätter* 18/19 (35): 64–70, (54): 96–97, (55): 109–111, (56): 148–153.

Stricker, W. 1977. Die bedrohten, verschollenen und ausgestorbenen Arten der Berliner Flora (I). *Berliner Naturschutzblätter* 21 (60): 266–272.

Strom, Elizabeth. 2001. *Building the New Berlin. The Politics of Urban Development in Germany's Capital City.* Lanham.

Stürmer, Rainer. 1991. *Freiflächenpolitik in Berlin in der Weimarer Republik.* Berlin Verlag.

Sucker, Ulrich. 2002. *Das Kaiser-Wilhelm-Institut für Biologie.* Steiner.

Sukopp, Herbert. 1958. Vergleichende Untersuchungen der Vegetation Berliner Moore unter besonderer Berücksichtigung der anthropogenen Veränderungen. Doctoral thesis, Freie Universität Berlin.

Sukopp, H. 1962. Das Naturschutzgebiet Teufelsbruch in Berlin-Spandau. *Sitzungsberichte der Gesellschaft der Naturforschenden Freunde zu Berlin* 2: 38–49.

Sukopp, H. 1966. Verluste der Berliner Flora während der letzten hundert Jahre. *Sitzungsberichte der Gesellschaft der Naturforschenden Freunde zu Berlin* 6: 126–136.

Sukopp, H. 1968. Der Einfluss des Menschen auf die Vegetation und zur Terminologie anthropogener Vegetationstypen. Bericht über das 7. Internationale Symposion in Stolzenau/Weser 1963 der Vereinigung für Vegetationskunde. In *Pflanzensoziologie und Landschaftsökologie. Bericht über das 7. Internationale Symposium in Stolzenau/Weser der Internationalen Vereinigung für Vegetationskunde,* ed. R. Tüxen. Junk.

Sukopp, H. 1970. Charakteristik und Bewertung der Naturschutzgebiete in Berlin (West). *Natur und Landschaft* 45: 133–139.

Sukopp, H. 1973. *Das Naturschutzgebiet Pfaueninsel in Berlin-Wannsee (1. Teil). Wissenschaftliche Grundlagenuntersuchungen in Berliner Natur- und Landschaftsschutzgebieten*. Technische Universität Berlin.

Sukopp, H. 1979. Naturschutz in der Großstadt. In *Ökologie Forum 1979. Schutz der Tier- und Pflanzenwelt sowie natürlicher Lebensräume*. Hamburg: Behörde für Bezirksangelegenheiten, Naturschutz und Umweltgestaltung.

Sukopp, H., and R. Böcker. 1975. *Das Naturschutzgebiet Grosses Fenn in der Forst Düppel, Wissenschaftliche Grundlagenuntersuchungen in Berliner Natur- und Landschaftsschutzgebieten*. Berlin: Institut für Ökologie.

Sukopp, H., P. Brahe, and W. Seidling. 1986. *Das Naturschutzgebiet Pfaueninsel (2. Teil). Wissenschaftliche Grundlagenuntersuchungen in Berliner Natur- und Landschaftsschutzgebieten*. I.A. des Senators für Stadtentwicklung und Umweltschutz, Oberste Naturschutzbehörde.

Sukopp, H., H. Dapper, S. Zimmermann-Jaeger, A. De Santo-Virzo, and R. Bornkamm. 1971. Beiträge zur Ökologie von *Chenopodium botrys* L. *Verhandlungen des Botanischen Vereins der Provinz Brandenburg* 108: 3–74.

Sukopp, H., W. Kunick, and C. Schneider. 1979. Biotopkartierung in der Stadt. *Natur und Landschaft* 3: 66–68.

Sukopp, H., and Martin Launhardt. 1985. Spontanvegetation im Berliner Stadtgebiet und auf der Bundesgartenschau. *Gartenpraxis* 5: 16–18.

Sukopp, H., and C. Schneider. 1979. Zur Geschichte der ökologischen Wissenschaften in Berlin. In *Wissenschaft und Gesellschaft*, ed. R. Rürup. Springer.

Sukopp, Herbert. 1957. Bericht über die Tätigkeit des Dahlemer Botanischen Vereins in den Jahren 1947 bis 1955. *Verhandlungen des Botanischen Vereins der Provinz Brandenburg* 83–97: 5–10.

Sukopp, Herbert. 1957. Verzeichnis von Neufunden höherer Pflanzen aus der Mark Brandenburg und angrenzenden Gebieten. *Verhandlungen des Botanischen Vereins der Provinz Brandenburg* 83–97: 31–40.

Sukopp, Herbert. 1958. *Denkschrift des Botanischen Vereins der Provinz Brandenburg über den Schutz von Mooren im westlichen Berliner Stadtgebiet*. Berlin: Botanischer Verein der Provinz Brandenburg.

Sukopp, Herbert. 1960. Die Vegetation des Naturschutzgebietes "Lichterfelder Schloßpark." *Berliner Naturschutzblätter* 4 (12): 224–227.

Sukopp, Herbert. 1961. Naturschutzgebiet Grunewaldsee im Bezirk Wilmersdorf. *Berliner Naturschutzblätter* 5 (13): 278.

Sukopp, Herbert. 1962. Aufruf zur Mitarbeit an der "Flora von Berlin." *Berliner Naturschutzblätter* 6 (16): 335–340.

Sukopp, Herbert. 1963. *Die Ufervegetation der Havel*. Berlin: Senator für Bau- und Wohnungswesen.

Sukopp, Herbert. 1967. Bericht über die Tätigkeit des Dahlemer Botanischen Vereins in den Jahren 1963 bis 1966. *Verhandlungen des Botanischen Vereins der Provinz Brandenburg* 104: 5–10.

Sukopp, Herbert. 1969. Der Einfluss des Menschen auf die Vegetation. *Vegetatio: Acta Geobotanica* 17: 360–371.

Sukopp, Herbert. 1969. Veränderungen des Röhrichtbestandes der Berliner Havel 1962–1967. *Berliner Naturschutzblätter* 13 (37): 303–313, (38): 332–344.

Sukopp, Herbert. 1971. Beiträge zur Ökologie von *Chenopodium botrys* L. I. Verbreitung und Vergesellschaftung. *Verhandlungen des Botanischen Vereins der Provinz Brandenburg* 108: 3–74.

Sukopp, Herbert. 1971. Bewertung und Auswahl von Naturschutzgebieten. *Schriftenreihe für Landschaftspflege und Naturschutz* 6: 183–194.

Sukopp, Herbert. 1972. Grundzüge eines Programmes für den Schutz von Pflanzenarten in der Bundesrepublik Deutschland. *Schriftenreihe für Landschaftspflege und Naturschutz* 7: 67–79.

Sukopp, Herbert. 1973. Die Großstadt als Gegenstand ökologischer Forschung. *Schriften z. Verbreitung naturwissenschaftlicher Kenntnisse* 113: 90–140.

Sukopp, Herbert. 1974. "Rote Liste" der in der Bundesprepublik Deutschland gefährdeten Arten von Farn- und Blütenpflanzen (1. Fassung). *Natur und Landschaft* 49 (12): 315–322.

Sukopp, Herbert. 1976. Zur ökologischen Beurteilung von Standorten für ein Kraftwerk in Berlin-Spandau. *Berliner Naturschutzblätter* 20: 166–170.

Sukopp, Herbert. 1977. Entwicklung und Aufgaben der Projektgruppe Ökologie und Umweltforschung. *Zeitschrift der TU Berlin* 9 (2/3): 279–322.

Sukopp, Herbert. 1979. Rodung in Gatow/Flughafen. Landesbeauftragter für Naturschutz und Landespflege. Pressemitteilung vom 9. November 1979. *Berliner Naturschutzblätter* 23 (68): 548–549.

Sukopp, Herbert. 1982. Untitled. In *Stellungsnahmen zum IBA-Neubaugebiet*, ed. Senatsverwaltung für Stadtentwicklung und Umweltschutz. Berlin: Senator für Stadtentwicklung und Umweltschutz.

Sukopp, Herbert. 1984. *Ökologisches Gutachten Rehberge Berlin, Landschaftsentwicklung und Umweltforschung*. Technische Universität Berlin.

Sukopp, Herbert. 1990. Ökologische Grundlagen für die Stadtplanung und Stadterneuerung. In Ergebnisse der 10. Sitzung der Arbeitsgruppe. In *Biotopkartierung im besiedlten Bereich*, ed. H. J. Condert. Courier Forschungs-Institut Senckenberg.

Sukopp, Herbert, ed. 1982. *Beiträge zur Stadtökologie von Berlin (West): Exkursionsführer für das 2. Europäische Ökologische Symposium im September 1980.* Technische Universität Berlin.

Sukopp, Herbert, ed. 1990. *Stadtökologie. Das Beispiel Berlin.* Reimer.

Sukopp, Herbert, Klaus Anders, Hartmut Bierbach, Arthur Brande, Hans-Peter Blume, Hinrich Elvers, Manfred Horbert, et al. 1979. *Ökologisches Gutachten über die Auswirkungen von Bau und Betrieb der BAB Berlin (West) auf den Großen Tiergarten.*

Sukopp, Herbert, and Axel Auhagen. 1979. Die Naturschutzgebiete Großer Rohrpfuhl und Kleiner Rohrpfuhl im Stadtforst Berlin Spandau, Teil 1. *Sitzungsberichte der Gesellschaft Naturforschender Freunde Berlin* 19: 93–170.

Sukopp, Herbert, and R. Böcker. 1971. *Das Naturschutzgebiet Barsee mit der Saubucht. Wissenschaftliche Grundlagenuntersuchungen in Berliner Natur- und Landschaftsschutzgebieten.* Technische Universität Berlin.

Sukopp, Herbert, and H. Elvers. 1982. *Rote Listen der gefährdeten Pflanzen und Tiere in Berlin (West), Landschaftsentwicklung und Umweltforschung 11.* Technische Universität Berlin.

Sukopp, Herbert, Hinrich Elvers, and H. Matthes, eds. 1982. Studies on Urban Ecology of Berlin (West)—Animals in the Urban Environment. In *Proceedings of the Symposium on the Occasion of the 60th Anniversary of the Institute of Zoology of the Polish Academy of Sciences Warszawa-Jablonna 22–24 October 1979.*

Sukopp, Herbert, and G.-H. Köster. 1970. *Das Naturschutzgebiet Grunewaldsee (südlicher Teil).* Technische Universität Berlin.

Sukopp, Herbert, Wolfram Kunick, and Christian Schneider. 1979. Biotopkartierung in der Stadt. Ergebnisse der ersten Sitzung der Arbeitsgruppe Biotopkartierung im besiedelten Bereich. *Natur und Landschaft* 54 (3): 66–68.

Sukopp, Herbert, and Barbara Markstein. 1984. Möglichkeiten und Grenzen des Ausgleiches von Eingriffen in den Naturhaushalt, dargestellt am Beispiel der Pflanzenwelt urban-industrieller Standorte. *Laufener Seminarbeiträge* 9 (83): 30–38.

Sukopp, Herbert, Barbara Markstein, and Ludwig Trepl. 1975. Röhrichte unter intensivem Großstadteinfluß. *Beiträge zur naturkundlichen Forschung in Südwestdeutschland* 34: 371–385.

Sukopp, Herbert, and A. Straus. 1967. 1968. Das Naturschutzgebiet Pfaueninsel in Berlin-Wannsee. I. Beiträge zur Landschafts- und Florengeschichte. Wissenschaftliche Grundlagenunteruchungen in Berliner Natur und Landschaftsschutzgebieten. *Sitzungsberichte der Gesellschaft Naturforschender Freunde zu Berlin* 8: 93–129.

Sukopp, Herbert, Werner Trautmann, and Jörg Schaller. 1979. Biotopkartierung in der Bundesrepublik Deutschland. *Natur und Landschaft* 54 (3): 63–65.

Sukopp, Herbert, and Rüdiger Wittig, eds. 1993. *Stadtökologie*. Fischer.

Szamatolski, Clemens. 1974. Die realisierte Grün- und Erholungsflächenpolitik in West Berlin 1949–1971 und ihre Perspektiven bis 1980. *Das Gartenamt* 23 (8): 449–459.

Tacacs, David. 1996. *The Idea of Biodiversity: Philosophies of Paradise.* Johns Hopkins University Press.

Tarr, Joel A. 2002. The Metabolism of the Industrical City. The Case of Pittsburgh. *Journal of Urban History* 28: 511–545.

Taut, Max. 1946. *Berlin im Aufbau*. Aufbau-Verlag.

Taylor, Peter. 1988. Technocratic Optimism, H. T. Odum, and the Partial Transformation of Ecological Metaphor after World War II. *Journal of the History of Biology* 21 (2): 213–244.

Technische Universität Berlin. 1972. Entwicklungsplan des Fachbereiches Landschaftsentwicklung (FB 14) für die Jahre 1972–1975.

Thellung, A. 1918. Zur Terminologie der Adventiv- und Ruderalfloristik. *Allgemeine botanische Zeitschrift für Systematik, Floristik, Pflanzengeographie etc.* 24: 36–42.

Thielke, G. 1982. Grundzüge eines Artenschutzprogramms am Beispiel eines Vogelschutzprogramms für die BRD. In *Artenschutz*, ed. G. Olschowi. Bonn–Bad Godesberg: Bundesanstalt für Vegetationskunde, Naturschutz und Landschaftspflege.

Till, Karen E. 2005. *The New Berlin: Memory, Politics, Place.* University of Minnesota Press.

Tischer, Stefan. 1996. Berlin-Adlershof. *Garten + Landschaft* 7: 30–34.

Tischler, Wolfgang. 1992. *Ein Zeitbild vom Werden der Ökologie*. Fischer.

Tomasek, Wolfgang. 1978. Über Beziehungen zwischen Landschaftsplanung und Ökologie. *Natur und Landschaft* 51 (11): 309–310.

Trepl, Ludwig. 1983. Ökologie—eine grüne Leitwissenschaft? Über Grenzen und Perspektiven einer modischen Disziplin. *Kursbuch* 74: 6–27.

Turner, Stephen. 2000. What Are Disciplines? And How Is Interdisciplinarity Different? In *Practicing Interdisciplinarity*, ed. P. Weingart and N. Stehr. University of Toronto Press.

Tüxen, Reinhold. 1956. Die heutige potentielle natürliche Vegetation als Gegenstand der Vegetationskartierung. *Angewandte Pflanzensoziologie* 13: 5–42.

Uekötter, Frank. 2004. *Naturschutz im Aufbruch. Geschichte des Naturschutzes in Nordrhein-Westfalen.* Campus.

Uekötter, Frank. 2006. *The Green and the Brown: A History of Conservation in Nazi Germany*. Cambridge University Press.

Uekötter, Frank. 2007. Native Plants: A Nazi Obsession? *Landscape Research* 32 (3): 379–383.

Umweltschutzforum. 1973. *Forschen für die Umwelt*. Special issue dated April 8, 1973.

van den Belt, Henk. 1993. Networking Nature, or Serengeti behind the Dikes. *History and Technology* 20 (3): 311–333.

Vetter, Jeremy. 2010. Capitalizing on Grass: The Science of Agrostology and the Sustainability of Ranching in the American West. *Science as Culture* 19 (4): 483–507.

Wächter, Monika. 2003. *Die Stadt: umweltbelastetes System oder wertvoller Lebensraum? Zur Geschichte, Theorie und Praxis stadtökologischer Forschung in Deutschland UFZ-Bericht*. Umweltforschungszentrum Leipzig-Halle.

Wagner, Martin. 1915. Das sanitäre Grün der Städte. Thesis, Berlin.

Wagner-Conzelmann, Sandra. 2007. *Die Interbau 1957 in Berlin. Stadt von heute— Stadt von morgen*. Imhof.

Wahnschaffe, F., P. Graebner, and R. von Hanstein, eds. 1907. *Der Grunewald bei Berlin. Seine Geologie, Flora und Fauna*, volume 1. Fischer.

Wahnschaffe, F., P. Graebner, and R. von Hanstein, eds. 1912. *Der Grunewald bei Berlin. Seine Geologie, Flora und Fauna*, volume 2. Fischer.

Wakeman, Rosemary. 2000. Dreaming the New Atlantis: Science and the Planning of Technopolis, 1955–1985. In *Science and the City (Osiris* 18), ed. S. Dierig, J. Lachmund, and J. A. Mendelsohn. University of Chicago Press.

Waterton, Claire. 2002. From Field to Fantasy: Classifying Nature, Constructing Europe. *Social Studies of Science* 32 (2): 177–204.

Waterton, Claire. 2003. Performing the Classification of Nature. In *Nature Performed: Environment, Culture and Performance*, ed. B. Szerszynski, W. Heim, and C. Waterton. Blackwell.

Wedek, Horst. 1973. Zur Bedeutung ökologischer Grundlagen für das Gebiet der Landespflege. *Natur und Landschaft* 48 (1): 23.

Weigmann, Gerd, Hans-Peter Blume, Hermann Mattes, and Herbert Sukopp. 1981. Ökologie im Hochschulunterricht. In *Didaktik der Ökologie*, ed. W. Riedel and G. Trommer. Daulis.

Weiß, Heinrich. 1978. Die nächsten Aufgaben des Natur- und Landschaftsschutzes. *Berliner Naturschutzblätter* 22 (64): 381–384.

Weiß, Heinrich. 1987. Der Naturhaushalt als Argument. *Berliner Naturschutzblätter* 13 (3): 75–83.

Weiß, Heinrich. 1989. Naturschutz und Politik. *Grünstift* 6: 42–46.

Weiß, Heinrich. 1997. Von der Naturdenkmalpflege zur Naturschutzpolitik. 75 Jahre Volksbund Naturschutz e.v. *Berliner Naturschutzblätter* 41 (2): 711–736.

Welter, Volker M. 2002. *Biopolis: Patrick Geddes and the City of Life.* MIT Press.

Wendland, Folkwin. 1993. *Der Grosse Tiergarten in Berlin.* Mann.

Wendland, Victor. 1971. *Die Wirbeltiere Westberlins.* Dunker & Humblot.

Werner, Frank. 1969. *Stadtplanung Berlin: 1900–1950.* Kiepert.

Wettengel, Michael. 1993. Staat und Naturschutz 1903–1945. Zur Geschichte der Naturdenkmalpflege in Preussen. *Historische Zeitschrift* 257 (2): 355–399.

Whyte, Anne. 1985. Ecological Approaches to Urban Systems: Retrospect and Prospect. *Nature and Resources* 21 (1): 13–20.

Willdenow, C. L. 1787. *Flora Berolinensis Prodromus.* Berlin.

Williams, John Alexander. 2007. *Turning to Nature in Germany: Hiking, Nudism, and Conservation, 1900–1940.* Stanford University Press.

Williams, Raymond. 1983. *Keywords.* Fontana.

Wilson, Jeffrey K. 2006. Environmental Protest in Wilhelmine Berlin: The Campaign to Serve the Grunewald. *Bulletin of the German Historical Institute* Supplement 3: 9–25.

Wimmer, Clemens A. 1992. *Parks und Gärten in Berlin und Postdam.* Nicolay.

Wise, Norton. 1999. Architectures for Steam. In *The Architecture of Science*, ed. G. Peter and E. Thomson. MIT Press.

Witte, Fritz. 1952. 3 Jahre "Hilfe durch Grün" im Aufbau Berlins. *Garten + Landschaft* 62 (9): 4–7.

Witte, Fritz. 1960. Grünplanung und Natur- und Landschaftsschutz in Berlin. *Das Gartenamt* 9: 216–218.

Wittig, Rüdiger, Dagmar Diesing, and Michael Gödde. 1985. Urbanophob— Urbanoneutral—Urbanophil. Das Verhalten der Arten gegenüber dem Lebensraum Stadt. *Flora* 177: 265–282.

Wolman, A. 1965. The Metabolism of Cities. *Scientific American* 213 (3): 179–188.

Yearley, Steven. 1991. *The Green Case.* HarperCollins.

Zacharias, Frank. 1972. *Blühphaseneintritt an Straßenbäumen (insbesondere Tilia x euchlora Koch) und Temperaturverteilung in Westberlin.* Freie Universität Berlin.

Index